ENERGY, THE SUBTLE CONCEPT

Nicolas Poussin, *A Dance to the Music of Time*, c. 1640 (By kind permission of the Trustees of the Wallace Collection, London).

Energy, the Subtle Concept

The discovery of Feynman's blocks from Leibniz to Einstein

JENNIFER COOPERSMITH

Revised Edition

OXFORD
UNIVERSITY PRESS

OXFORD
UNIVERSITY PRESS

Great Clarendon Street, Oxford, OX2 6DP,
United Kingdom

Oxford University Press is a department of the University of Oxford.
It furthers the University's objective of excellence in research, scholarship,
and education by publishing worldwide. Oxford is a registered trade mark of
Oxford University Press in the UK and in certain other countries

First Edition published in 2010
Revised Edition published in 2015

Published in the United States of America by Oxford University Press
198 Madison Avenue, New York, NY 10016, United States of America

British Library Cataloguing in Publication Data
Data available

Library of Congress Control Number: 2015930102

ISBN 978-0-19-871674-7

To

Bertie Coopersmith

and

Murray Peake,

sages both

Preface to the Revised Edition

This book is both a history of the emergence of the concept of energy and an explanation of energy through that history. As it is primarily a book of explanation, the historical coverage is not exhaustive, and also 'energy' is brought in ahead of its final discovery, wherever an explanation in modern terms would be an aid to understanding. The level is popular science but hard thinking is required—we might call it serious popular science. The facts do not, in general, speak for themselves, and the aim is not only to explain, but, most of all, to imbue the reader with a sense of awe—over these very facts, and the genius of their discoverers.

This second edition is more comprehensive in its coverage, especially in Chapter 17, 'A Forward Look', but at the same time it is more streamlined in order to better bring out the logic of the arguments—arguments that will lead to the concept of energy. The expertise gained while researching *The Lazy Universe: the Principle of Least Action Explained* (Oxford University Press, in preparation) has led to some changes in Chapter 7, Part IV, and Chapter 13, 'Hamilton', and to some new material in Chapter 18, 'Impossible Things, Difficult Things'.

'Energy' is a huge topic, and to keep the book at a reasonable length, and in one volume, energy in the current era has not been considered; in fact, the aim has been to cover material, and to provide insights, *not* easily found by a computer search. Likewise, the amount of biographical detail is mostly in inverse proportion to the fame of the given scientist.

Some 'housekeeping':

Quotation marks appear all over the place, sometimes in thickets, and with little attempt at consistency. This is because as new ideas arrive, new words hover over them, and it takes a few decades for the right word to land. (There are also many commas—this is the OUP house style.)

Mathematical notation: 'height $\propto$ age' means 'height is proportional to age', '$<x>$' means 'the average value of x', 'Δx' means 'a small increment in x', '$\int E \mathrm{d}t$' is the 'integral of E with respect to t' or 'the integral of E over t', or, less often, 'the integral of E through t', and is roughly equivalent to a summation of the separate increments, $E_1 \mathrm{d}t + E_2 \mathrm{d}t + E_3 \mathrm{d}t + \ldots$ Vectors are quantities that have a direction as well as a magnitude (a bit like an arrow, of length 40 cm, pointing due west). They are indicated in bold;

for example **v** is velocity, whereas v is just the magnitude of velocity, also known as speed.

The history has been given in roughly chronological order, but 'heavy' chapters on theoretical developments are interspersed with 'lighter' chapters on experiments and phenomenology. Chapters 7 (the last section), 13, 17, and 18 assume some background knowledge in the physical sciences: readers with this background will enjoy the broad sweep across the whole of physics (Chapter 17), and other readers should not be discouraged as much knowledge can seep in by osmosis. Also, all chapters contain gentle introductions, summaries, and occasional discursive sections (for example, on interactions between physicists, energy in the public domain, and global warming). The summaries were variously called conclusions, summary, resumé, review, comments, remarks, and so on, but the copy editor swept these aside and insisted on a uniform descriptor—'overview'. Just one 'Remarks' has survived, in the middle of Chapter 8 (the end of the section on Watt), as the remarks really are . . . remarkable—there is even a mention of sex.

Acknowledgements

Six marvellous books have guided me and been my constant companions: Donald Cardwell's *From Watt to Clausius: the Rise of Thermodynamics in the Early Industrial Age*, Cornelius Lanczos' *The Variational Principles of Mechanics*, Richard Feynman's *Lectures on Physics*, Richard Westfall's *Force in Newton's Physics*, Charles Gillispie's *The Edge of Objectivity*, and Brian Pippard's *The Elements of Classical Thermodynamics*.

In the case of the last two books, I have been privileged to have a correspondence with the authors. Both have been extremely enthusiastic and encouraging (Professor Sir Brian Pippard died last year). Professor Gillispie (Emeritus Professor at Princeton and founding father of the discipline of the history of science) has read many of my draft chapters and made invaluable suggestions. Professor Paul Davies (physicist, science writer, and founder and director of 'Beyond, Center for Fundamental Concepts in Science'), and Emeritus Professor Rod Home (Department of History and Philosophy of Science, Melbourne University) read some early draft chapters and were likewise very encouraging.

I would like to thank past colleagues and friends at King's College London, TRIUMF, UBC, Logica SDS, and the Open University. I would especially like to thank the Open University students (London, Winchester, and Oxford, 1986–96) who asked all those difficult-to-answer questions.

My present colleagues at La Trobe University, Bendigo Campus, and Swinburne University of Technology in Melbourne have been very supportive. In particular, I would like to thank the late Rob Glaisher, Katherine Legge, Glenys Shirley, John Schutz, Mal Haysom, Joe Petrolito, John Russell, Glen Mackie, and Sarah Maddison. For special tasks: Mal Haysom took the cover photos, Andrew Kilpatrick was tireless in drawing diagrams, even randomizing the position of dot-molecules by hand (mind you, I assisted with his marking-avoidance strategies), Sabine Wilkens translated Euler and trawled through Helmholtz's endless accounts of indigestion, and Glenys Shirley helped with anything and everything, and with characteristic good humour.

I thank the University of La Trobe, Faculty of Science, Technology and Engineering, Bendigo Campus, for providing me with an office,

computing and library facilities, and the intangible but crucially important ambience of learning, research, and cooperation.

The library staff at La Trobe University, Bendigo showed great for-
bearance in my constant requests for inter-library loans and my loss
of magnetic belts, late returns, and so on. (Having previously lived a
stone's throw from the Blackwell bookshop and the Bodleian in Oxford,
I naturally didn't embark on this project until I had moved to central
Victoria.)

I thank Howard Doherty, of Doherty's Garage in Bendigo, for interesting conversations about cars, their design, and their impact on society.

Thank you to Wikimedia Commons and Google Books for making
source documents available online. In material reproduced in this book,
every reasonable effort has been made to contact the original copyright
holders. Any omissions will be rectified in future printings if notice is
given to the publisher.

I am unusually fortunate in having not one but two sages to call upon
in my family (I am daughter of one and married to the other).

My father (a physicist-turned-programmer) is a deep and highly
original thinker, always aiming to get back to the essentials. When I
asked him where my grandfather had come from in Russia and what
the family name was (before 'Coopersmith' was adopted upon arrival
in South Africa around 1910), he replied 'I don't know. Anyway it's just
contingent.'

My husband, Murray, has a profound understanding of mathematics
and physics (he was a prize-winning physics student and applied mathematician before being drawn down that well-trodden path of computer
programming). His compendious knowledge of pre-twentieth-century
arts and crafts has also been useful in the writing of this book. After one
of his weekday 'Murraycles' (for example, being instrumental in a quiet
revolution in the calculation of risk in banking), he turns to tying knots,
grinding wheat, making rope, bread, tofu, and soap; restoring gas lamps,
kerosene burners, sextants, and clocks; and calculating lunar motions
using Stumpf's method. He is an unassuming polymath, so why switch
on the computer or find the calculator when you can always ask 'Mur-
raypaedia' or get him to do a back-of-the-envelope calculation with the
appropriate series approximation? His large collection of books ranges
from *Spruce-root Basketry of the Tlinglit* to *Galactic Dynamics*, and he has
a knack for swiftly finding the very book needed to solve one's problem.
Together, we usually manage to complete Araucaria. The fact that he is
also a teddy bear of a man is just the icing on the cake.

My mother has frequently held the fort (or fortress, as our Dutch uncle calls it) with inimitable Granny-ish style and vigour. She is nevertheless impervious to my constant exhortations: 'You don't need to practice worrying—when you need to worry you'll know how'.

I would like to thank the science editor Sönke Adlung, at OUP, and the copy editor, Geoffrey Palmer. Also Jan Wijninckx (the Dutch uncle) was very generous with his time and software expertise and saved me from some Word-wracking moments.

Sustenance was provided at the congenial Good Loaf Bakery, Bendigo; they unwittingly assisted in the transformation 'BB' ('bread-to-book') or 'TTT' ('toast-to-text').

Many thanks also to the 'patchwork quilt' of childcare, most especially Helen and Richard Jordan and Clare and Phil Robertson.

Finally, I thank my children, David, Rachel, and Deborah, for helping in all the usual ways; in other words, without their efforts this book could have been completed in half the time. Seriously, though, they have been very appreciative of this enterprise and provided a wonderful selection of live and recorded music to write books to. Deborah conceived the design of the cover and helped with copy editing.

Contents

List of Illustrations

Praise for the First Edition

'In clear and engaging prose, Coopersmith shows how the modern understanding of energy was formulated, moving from the first documented discussions of simple machines and perpetual motion in ancient Greece, to the work of Gottfried Leibniz and other 17th-century thinkers, to Einstein's theory of relativity and beyond . . . *Energy, the Subtle Concept* is a fascinating read; both physicists and nonphysicists who want to learn more about the history of energy will enjoy it'.
Lisa Crystal, *Physics Today*

'The conservation of energy is arguably the most important law in physics. But what exactly is being conserved? Are some forms of energy more fundamental than others? You will have to read the book to find out. Coopersmith sets out to answer such questions and to explain the concept of energy through the history of its discovery. This is neither a straightforward narrative nor one for the faint-hearted. Those not put off by the odd bit of mathematics will be well-rewarded by dipping into this book'.
Manjit Kumar, *New Scientist*

'The work is full of surprises, and some illuminating aperçus. It makes one think about the subject in a new way—the connections made with dynamics, Hamilton and Lagrange are germane, and one never sees these in books on thermodynamics'.
Sir Aaron Klug, Nobel laureate, President of the Royal Society 1995–2000

'This is a work of physics in substance and history in form. *Energy, the Subtle Concept* is as much concerned with physicists as with physics. Its scientific interest is matched by human interest. Jennifer Coopersmith deftly brings to life the people who made the science throughout its history'.
Charles C. Gillispie, Professor of History of Science Emeritus, Princeton University

'The more I read this book, the more difficult it was to put it down . . . [It] has a fascinating story to tell about the development of our understanding of energy as a physical quantity . . . '
Matt Chorley, *Popular Science*

'I am pleased to heartily recommend Coopersmith's readable, enjoyable, and largely nonmathematical yet profound account of the development of an important physical concept—energy. With a vein of humour running throughout, it deals with an enormous compass of important topics seldom found elsewhere at this level. It should be of great interest and utility to students, both undergraduate and graduate, historians of science, and anyone interested in the concepts of energy and their evolution through time'.
George B. Kauffman, *Chemical and Engineering News*

'This book makes me proud to be a physicist, for two reasons. First it is a tale of the giants of the past who contributed to our present understanding of energy, people whose astonishing intuition took them from gossamer clues to the understanding we have today of one of the most basic explanatory concepts in physics. We've had some pretty good players in our team. More than this—and this is the second reason—this is a story as much about invention as discovery . . . I am sure all physicists would enjoy this book and indeed learn from it'.
Australian Physics

'[B]eautifully written text . . . Throughout, the book is sprinkled with anecdotes and, most importantly, insightful commentary, with a plethora of figures that assist the reader in digesting the concepts detailed'.
Jay Wadhawan, University of Hull

1

Introduction: Feynman's Blocks

Ask a physicist what physics is all about and she'll reply that it's something to do with the study of matter and energy. 'Matter' is dispensed with quite swiftly—it's stuff, substance, what things are made from. But 'energy' is a much more difficult idea. The physicist may mumble something about energy having many different forms, and about how it is convertible from one form to another, but its total value must always remain the same. She'll look grave and say that this law, the law of the conservation of energy, is a very important law—perhaps the most important law in all of physics.

But what, exactly, is being conserved? Are some forms of energy more fundamental than others? What is the link between energy, space, time, and matter? The various forms and their formulae are so seemingly unrelated—is there some underlying essence* of energy?

The aim of this book is to answer these questions and to *explain* the concept of energy *through the history* of its discovery. The reason behind the peculiarly anthropocentric quality of the concepts ('work', 'machine', 'engine', and 'efficiency') then becomes clear. More than this, the history shows that great philosophical and cultural—intellectual—revolutions were required, and these were every bit as paradoxical, as discomforting, and as profound as those that occurred with the arrival of quantum mechanics and Einstein's Relativity. The long list of revolutions, before the twentieth century, includes: the physical world must be looked at and measured; mathematics is the language of physics; time is a physical parameter, and can be compared with distance; the vacuum exists; perpetual motion—of a continually acting, isolated, machine—must be vetoed; perpetual motion—of an isolated body—must be sanctioned; the latter must be vetoed, *in practice*; nature must conserve her resources;

---

* We'll find that the answer is NO—there is no 'essence of energy'.

nature must be economical with her resources; action-at-a-distance is absurd but 'true'; the world is cyclical and deterministic; the world is not cyclical but progressive; most subtle, weightless fluids are redundant; light is particulate; light is wave-like; Newton's Third Law isn't always true; physics has probabilistic aspects; light and matter can be treated by one mechanics; electricity, magnetism, and light are all manifestations of the same thing; the electromagnetic field really exists, and electromagnetic waves can travel through empty space.

Also, the historical sequence of events is surprising. The steam age was well under way—steam locomotives pulling trains and steam engines powering industry—decades before 'energy' was discovered and its conservation stated as the First Law of Thermodynamics. Also, the Second Law trumped the First Law by being discovered first; then came the Zeroth Law, and the Third Law might not be a law at all.

Finally, the historical approach is hugely entertaining. In fact, the history of scientific ideas is the ultimate 'human interest' story; and it is a ceaseless wonder that our universal, objective science comes out of human—sometimes all too human—enquiry.

This is a tale[1] of persecuted genius and of royal patronage, of ivory-towered professors and lowly laboratory assistants, of social climbers and other-worldly dreamers, of the richest man in all Ireland, the richest man in all England, and one who had to make a loaf of bread last a week. It includes feuding families and prodigal sons, the Thomson and Thompson 'twins', a foundling, a Lord, revolutionary aristocrats, entrepreneurs and industrialists, clerics, lawyers, academics, engineers, *savants*, doctors, pharmacists, diplomats, a soldier, a teacher, a spy, a taxman, and a brewer. Some were lauded and became grandees of science; others only received recognition posthumously. A few were persecuted (under house arrest, imprisoned, one even guillotined, another lampooned) while others carried on through trying domestic circumstances (for example, the mathematician Euler, old and blind, did the calculations in his head, surrounded by a horde of grandchildren, all possessions lost in a house-fire, and yet maintained his work-rate of one paper a week for 60 years). There was a ladies' man, a pathologically shy man, a gregarious American, and a very quiet American. Yes, they were all men, although two wives and a mistress put in an appearance (one wife actually appears twice, married first to Lavoisier and then to Count Rumford).

There were English eccentrics and gentlemen-scientists; Scottish 'Humeans' and, later, Scottish Presbyterians; French *philosophes* and, post-Revolution, French professional scientists; and German Romantics then materialists and, later, positivists. Some were great travellers and others travelled only in their minds (Newton never even visited Oxford, let alone other countries). However, the community of scientists was small enough that all could know each other. Communication was obviously slow (for example, by stagecoach in the seventeenth and eighteenth centuries) yet, apart from one glaring exception (the Scottish physicist, P.G. Tait), there was always a spirit of co-operation between scientists of different nationalities, even when wars were raging.

Some were what we would now call philosophers (Descartes and Leibniz), applied mathematicians (Euler), mathematicians (Lagrange and Hamilton), chemists (Boyle, Lavoisier, Scheele, Black, and Davy), and engineers (Watt and Carnot), while others were true physicists (Newton, Daniel Bernoulli, and Joule). Some were lovelorn and poetical (Hamilton and Davy) or dour and preoccupied with earning a living (Watt), and some were universal geniuses (Leibniz and Young). All were very hard-working.

There seems to be a conspiracy of silence surrounding certain topics, such as: Why are there two main types of energy, kinetic and potential, and is one more fundamental than the other? Which is more important, force or energy? Where does the kinetic energy go to when we change our viewpoint? Why is the difference between the kinetic and potential energies significant? Why does kinetic energy have the form $\frac{1}{2}mv^2$? What is 'action'? Many of these ideas came through in the century from Descartes to the French Revolution, the 'only period of cosmic thinking in the entire history of Europe since the time of the Greeks'.[2] At other times, cosmic thinking did continue, but it was carried out by individuals rather than by society as a whole: from Galileo's deep insights, culminating in Einstein's equivalence between mass and energy.

Other special questions include: why the steam engine was discovered at just one time and in just one place; what heat is; what temperature is; why temperature is more fundamental than pressure; what 'the Hamiltonian' is; and a link between the Second Law of Thermodynamics and global warming.

Just how slippery the concept of energy is has been captured by Richard Feynman in an allegory in his Lectures on Physics.[3] (If Feynman says the concept of energy is difficult, then you know it really *is* difficult.) Feynman imagines a child, 'Dennis the Menace', playing with blocks. Dennis' mother discovers a rule—that there is always the same number of blocks, say 42, after every play-session. However, one day she finds only 39—but after some effort finds one under the rug, and two under the couch (she must look carefully); and another day she finds only 40—and eventually discovers that two had been thrown out of the window. On yet another occasion there were 45—it turned out that a friend, Bruce, had left three of *his* blocks behind (she must be thorough, and alert to different possibilities).

Banning Bruce and closing the window, everything is going well, until one day there are only 39 blocks. The mother, being ingenious, finds that the locked toy box, which weighs 16 ounces when empty, now weighs more. She also knows that a block weighs 3 ounces, and so she discovers the following rule:

number of blocks seen + [weight of box −16 oz]/3 oz = constant

On another occasion the water in the bath has risen to a new level but is so dirty (!) that she can't see through it. She finds a new term that she must add to the left-hand side of the above equation: it is [height of water − 6 inches]/0.25 inch.

As Dennis finds more and more places in which to hide the blocks, so his mother's equation becomes more and more complicated, adding together a longer and longer list of seemingly unrelated formulae. The complex sum of formulae is a quantity *that has to be computed*, but that always adds to the same final value.

Feynman means us to draw a comparison between Dennis' blocks and energy: both sum to a constant value when a closed room (isolated system) is inspected; both require the summing of ever-more-complicated mathematical terms. But Feynman concludes with a remark that strikes at the heart of the slipperiness of energy:

What is the analogy of this to the conservation of energy? The most remarkable aspect that must be abstracted from this picture is that *there are no blocks*.[4]

Something is being conserved—but what? There are some other lessons to be learned from Feynman's allegory: it is necessary to examine a *system*, with unambiguous boundaries; mathematization and quantification are essential. In fact, Feynman shows us that it is *only* by uncovering the individual mathematical formulae that energy will be uncovered. We shall, therefore, track the emergence of these formulae—the nonexistent 'blocks of energy'—through the history of their discovery.

2

The Quest for Perpetually Acting Machines

A mathematical description of nature, leading to the eventual discovery of energy, emerged slowly, over millennia, and was prompted, more than anything else,[1] by one endeavour—the construction of devices. These were tools, simple machines, and 'engines', discovered independently, again and again, all over the world.[2]

The first known *quantitative* analysis was the description of the lever, by the followers of Aristotle. These were the Peripatetics, so-called because they wandered around the gardens of the Lyceum in Athens while carrying on with their discussions. In the *Mechanica*,[3] of around 300 BC, they posed the question:

Why is it that small forces can move great weights by means of a lever?

and answered:

the ratio of the weight moved, to the weight moving it, is the inverse ratio of the distances from the centre . . . the point further from the centre describes the greater circle, so that by use of the same force, when the motive force is further from the lever, it will cause a greater movement.

This sounds quite modern until we read:

The original cause of all such phenomena is the circle. It is quite natural that this should be so; for there is nothing strange in a lesser marvel being caused by a greater marvel, and it is a very great marvel that contraries should be present together, and the circle is made up of contraries.[4]

The most famous proof of the law of the lever is that due to Archimedes (in around 250 BC, in Syracuse). He determined the location of the centre of gravity when the lever was in equilibrium. (Apparently, 'centre of gravity' was a self-evident idea, needing no justification.) Legend has it (or, rather, Pappus of Alexandria, *c.* AD 340) that Archimedes said 'Give me a place to stand on, and I will move the Earth.'

The enigmatic Hero of Alexandria lived around AD 60, and devised many ingenious machines and 'toys', principally for use in temples and as spectacles for an audience. One such toy, the *aeolipile*, was a sort of kettle, with two bent spouts, and supported on mounts that allowed the kettle to keep on turning as the steam escaped (Fig. 2.1). This is possibly the first steam engine in history (the Chinese used steam in antiquity, but in laundry presses rather than as a source of power[5]). Whether or not it actually was the first, it's certainly true that steam-driven devices are the one exception to the rule: they were *not* discovered independently, again and again, all over the world (in fact, they emerged again only 1,600 years later, in Britain, and were brought to perfection by just two individuals, Newcomen and Watt— the reason why the steam engine is special in this respect is explained in Chapter 8). Hero listed the five simple machines as the wheel and axle, the lever, the pulley, the wedge, and the inclined plane. He understood that all were variations on one elemental machine, and that all obeyed the fundamental law: 'Force is to force as time is to time, inversely'.[6]

Not much happened (as regards the mathematization of machines) for well over 1,000 years. Around AD 1250, Jordanus de Nemore derived the law of the lever again, and was the first to use the concept of 'virtual displacements' (to be put to important use by Johann Bernoulli in the eighteenth century; see Chapter 7).

The semi-legendary Chinese figure, Ko Yu, is credited with the invention of the wheelbarrow around AD 1 (AD will be assumed from now on). Later, these were often sail-powered (a wheelbarrow with a sail was a popular image for China in eighteenth-century Europe). A window in Chartres Cathedral, dated to 1225, has the earliest depiction of a wheelbarrow in use in the West.

The first water-wheel was found in China around 200 BC. The development of gears about this time enabled the wheel to be ox-powered and used for irrigation. In Illyria (Albania), in around 100 BC, water-powered mills were used for grinding corn. By 1086 in England, the Domesday book listed 5,624 water-wheel mills south of the River Trent, or one for every 400 people. The earliest known windmills were used in Persia (Iran) in around 600. They had a vertical shaft and horizontal sails, and were used to grind grain. Windmills—for draining the land— were found in Holland from around 1400.

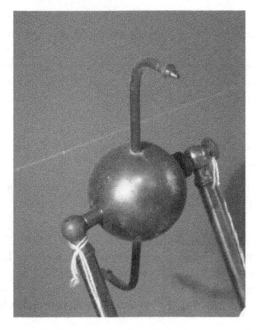

Fig. 2.1 A photograph of a reconstruction of Hero's steam toy (courtesy of the Museum of the History of Science, Oxford).

But what happened when the rivers dried up and the wind didn't blow? Since ancient times there have been continuing attempts to make a machine that, once started, would run forever—a perpetual motion machine (or *perpetuum mobile* in Latin, the language of culture in Europe since Roman times). The idea of a perpetually acting device appears to have originated in India, where a perpetually rotating wheel had religious significance, symbolizing eternal cycles such as the cycle of reincarnation. (Wheel symbols often appear in Indian temples.) The first mention of a perpetual motion machine occurs in the Brahmasphutassidhanta, a Sanskrit text by the Indian mathematician and astronomer Brahmagupta in 624. He writes:

Make a wheel of light timber, with uniformly hollow spokes at equal intervals. Fill each spoke up to half with mercury and seal its opening situated in the rim. Set up the wheel so that its axle rests horizontally on two supports. Then the mercury runs upwards [in some] hollow spaces, and downwards [in some others, as a result of which] the wheel rotates automatically forever.[7]

A subsequent Sanskrit text, by the Indian astronomer Lalla in 748, describes a similar wheel, and around 1150, the Indian mathematician and astronomer Bhaskara II describes this sort of wheel again: 'This machine rotates with great power because the mercury at one side of the axle is closer than at the other.' (Bhaskara II is also famous as the first mathematician to define dividing by zero as leading to infinity.)

Such overbalancing wheels were to prove one of the most popular methods in the attempt to achieve perpetual 'motion'. The French architect, de Honecourt, tried it in 1235 (see Fig. 2.2):

Now there are major disputes about how to make a wheel turn of itself: here is a way one could do this by means of an uneven number of mallets or by quicksilver.[8]

Edward Somerset, the Marquis of Worcester, (*c*.1601–67), famously made an overbalancing wheel while held in the Tower of London, after the beheading of Charles I. The wheel was made on a huge scale ('fourteen foot over, and forty weights of fifty pounds apiece'[9]) and probably kept going for some considerable time by acting as a flywheel. Charles

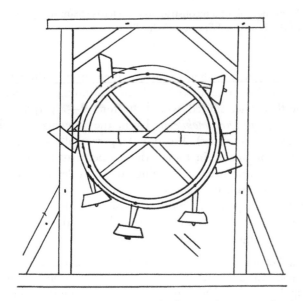

Fig. 2.2 de Honecourt's overbalancing wheel, 1235 (courtesy of Wikimedia Commons).

II was sufficiently impressed to release the Marquis, who subsequently developed his 'water-commanding engine', the very first steam engine (Chapter 4).

As remarked earlier, sometimes the rivers dried up and sometimes the wind didn't blow. Inventors then wondered if a fixed supply of water or air could drive a machine in a closed-cycle operation. Robert Fludd (1574–1637), physician and mystic, proposed a closed-cycle water mill in 1618. More obviously problematic was Zimara's self-blowing windmill from a hundred years earlier (1518). The artist's impression[10] shown in Fig. 2.3 is in a style appropriate to the age—could the vanes of the windmill ever hope to operate those huge bellows?

Fig. 2.3 Zimara's self-blowing windmill, 1518 (artist, Lee Potterveld).

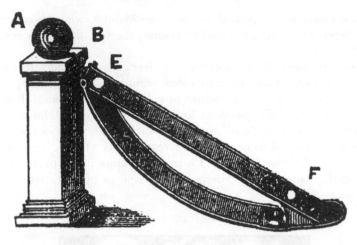

Fig. 2.4 Bishop Wilkins' magnetic perpetual motion, 1648, *Perpetuum Mobile, or the Search for Self-motive Power during the 17th, 18th and 19th Centuries,* by Henry Dircks (1861).

John Wilkins (1614–72), Bishop of Chester and an early official of the Royal Society (founded in London in 1660), proposed three sources that might lead to perpetual motion: 'Chymical Extractions', 'Magnetical Virtues' (Fig. 2.4), and 'the Natural Affection of Gravity'.[11] Wilkins was part of the seventeenth-century scientific establishment—clearly such ideas were not considered far-fetched.

Another proposed source of perpetual motion was capillary action, as in Boyle's, and also Papin's, 'goblet' in the seventeenth century (Fig. 2.5). (Robert Boyle was the famous 'chemyst' and discoverer of Boyle's Law, while Papin was famous for his pressure cooker or 'digester'; Chapter 4.)

Zonca's perpetually siphoning closed-cycle mill of 1656 (Fig. 2.6) was a similar idea.

Johann Bernoulli (1667–1748), of the Bernoulli family of mathematicians (see Chapter 7), proposed perpetual motion using a sophisticated but flawed hydrostatic analysis of a system of two fluids. Sir William Congreve (1772–1828), politician and inventor of the Congreve rocket, was amongst a number of perpetual motionists in the eighteenth century who utilized sponges and osmosis.

By the eighteenth century more and more charlatans and fraudsters were joining the ranks of genuine attempts at perpetual motion. In the

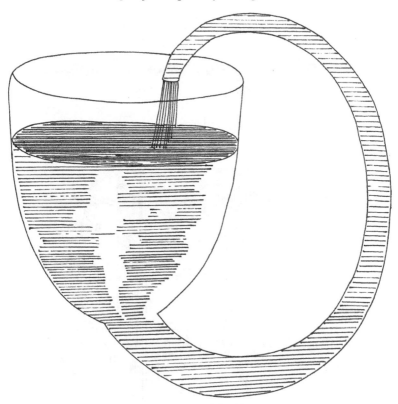

Fig. 2.5 Boyle's and Papin's 'goblet' (after Dircks 1861, as in Fig. 2.4 above).

tale of Orffyreus, a pseudonym for the German entrepreneur Bessler (1680–1745), a perpetually moving wheel was set up for display in Hesse Castle. However, the interior mechanism of the wheel was concealed, and the wheel itself kept in a locked room with the Landgrave of Hesse's seal upon the door. The Landgrave allowed the Dutch natural philosopher Willem s'Gravesande (1688–1742; Chapter 7) to inspect the wheel—but without opening up the mechanism. s'Gravesande subsequently sent a report of his inspection to Isaac Newton. It is not known whether or not Newton was impressed, but Orffyreus was so outraged that he smashed the wheel to bits, and apparently scrawled a message on the wall saying that he was driven to wreck his wheel by s'Gravesande's impertinence.

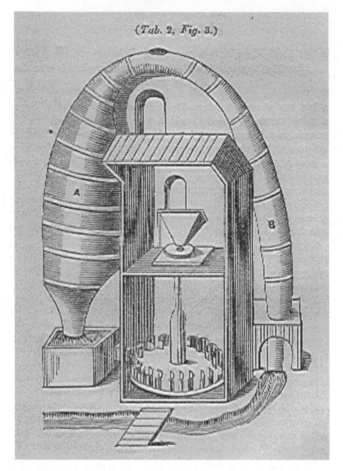

Fig. 2.6 Zonca's siphon, in *Nuovo Teatro di Machine*, Padua, 1656 (from Dircks 1861, as in Fig. 2.4 above).

Newton justified his Third Law of Motion (Chapter 3) by the absurdity of perpetual motion, yet, in his youthful entries in his 'Wastebook', he made a sketch of a perpetually turning wheel powered by gravity. The wheel was half covered by a 'gravity shield': when gravity rained down on the exposed half, that half got pushed downwards while the other half was dragged upwards, and the whole wheel began to turn. This *could* have worked if only a gravity shield could have been found . . .

Not one of the above machines was convincing, and by the middle of the eighteenth century it was becoming clear that all such attempts were doomed. In 1775, the French Royal Academy of Sciences said 'Non!' and banned further submissions, putting perpetual motion in the same category as squaring the circle, trisecting the angle, or doubling the cube (concurrent unresolved problems).

While there were always seekers after perpetual motion, there were, at the same time, always others who knew this quest was vain. Most interestingly, the *impossibility* of perpetual motion began to be used as the basis for a variety of proofs (proofs that, with considerable hindsight, all had a bearing on the conservation of energy).

The first such proof was that due to the 'geometer' Simon Stevin (1548–1620) of Bruges, in Flanders. Stevin considered a chain of uniform spheres* draped over two inclined planes (Fig. 2.7). As perpetual motion is 'absurd', he argued that the chain doesn't cycle round of its own accord, and everything must be in balance. Stevin's remarkable and simple conclusion?—for balance, the length of chain along a given side

Fig. 2.7 Stevin's 'wreath of spheres', from *De Beghinselen der Weeghconst*, 1586 (courtesy of the National Library of Australia, ANL).

---

* Think of the chain of ball-bearings used to open a blind. Also, in Stevin's idealized thought experiment, the chain never catches on the pointy bits.

must be equal to the length of that side. (Note that the chain mustn't be bunched up or stretched out; also, the freely hanging curve of chain at the bottom can be ignored—it already is in balance.) Stevin knew that he had made a long-lasting contribution to knowledge, and he had his 'wreath of spheres' as the frontispiece of his book[12] (rather than commemorating some of his other achievements, such as inventing decimal notation and the land-yacht).

A proof of the impossibility of perpetual motion will be used again, to astounding effect, by Sadi Carnot, some 240 years into the future (Chapter 12).

Overview

Perpetual motion—a perpetually acting machine—had been attempted for so many years, and in so many ways, but was never successful: did this mean that something was being conserved? This was a question that could barely be put—let alone answered—and was not finally resolved until the riddle of energy was solved in the middle of the nineteenth century. However, the sheer variety of ways in which perpetual motion was sought, and was failing, was in itself illuminating (overbalancing wheels, chemical, magnetic, and adhesive attractions, gravity shields, and so on). If anything was being conserved, then it was something universal, hard to define, and that crossed categories. Also, it is no accident that the first 'block of energy' was discovered in the same century, the seventeenth century, when the fervent quest for perpetual motion was at its height, and when the condemnation of it was brought in by the Scientific Revolution. We continue with this seventeenth-century tale, and the first 'block', in the next chapter.

3

Vis viva, the First 'Block' of Energy

Galileo

Archimedes had looked at the lever as a problem in statics. Two thousand years later, Stevin introduced motion only to ban it. Who could nowadays look at a lever and not see the motion of the parts, and what child could look at an inclined plane and not think of motion down it? Galileo Galilei (1564–1642) was one who looked at inclined planes afresh—and brought in *motion*, mathematically, for the first time.

It's not clear whether Galileo really did carry out the legendary experiments, dropping cannon balls from the Leaning Tower of Pisa (he may just have done this as a demonstration—a common practice at this time). What he *did* do was to realize that Stevin's 'wreath of spheres' experiment (Chapter 2) had relevance to the dropping balls experiment—the inclined plane slowed everything down so that the free-fall process was brought within experimental reach.

Let's be awe-struck and list[1] some of the mind-shifts that Galileo-brought in. First, the famous one:

(1) Bodies of different weights all fall at the same rate (have a constant acceleration).

Then:

(2) Looking at a process, at *motion*.

(3) Not just looking, but *measuring*.

(4) Recognizing that an experiment is always idealized, to some extent.[2]

The fifth mind-shift was even more momentous than the first four:

(5) 'Time' is a physical parameter.

Yes, time could be marked out by water-clocks, candle-clocks, and sun-
dials; yes, the seven ages of man were known about; yes, the acorn takes
time to grow into an oak; yes, the priest's sermon dragged on (in time) so
that Galileo's attention was diverted towards the swinging of the church
lamp—and so on. But that time could be put into a mathematical relationship, that it could be brought into comparison with distances travelled, that it was a 'dimension in physics'—this was new.[3]

How hard this step was is evidenced by the fact that, in his free-fall
(inclined-plane) experiments, Galileo tried and tried again for *over 20
years*[4] to find a link between one distance and another distance before
he could bring himself to introduce time. When he did, he eventually
found that, for a body dropped from rest:

the spaces run through in any times . . . are as the squares of those times.[5]

From this came also $v^2 \propto h$, a relationship* that would be remembered
by Huygens, and be important in the story of energy. (Meanwhile, Galileo joked that the Aristotelians were looking at free fall in a totally different light:[6]

[the Aristotelian] must believe that if a dead cat falls out of a window, a live one
cannot possibly fall too, since it is not a proper thing for a corpse to share in
qualities that are suitable for the living.

A cat paradox preceding Schrödinger's?

Incidentally, how did Galileo measure the time of fall? A water-clock
would not have been suitable for such short time intervals. The Galileo scholar, Stillman Drake, suggests an alternative method—singing.[7]
It is possible to sing to a constant rhythm, achieving a timing precision of 1/64th of a second. Drake suggests that gut lute strings (Galileo
was an accomplished lute-player) were tied around the inclined plane,
and as the ball rolled down it would make a noise when passing over
each string. With repeated trials, the position of the strings could be adjusted until the soundings coincided with the regular beat of a song. This
method had the advantage of high precision, but had the disadvantage
of linking speed with distance rather than with time of fall. But eventually the speed versus *time* relationship in uniform acceleration was to
emerge. Drake has it that this was the first time in the history of science

---

* v is the final speed and h is the vertical distance fallen. The symbol '$\propto$' means 'proportional to'.

that experiments were carried out in order to discover a quantitative relationship rather than just to demonstrate a known effect.

Mind you, Galileo was expecting the result that he finally determined; he had long realized that the Aristotelian claim that heavy bodies
fall faster than light ones was wrong. First, he argued it from reason:
how could a heavy composite body fall faster than its parts? Second,
he argued it from experience: he remembered a hailstorm he had witnessed as a youth—large and small hailstones all arrived at the ground
at the same time. So Galileo realized that, on both rational and experimental grounds, the acceleration was the same for bodies of different
weights—or was it? The final unravelling required the appreciation that
no experiment is perfect. In his book *Two New Sciences*,[8] Galileo answers
criticisms from an Aristotelian:

Aristotle says, 'A hundred-pound iron ball falling from the height of a hundred
braccia [one braccia is 21–22 inches] hits the ground before one of just one
pound has descended a single braccio'. I say that they arrive at the same time.
You find, on making the experiment, that the larger anticipates the smaller by
two inches . . . And now you want to hide, behind those two inches, the ninety-
nine braccia of Aristotle.

Let's pause and look briefly at Galileo the man.[9] Born Galileo Galilei (1564–1642) in Pisa, Italy, there are few indications that he was
one of the most advanced thinkers of all time: he was happily 'married', an attentive and affectionate father, played the lute, appreciated wine, and saw the commercial potential of his discoveries (the
telescope and a 'geometric and military compass'). His books demonstrate a beautiful literary style, a sense of humour, and an enormous
talent for polemical writing and explaining things simply. As regards
his famous breach with the Catholic Church, it seems that the bishops
and cardinals were bending over backwards to accommodate Galileo's
views, if only he would just keep quiet and not make waves.[10] However, he kept on and on, unable to accept that the church couldn't
be won round by reason. He spent the last few years of his life under
house arrest at his villa near Florence, blind but continuing to write
(the *Two New Sciences* was smuggled out to Holland and published by
the Elsevier family in 1638). No one knows whether he really mut-
tered the words 'and yet it [the Earth] moves' under his breath at his
recantation trial.

Galileo's introduction of 'time' was obviously crucial to the development of the concept of energy—how could something be seen to be

conserved before time had been discovered? But we still haven't finished with our list of Galileo's extraordinary mind-shifts:

(6) The heavens have imperfections (the Moon has craters, and the Sun has spots and rotates).

(7) The Earth moves and, in fact, orbits the Sun (supported by the phases of Venus, and Jupiter's moons, further discoveries of Galileo).

(8) The 'fixed' stars are extremely far away (follows from (7)).

(9) Physics is 'written in the language of mathematics'.[11]

Finally, and possibly Galileo's single most important contribution to physics:

(10) Motion is relative.

Taking the example of the inclined plane yet again, Galileo asked what would happen if the plane was made shallower and shallower. He answers:

a body with all external and accidental impediments removed travels along an inclined plane with greater and greater slowness according as the inclination [of the plane] is less.[12]

What happened when the plane was horizontal? The answer is given in the dialogue between a common man (Simplicio) and Galileo (Salvatio):[13]

> SIMP: I cannot see any cause for acceleration or deceleration, there being no slope upward or downward.
> SALV: Exactly so. But if there is no cause for the ball's retardation, there ought to be still less for its coming to rest; so how far would you have the ball continuing to move?
> SIMP: As far as the extension of the surface continued without rising or falling.
> SALV: Then if such a space were unbounded, the motion on it would likewise be boundless? That is, perpetual?
> SIMP: It seems so to me, if the movable body were of durable material.

So here we have to come an astonishing and very non-Aristotelian result—motion on a horizontal surface, free from 'impediments' such as friction or a body that breaks up, continues forever (perpetually) and without need of any empathies, forces, or angels to push it along. Perpetual motion, *of a machine*, was looking less and less likely, but perpetual motion of an unimpeded body was a *requirement*. The English philosopher Hobbes was so bowled over that he came to visit Galileo at his villa in 1636, and made Galilean relativity the core of his philosophy.

There was so much that could be gleaned from this result. As the inclined plane became shallower, the acceleration was reduced until eventually it was zero: but the final speed wasn't zero—what was it? From $v^2 \propto h$ (see above), a body on a horizontal plane has a constant speed dependent only on the height from which it was released. But the body itself has no 'memory' of this height, and so—it (the height or speed) doesn't matter. To restate this less obscurely: there is no difference between one speed and another; speed cannot be determined absolutely; only motion *between* bodies is important:[14]

If, from the cargo in [a] ship, a sack were shifted from a chest one single inch, this alone would be more of a movement for it than the two thousand mile journey [from Venice to Aleppo] made by all of them together.

Even the state of rest isn't special in any way:[15]

Shut yourself up with some friend in the cabin below decks on some large ship, and have with you there some flies, butterflies and other small flying animals. Have a large bowl of water with some fish in it; hang up a bottle that empties drop by drop into a wide vessel beneath it. With the ship standing still, observe carefully how the little animals fly with equal speed to all sides of the cabin. The fish swim indifferently in all directions; the drops fall into the vessel beneath; and in throwing something to your friend, you need throw it no more strongly in one direction than another, the distances being equal; jumping with your feet together, you pass equal spaces in every direction . . . [then] have the ship proceed with any speed you like, so long as the motion is uniform and not fluctuating this way and that. You will discover not the least change in all the effects named, nor could you tell from any of them whether the ship was moving or standing still.

In other words, only *relative* motion is important and can be determined *absolutely*. This came to be known as the Principle of Galilean Relativity, and could be said to mark the start of physics. It was the very beginning of a line continuing with Descartes, Huygens, and Newton, and culminating with Einstein's Principle of Relativity (see Chapter 17). Needless to say, it will be crucial in the story of energy.

Galileo's principle of the relativity of motion gave very good ammunition for his arguments in favour of a moving Earth: the Earth *does* move, he said, but—*we are not aware of it*.

There is, of course, one way to tell whether a ship is moving (relative to something else)—look out of a porthole and watch the view going by.

For the Earth, watching the view meant looking at the stars—they were seen to rotate about the pole star every day. But was it the Earth that was spinning or were all the stars rotating in unison? Simplicity argued the case for a spinning Earth, but Galileo also considered the problem from an intriguing angle (which had a future connection with energy)—the enormous 'power' that would be needed to have the stellar sphere—so far away, so many and such large bodies—moved instead of the Earth;[16] 'Tiring is more to be feared for the stellar sphere than the terrestrial globe'.

We have not yet mentioned the *direction* of motion. The perpetually moving body moves not only at constant speed but also in a fixed direction. Here, Galileo betrayed an Aristotelian legacy. For him, the 'horizontal' plane was only a small part of the great circle going around the centre of the Earth. He appreciated that, by mathematical correctness, this plane was only horizontal at one point—the point of contact with the Earth's surface. Motion away from this point was like going uphill, and there would be a resistance to such motion. However, this effect could be lumped together with friction and other 'impediments to motion'. For both Aristotle and Galileo, all 'natural' motion (unforced—not due to cannons, crossbows, or muscles etc.) was either straight down towards a 'gravitating' centre, or it was circular and equidistant from such a centre. A straight line extending to infinity out in space, or a circle not centred on the Earth or other 'centre', was unnatural, unthinkable.

The relativity of motion incorporated yet another mind-shift:

(11) The system is important.

For example, the ship and all its cargo constitutes a system, and the shoreline is outside this system. Galileo distinguished between the cases of a body dropped from a tower and a body dropped from the mast of a ship. The position of the body relative to the bottom of the tower, and the bottom of the mast, was not the same. In the former case (system), both the tower and the surrounding air were carried together (on the moving Earth); in the latter case, the boat had an additional motion *through* the air.

The system can be in Venice or Aleppo, or moving uniformly between the two—it doesn't make any difference. Also, the system can be scaled up or down and it doesn't make any difference. But, for any alteration, one must be careful to apply it to the *whole* system. For example, Galileo considered the case of a bone in a small animal and in a giant (Fig. 3.1), and noted that they would *not* have the same proportions.[17]

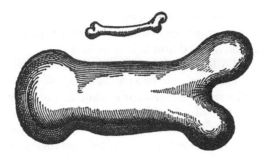

Fig. 3.1 Bone in a small and a large animal, from Galileo's *Two New Sciences*, 1638 (with the permission of Dover Publications, Inc.).

What constitutes the total system is important—and not always easy to identify. This will be a recurring theme in the identification of energy. (We remember how, in Feynman's allegory, knowing what constitutes the system is important.)

Let's now consider Galileo's work in a totally different arena—the machine. (Galileo was frequently consulted by his patron, the Grand Duke of Tuscany, about schemes put forward by inventors of pumps, milling machines, and olive presses.) Galileo understood that a perpetually acting machine was an impossibility: 'nature cannot be . . . defrauded by art'.[18] However, in considering percussion the situation was not so clear, and Galileo admitted to 'dark corners which lie almost beyond the reach of human imagination'.[19] The trouble arose in comparisons between the force of a heavy dead weight and the 'energy and immense force [*energia e forza immensa*]'[20] of a much lighter but moving weight, such as the blows of a hammer. The hammer could act as a pile-driver, whereas, in gently removing and replacing the heavy dead weight, nothing happened. This problem had to wait for the advent of Newton and his mechanics for its resolution.

Descartes

René Descartes (1596–1650) is hardly ever given his due in histories of physics. Somehow his famous 'Cogito ergo sum' ('I think, therefore I am') has resulted in his being hijacked by the philosophers. However, Descartes was just as motivated to develop his theory of the physical world (*res extensa*) as his theory of the mental world (*res cogitans*). These two domains—extension and thought—were the only ones Descartes' 'method of doubt' allowed him to be certain about.

Descartes had a life-changing experience that made it clear to him what his mission in life was: to be a philosopher, and to construct a complete system of knowledge. Apparently, while shut up in an 'oven' (a sort of boiler-room), he had a day of solitary thought followed by a night of restless dreams. It became clear to him what the correct path was: he had to start from no prior knowledge, no beliefs, and then, as it were, re-peel the onion, admitting only those layers of knowledge, one by one, about which there could not be any doubt. Another anecdote about Descartes tells of his antipathy to getting up early. He never started work before 11 in the morning. When he went to Sweden to tutor Queen Christina in 1650, she insisted that lessons start at 5 a.m. This, combined with the harsh Swedish climate, was too much for Descartes, and he caught a chill and died.[21]

If the physical domain was characterized only by the property of extension, where did matter come in? Descartes answered this by considering that the chief property of matter *was* extension. He also argued this the other way around: if there was no matter, then there was no extension—in other words, empty space was an impossibility. This seems very strange to us, even more so when we recall that Descartes invented the way of representing space that is still used today—the Cartesian coordinate system. Perhaps because he *had* invented it he couldn't believe it corresponded to something real.

The straight axes of Descartes' coordinate system extended to infinity. For the first time, the universe was of infinite extent. The realm of the stars in Copernicus' celestial system was much further away than it had been in Ptolemy's, but all the celestial bodies still moved in *finite* perfect circles. This was also true in Galileo's universe. But Descartes had a more abstract concept of motion altogether; it could continue indefinitely. But despite his infinite universe, Descartes still thought that only *local* motion made any sense.[22] Also, he didn't accept Galileo's division of motion into natural and unnatural (forced) kinds, and replaced them by a single, more general, abstract kind of motion: rectilinear (in a straight line), unaccelerated, unbounded, and unconnected with any centres of activity. This intellectual leap was all the more impressive in view of the fact that such rectilinear motion was never to be observed in practice: Descartes' universe was completely filled with matter (a plenum), and so unobstructed rectilinear motion could never actually take place.

The infinity also operated in reverse: matter was infinitely divisible—the smaller parts could always be subdivided into even smaller parts. Space was completely filled with this continuum of matter, and so atoms didn't exist. This was consistent with Descartes' banishing of the void (if atoms existed, then there would have to be empty space between those atoms). Descartes did, however, posit three types of matter.[23] The first element (fire) was 'the most subtle and penetrating liquid'; the second (the element of air) was also a subtle fluid; the third (the element of earth) was made from bigger, slower particles.

Common to all these types was the requirement that all matter was inert, had extension, and could be in motion. However, as space was already completely full, this motion could only occur through some matter displacing other matter in a continuous loop. This loop or vortex motion was made to account for the orbits of the planets, and for gravity at the Earth's surface. The trouble was, it didn't reproduce Galileo's free-fall results.

However, the important point to notice is the leap made by Descartes in going from matter moving in closed whirlpools, bordering and constrained by other whirlpools, to that motion which matter *would* have if it wasn't so constrained. Descartes considered this hypothetical, 'free' motion (we now say 'inertial') to be of constant speed and rectilinear. His first two laws of nature were:[24]

That each thing remains in the state in which it is, so long as nothing changes it.

That every body which moves tends to continue its motion in a straight line.

He appreciated that the telling question was not what will keep a body in motion, but what could cause that motion to cease:[25]

we are freed of the difficulty in which the Schoolmen [the Scholastics] find themselves when they seek to explain why a stone continues to move for a while after leaving the hand of the thrower. For we should rather be asked why it does not continue to move forever.

Thus, Descartes appreciated that motion was relative, and that a state of rest had no special status compared to a state of motion. He may well have exploited this relativity in order to conceal his adoption of the Copernican system, with its requirement for a moving Earth (Descartes had taken careful note of Galileo's troubles with the church):

The Earth is at rest relative to the invisible fluid matter [Descartes' second element] that surrounds it. Even though this fluid *does* move [around the Sun] and carries the Earth with it, the Earth does not move any more than do those who as they sleep are carried in a boat from Calais to Dover.[26]

The original source of all matter and motion was God. As God and His creations were immutable, then the total quantity of matter and the total 'quantity of motion' could not change, although particles could acquire or lose motion after colliding with other particles. Indeed, this was the fundamental mechanism of change in Descartes' universe. All physical phenomena (for example, gravity and magnetism) and all qualities (colour and smell) were to be explained solely in terms of *collisions between bodies*. No occult forces, attractions, empathies and antipathies, action at a distance, and so forth, were to be admitted. This was the seventeenth-century's 'mechanical philosophy': it had many followers (such as Boyle and Gassendi), but Descartes was its chief exponent.

The 'quantity of motion' was defined by Descartes as the 'mass' of a body times its 'velocity'. 'Velocity' was what we would now call speed, as the body's direction—'determination'—was not considered relevant. 'Mass' was an umbrella term including the size and shape of the body. Mindful of these contemporary distinctions, we can say that 'quantity of motion' equals mv.

Now, as we have said, the mechanical philosophy had it that collisions between bodies accounted for every physical process in the world. A theory of collisions was therefore of paramount importance, and this was one of the few scenarios where Descartes attempted a quantitative treatment. His guiding principle was that the total mv before collision should equal the total mv after collision. This reminds us of the modern law of conservation of momentum—but Descartes' physics was such that his 'quantity of motion' had always to be a *positive* quantity; in other words, mv was a pure magnitude, and there was no change in sign following a body's reversal of direction.

(As there were no forces causing bodies to change direction, Descartes had another solution: the redirections occurred in the pineal gland, sited in the brain. *Only human brains could do this; all other animals had no minds and were automata.* There was much correspondence on this topic between Descartes, Gassendi, and Hobbes.)

In the case of elastic or completely inelastic collisions between identical bodies, Descartes' rules gave identical results to the accepted ones. However, in most cases his rules did not agree with experiment. The most glaring mismatch occurred in the case of a collision between a small body moving at constant speed towards a larger body at rest (rule 4). Descartes maintained that the larger body could never be moved by the smaller one, no matter how rapidly this smaller one approached. Of course, Descartes *did* realize that this didn't work out in practice and so he developed the perfect fudge: his rules were only for the collisions between two bodies in a vacuum—but the vacuum didn't exist. In reality (argued Descartes), there were innumerable collisions between gross bodies and the matter making up the invisible fluid medium (Descartes' second category of matter), and these collisions would be just such as to account for the differences between theory and experiment . . . (Descartes tried to use a similar argument to explain why his vortex theory of gravity didn't accord with Galileo's constant acceleration of freely falling bodies.)

As well as his collision theory, Descartes also gave a new analysis of another problem that would have great relevance to energy: simple machines (the lever, pulley, and inclined plane). He started with:[27]

the effect must always be proportional to the action that is necessary to produce it . . .

and saw that, despite their various methods of working, the 'action' of all the simple machines could be generalized to just one case, the weight-raising machine:

the action . . . is called the force by which a weight can be raised, whether the action comes from a man or a spring or another weight etc.

Finally, he was able to propose a quantitative measure for such action:

The contrivance of all these machines is based on one principle alone, which is that the same force which can raise a weight, for example, of a 100 pounds to a height of two feet, can also raise one of 200 pounds to the height of one foot, or one of 400 to the height of a half foot, and so [on] with others.

Thus Descartes' measure for action is identical to our measure for work (against gravity), and as such it is the very start of our quest for the concept of energy.

Descartes' legacy was almost as boundless as his universe: the merging of geometry and algebra, the mechanical philosophy, in which animistic and occult influences were given short shrift, the generalizing of the concept of 'free' motion, and, of most particular importance to our theme, the emergence of two measures—the 'quantity of motion' (roughly mass × speed) and the 'action' of simple machines (weight × height). However, Descartes' world-view was empirical only to the extent that the facts of nature were to be taken as a given—there was no need to actually carry out experiments.* Physical knowledge could be built up by reason alone ('my physics is nothing but geometry',[28] he once wrote), and ultimately his physics was flawed in this respect.

Huygens

Christiaan Huygens (1629–95) had the misfortune to fall between Galileo and Descartes on the one hand and Newton and Leibniz on the other—his greatness would otherwise have been more readily apparent. Today, Huygens is famous to physicists for his wave theory of light, to astronomers for discovering the rings of Saturn, and to clockmakers (maybe) for his magnum opus, the *Horologium oscillatorium*.[29]

Huygens was born in The Hague, into a cultured family of Dutch diplomats serving the House of Orange. Descartes was a frequent visitor, and Huygens was brought up immersed in the prevailing Cartesian orthodoxy. His tutor was therefore shocked to find that Huygens was working on some errors in Descartes' theory of collisions.

The results of Huygens' studies on collisions were collected into a treatise, *De motu corporum ex percussione*, in 1656, while Huygens was working at the recently formed Académie Royale des Sciences in Paris. The most important theorems from this work were presented, in 1668, to the also recently formed Royal Society in London, in response to a call for papers specifically on this topic. John Wallis (professor of mathematics at the University of Oxford) and Sir Christopher Wren (the mathematician and architect) were the other respondents.

Huygens visited England and, apparently, he and Wallis shared a long stagecoach journey—one can only speculate as to the conversation, and wonder in which language it was conducted. Despite frequent wars between France, Holland, and England during the latter half of the seventeenth century, a free interchange of ideas continued between the natural philosophers of the day. However, Huygens was guarded about his work, and often coded his results as anagrams or numerical ciphers. He lived at Louis XIV's expense in an apartment at the Bibliothèque Royale. In later years, while Huygens was on a visit to The Hague, his patron, the Minister Colbert, died, and the tide turned against protestants. Huygens didn't return to Paris, but stayed on in Holland for the rest of his days.

---

* He did do careful experiments in optics and animal dissections.

Getting back to the *De motu*, Huygens put forward a series of hypotheses and propositions, and set out to refute Descartes much in the manner of a proof by Euclid or Archimedes:[30]

Hypothesis I, A body will continue in a straight line with constant speed if its motion is not opposed.

Hypothesis II, Equal bodies approaching with equal speeds will rebound with their speed of approach.

Hypothesis III, For any collection of bodies, only their motion relative to one another makes any difference.

The first hypothesis reasserted the principle of free or unforced motion (what will, post-Newton, be called inertial motion), inherited from Galileo and Descartes. His second hypothesis appealed to inherent symmetries, and the third hypothesis reasserted Galileo's profound insight on the relativity of motion. Unlike Hypotheses I, II, and III, Hypothesis IV was not of universal scope, but Huygens found that he couldn't determine all possible outcomes unless this particular case was specified:

Hypothesis IV, A big body hitting a smaller one at rest will lose some of its motion to the smaller body.

Finally, in the fifth hypothesis, Huygens managed to generalize the scenario of the second hypothesis (equal bodies, equal speeds) to the case of unequal bodies:

Hypothesis V, Two bodies whatever [*quelconques*] collide: if one conserves all its motion then so does the other one.

This hypothesis stressed the idea of conservation rather than symmetry. Huygens was so convinced of its importance that in the original manuscript he underlined it with a wavy line.

Having established this base of hypotheses, Huygens had sufficient theoretical weapons in his armoury to begin his attack on Descartes. The rules of impact established by Descartes disagreed with experiment, the most conspicuous conflict arising from rule 4—that a fast-moving small body could never shift a larger body at rest. Huygens was of a different ilk to Descartes—no fudge could ever conceal an outright clash with observations. But Huygens' objections went deeper. The 'quantity of motion' had to maintain an absolute constant value, while Descartes still adhered to the relativity of motion; but (wondered Huygens) how could motion be both relative and absolute at the same time? There was no doubt in Huygens' mind: of the two principles, it was the relativity of motion that had to be retained.

Galileo had asserted that there is no absolute state of motion, and Huygens took this on board—quite literally: he envisaged a collision on board a smoothly coasting canal boat, and then imagined viewing this collision from this boat, and also from the banks of the canal. The frontispiece to the *De motu* (Fig. 3.2) shows the imagined set-up.

The experiment was an elastic collision between two pendulum bobs (spherical hard bodies, but not necessarily of the same size). Of course, it was idealized, and not as feasible as it looks. Apart from external impediments (air resistance, torsion in the cords, choppy waters), how would the bodies be given specific starting speeds, and how would the speeds after impact be measured? This was almost certainly a thought experiment rather than a real one. (Another setting Huygens could have chosen was 'billiard-ball collisions', as the game of billiards was popular at the time, and familiar to him.[31]) Nevertheless, Huygens' thought experiment marked a new epoch in physics: this was the first time that Galilean Relativity was put to a *quantitative* test.

On board the boat, a small body collides with a larger body at rest (this is different from Fig. 3.2; imagine one small pendulum bob and one large pendulum bob). If the small body has the same incoming speed, v, as the boat but in the opposite direction, then, for an observer on the bank, it will appear as a collision between the larger body moving

Fig. 3.2 Huygens' thought experiment in *De motu corporum ex percussione*, 1656 (published by Martin Nijhoff NV).

with speed, v, and the smaller body at rest. Descartes had different outcomes for these two viewpoints but Huygens insisted that, by Galileo's principle of relativity, the outcomes should be the *same* (there was, after all, only one collision, albeit imagined).

Huygens had managed to correct the error in Descartes, but he wanted to go further and discover the overriding principles governing all collisions. (He continued, using propositions rather than hypotheses.) First, he proposed that the relative speed of approach equals the relative speed of recession (Proposition IV), and then that an elastic collision will always be reversible (Proposition V). By Proposition VI, he was ready to propose that Descartes' 'quantity of motion' is not always conserved. Interestingly, although Huygens managed to show that direction was the ingredient missing from mv, he didn't elevate this to an axiom, a hypothesis, or even a proposition. He still thought of motion in Cartesian terms; in other words, it was for him always a *positive* quantity.

Huygens was still not satisfied that he had reached a general principle from which all the rules would follow—until he arrived at Proposition VIII:

for two bodies approaching each other with speeds inversely proportional to the magnitude of the body, each will rebound with a recession speed equal to its approach speed.[32]

(by 'magnitude', Huygens meant, roughly speaking, 'mass'). Huygens realized that he had at last found what he was looking for—a proposition that preserved the symmetry of Hypothesis II, but that was applicable to the more general case of unequal bodies and unequal speeds.

From today's perspective, all these hypotheses and propositions appear rather bloodless—no body ever moves because it has been *hit*. Huygens tried to avoid all mention of 'force' (*vis*) and instead stressed symmetries and conservation. It should be borne in mind that all this *precedes* Newton's force and the 'game plan' of Newtonian mechanics.

Why did Proposition VIII work? Huygens recognized that in the special scenario where the speeds are inversely proportional to the 'body-magnitudes', then the centre of gravity of the total system remains fixed. While the proposition singles out a special viewpoint (that in which the centre of gravity is at rest), the principle of relativity guarantees that any results found from *this* viewpoint can be applied to all other viewpoints.

The centre of gravity was a concept that had been known about since antiquity and, as we saw in Chapter 2, was taken by Archimedes to be self-evident. But why was Huygens so sure of the truth and importance of Proposition VIII? He remembered a result of Torricelli's (Chapter 4), that when two bodies are linked together and constrained to move only in the vertical plane, then no movement freely undertaken by these bodies can raise their common centre of gravity. Huygens generalized Torricelli's result to any number of bodies moving any which way, and realized that 'the common centre of gravity of bodies cannot be raised by that motion of the bodies which is generated by the gravity of those bodies themselves'.[33] In other words, levitation is impossible (you can't raise yourself off the ground by pulling up on your hair). So, the centre of gravity must remain at the same level but, more than that, its motion must be 'inertial' (must satisfy Hypothesis I). Huygens finally arrived at the following 'admirable law':[34]

ADMIRABLE LAW OF NATURE . . . It is that the common centre of gravity of two, or three, or however many bodies you wish, continues to move uniformly in the same direction, in a right line, before and after their impact.

In other words, whatever the internal details of the system, the mass acts as if it is one total mass concentrated at the centre of gravity; and this centre of gravity moves uniformly, perpetually, and sublimely unaware of what the constituent masses are up to.

But what is the relevance of prohibiting the centre of gravity from undergoing any change in *height* when Huygens was considering bodies colliding in a *horizontal* plane? Huygens took a leaf out of Galileo's book (the *Discourses on Two New Sciences*) where bodies are dropped from certain heights and then fall freely until deflected into horizontal motion. Now Galileo had found that v^2 is proportional to h. All that was needed was a way of converting the vertical motion into horizontal motion and vice versa. Huygens' frontispiece is suggestive—the bodies are attached to cords and can therefore be released from prescribed heights. In a footnote, Huygens has a tiny diagram that shows another possibility— elastic bands (or whatever the seventeenth-century equivalent would have been) at a 45-degree angle to the vertical, acting as deflectors.

Arguing the other way around: a body on the horizontal, moving uniformly with speed v, will, when deflected upwards, have its speed retarded until, finally, at the *same* height, h, the motion is exhausted (exactly all used up). Galileo had asserted this reversibility (assuming no impediments) and Huygens saw that this *must* be so; otherwise a 'perpetual motion' would result.

Thus, ultimately, it was the inadmissability of 'perpetual motion' that brought in the requirement that the centre of gravity maintains . . . a perpetual motion. This contradiction is resolved when it is understood that, in the first usage, 'perpetual motion' means a 'perpetually acting mechanism'. It was this impossibility of a perpetual mechanical *process* that Huygens saw as the real physical explanation behind all the rules of collision.

(Before we jump to conclusions ahead of the times—the seventeenth century—it is worth noting that Huygens still hadn't ruled out the possibility of a perpetual motion by *non*-mechanical means. For example, he wrote of a clock invented by Robert Hooke (1635–1703):

[a] clock invented by Monsieur Hook [*sic*] . . . moves with the aid of a loadstone: if it uses no other motor, this would be a type of perpetual motion, and it would be an admirable invention.[35])

So Huygens, who wanted to keep all considerations of forces out of the picture, had to allow gravity as the force that, in collisions, gave the bodies their initial speeds and soaked up their recoil speeds. (Even so, Huygens was never able to accept Newton's Law of Universal Gravitation because of the occult requirement for action-at-a-distance.) In this way, height came into a scenario that was totally on the horizontal. This seems like reasoning a long way round, but it had a very interesting by-product. For the height of the centre of gravity to remain constant, the mathematics showed that another quantity had to stay constant: the total mv^2 before and after the collision.[36] This was the first time that the formula 'mv^2' made an appearance in physics. Huygens merely noted that it was a conserved quantity, and acknowledged its consequent utility as a predictor of the final speeds after collision.

In Huygens' later work on circular motion (*De vis centrifuga*, 1659), he finally succumbed to the idea of force (*vis*), and coined the term 'centrifugal' to describe the tendency of a circling body to 'flee' the centre of rotation. He found that this centrifugal force increased with the speed of rotation, and decreased for a larger radius—but what about its dependence on mass? 'Mass' was still a perversely insubstantial thing, and Huygens always referred evasively to 'equal', 'unequal', or 'whatsoever' bodies ('corps quelconques').

Still in connection with centrifugal force, Huygens contrived another thought experiment: a homunculus sitting on the rim of a giant wheel, and holding a mass or bob on a cord. As the wheel rotates, the bob pulls away from the centre of rotation. Substituting greater and greater masses for the bob, Huygens noted that the centrifugal force increases proportionately (the bob pulls away more and more strongly). The parallel with gravity was not lost on Huygens.[37] Huygens' biographer, Crew, notes that 'This is the first time that a distinction between mass and weight appears in physics'[38]—yet another first for Huygens. That mass can exhibit weight merely because of its acceleration (e.g. a rotation), will have great significance in the hands of Einstein; and, even more significant, Einstein will famously show that mass and *energy* are similar sorts of things (Chapter 17).

To summarize, Huygens was the first to apply Galilean relativity in a quantitative way. He was also the first to bring out a distinction between mass and weight. Most crucial of all for energy, he showed that, in the case of collision experiments, a certain quantity—a mere algorithm ('calculatrix')—was conserved. This was the quantity mv^2.

Newton

In 1687, Isaac Newton (1643–1727) published his great work, *Philosophiae naturalis principia mathematica*[39] (*The Mathematical Principles of Natural Philosophy*), usually referred to as 'the *Principia*'. As claimed in the title, the book set out the founding principles of physics: *it is probably the most influential book ever written* (by a single author). Let us not celebrate Newton's achievements but, rather, comb his work for anything of relevance to 'energy'.

Newton starts off with defining mass for the first time. (Huygens, as we have seen, sidesteps the issue by referring to a body 'quelconque'.) He defines 'mass' as being equal to the volume of a body times its density ('density' and 'volume' are taken to be understood). Significantly, the definition is in the measure—a new approach, starting with Newton. He performs careful experiments (timing pendula with bobs made from hollowed-out spheres of gold, copper, silver, and wood) and finds that the weight of a body is proportional to its mass—also very significant, but that's another story, for the time being. Newton has it that mass and matter are the same, and asserts that all matter is made, ultimately, from infinitely hard atoms that are indivisible, impenetrable, and are capable of motion. The quantity of matter then boils down to a head-count of the atoms.

Masses or bodies move about in empty space—a complete departure from Descartes' plenum. (The young Newton bought Mr Cartes' *La geometrie* at a fair in Sturbridge, Cambridge, and tossed it aside until persuaded to have another look by his tutor.[40]) But Newton does follow

Cartesian ideals in a number of ways: the physical world is materialistic (as opposed to animistic or psychic), and matter is passive (Newton calls it inert) and unaffected by its own motion. This passivity or inertia is exemplified by the law of 'inertial' motion, brought in by Galileo, Descartes, and Huygens. Newton gives it prominence as his First Law of Motion:

LAW I: Every body perseveres in its state of rest, or of uniform motion [meaning at constant speed] in a right line, unless it is compelled to change that state by forces impressed thereon.[41]

Newton defines such a uniformly moving, constantly directed, body as having a '*vis insita*' or '*vis inertia*'—an 'internal force' that keeps the body moving. But, in contrast to Descartes, for Newton this is emphatically not the *true* force. First, it can be cancelled out by motion in the opposite direction (Newton sticks to his mathematical guns on this) and, more importantly, it is only ever a relative quantity, dependent on the observer's point of view.

Uniform motion is relative but one thing that can be determined absolutely is a difference between earlier and later motions, a *change* in motion. But what causes the change in motion? Newton looked at Huygens' bumper-car kinematics (excuse the anachronism), obeying all the symmetries of collisions, and realized that this would still never be *enough* to account for everything, for all the phenomena in the physical world. Huygens had ruled out 'perpetual motion', but what had he ruled *in*? Some new entity—force or '*vis*'—was required:

LAW II: The alteration of motion is ever proportional to the motive force impressed; and is made in the direction of the right line in which that force is impressed.[42]

('Alteration of motion' was ambiguous. It meant 'alteration of motion during a given time'—or *acceleration*.[43] However, in all Newton's examples and calculations, a certain time interval was assumed, either implicitly or explicitly, and so no errors arose.)

Thus, Newton's First and Second Laws, taken together, showed that if there was *no* change in motion, then there was *no* force; if there *was* a change in motion, then there *was* a force; and if there *was* a force, then there *would be* a change in motion.

For the first time, the force did not reside *in* the body but was external to it, impressed *on* the body. Also, and for the first time, the effect of this true force was not to maintain motion but to *change* it: even a decrease in the speed was the result of a force. Also, and again for the first time, a change in direction alone (i.e. no change in magnitude) required a force—so the 'natural' circular motion of Galileo was no longer unforced. The effect of the force was not merely qualitative but quantitative: the change in the body's motion was proportional to and in the same line as the impressed force. Finally, two or more forces were added by the 'parallelogram law' for the addition of directed quantities (this is given as Corollary I, immediately following the Laws of Motion).

The amazing thing about Newton's Second Law is not so much the invocation of an invisible, almost magical, 'force', but the fact that it has just one effect—to cause an acceleration (it could, for example, have left the motion unaltered but caused the mass to swell, or it could have caused a second-order change in the velocity, and so on). Also, this one effect is then to account for *everything* in the physical world (the shape and size of pebbles on the beach, the trajectory of one such pebble when it is thrown into the sea, a fly walking on the surface of water, the path of the Moon—and so on).

It's strange to say, but Newton cited Galileo as the source of the Second Law, presumably because of the latter's discovery of the constant acceleration of bodies in free fall. But the connection between acceleration and force is not always so intuitively obvious as it is in the case of free fall. Take the modern-day exemplar of Newtonian mechanics, the 'simple' case of billiard-ball collisions: where is the force, and where are the accelerations? The bodies are not seen to slow down before impact or to speed up afterwards. In fact, the only evidence of anything *happening* is a change of direction. But Newton, armed with intuition and with mathematics, understood that this too counts as acceleration.

This is the true import of the Laws: motion is relative, but *happenings* are absolute, and need a cause. The First Law defines the simplest case—one body, on its own, and nothing is happening (in other words, uniform motion doesn't count as a happening). Anything more complicated, such as a collision between bodies, constitutes 'something happening': all observers will agree that the collision takes place, even though they may disagree on speeds and directions.

An absolute acceleration opened the way for the possibility of an absolute force, as Newton required. However, the absolute acceleration seemed to rely on there being an absolute space that was itself absolutely unaccelerated. This problem was not lost on Leibniz (see below), who was all too ready to pick through Newton's masterwork and find fault with it. The physicist and philosopher Ernst Mach (1838–1916) also criticized Newton's absolute space and time. There was no absolution! Newton had managed to wrest something absolute from out of the shifting sands of relativity, and was then criticized for his need for an absolute space and time.

However, this turned out to be mostly a red herring—let's say, a red anchovy—as Newtonian mechanics, in distinction to Newton, does not, for the most part, require space and time to be absolute. This is because only *intervals* in space, and *intervals* in time, occur in the formulae—there are no absolute designations such as '14 October 1066' or 'the battlefield at Hastings'. It is for this reason that Newtonian mechanics is so incredibly successful, explaining everything from billiard balls to the trajectories of spaceships. It does begin to fail at very high speeds, but not because of those speeds *per se* but because a disallowed absolute—the speed of light—is being approached. (Later, in Einstein's Theory of Special Relativity, we shall find that even intervals (of space and of time) are not enough to save Newtonian mechanics, but a new 'spacetime' must be assumed. Even more startling, the subtle concept 'energy' survives, but there must be an equivalence between energy and mass; see Chapter 17.)

Another philosophical problem with the Second Law is that it appears to be circular—defining mass in terms of force, and force in terms of mass. (Actually, 'mass' isn't even mentioned—look back to the definition, given earlier—and we need to substitute 'rate of alteration of motion *of a body*' for 'alteration of motion'.) In fact, the Second Law would be circular—and devoid of empirical content—if it applied to only one case. However, it is a statement about the physical world, and it predicts the results of countless experiments, such as applying the same force to bodies of different mass, or to bodies of the same mass but made from different materials, or applying different forces to one body, and so on (we then only need to agree on the notion of 'sameness'—see below).

Mach argues that Newton's Second Law amounts to a definition of mass—as 'reciprocal acceleratability'. However, one can continually pass the buck round when it comes to the attribute 'most-fundamental-and-irreducible-entity'. It is true that 'mass', when accelerating, could take on some of the burden of 'force'—and we shall see that this *is* what happens in d'Alembert's Principle (Chapter 7) and in Einstein's Theory of General Relativity (Chapter 17). There is no doubt, however, that Newton's intention was to consider mass as inert, and to have all changes as being due to the newly introduced agency of activity—force.

Except in the one case of gravity, Newton did not specify the exact mathematical form of this new entity, force; but he did identify many examples, such as stretched springs, magnetic attractions, cohesion of bodies, centripetal forces, and electrostatic attraction,[44] as well as gravitational attraction. The *same* force occurs when the conditions are the same: for example, a given spring stretched to a given extent; attraction to a particular piece of loadstone; or a gravitational attraction between two specific bodies. (Newton commented that it was simplest—and therefore wisest—to assume that gravity was the same force for distant celestial bodies as it was on Earth.)

However, Newton did introduce some provisos: forces arise from some material source—disembodied forces, or attractions to a point in space, are not allowed (actually, this did occur on a few occasions, and Newton wasn't very happy about it); they have direction and magnitude (in modern language, they are vectors); multiple forces are added in the same way as vectors are added (Newton sticks to his mathematical guns again); and they are attractive or repulsive, and act along the line joining the bodies. Finally, if the force arises from a body, A, and acts on another body, B, then Newton's Third Law of Motion imposes the constraint that B must exert an equal force back on A:

LAW III: To every action there is always opposed an equal reaction; or, the mutual actions of two bodies upon each other are always equal, and directed to contrary parts.[45]

Newton goes on to give examples of this law:

If you press a stone with your finger, the finger is also pressed by the stone. If a horse draws a stone tied to a rope, the horse (if I may say so) will be equally drawn back towards the stone.[46]

The Third Law leads to the same results (Newton calls them corollaries) as Huygens' rules of collision: the total 'quantity of motion' (in modern terms, the momentum) is conserved, and the common centre of gravity is unaffected by the motions of the individual bodies. But if the finger presses the stone more strongly, the stone will press back more strongly. Newton is therefore asserting something more than just conservation of momentum; he is asserting an equilibrium between the 'action' and 're-action' forces. In the uniform and unchanging motion of the common centre of gravity, both Newton and Huygens saw the evidence of a larger truth—the impossibility of perpetual motion. In the Scholium, following the Laws of Motion, Newton gives an example to demonstrate this:

I made the experiment on the loadstone and iron. If these, placed apart in proper vessels, are made to float by one another in standing water, neither of them will propel the other; but, by being equally attracted, they will sustain each other's pressure, and rest at last in equilibrium.[47]

If the attraction of the loadstone on the iron had been greater than the attraction of the iron on the loadstone, then the iron would accelerate towards the loadstone '*in infinitum*', and an absurd perpetual motion would result.

Newton went on to apply this same *reductio ad absurdum* argument to a thought experiment in which the Earth is partitioned into two unequal parts, and where the action of one part must equal the reaction of the other part. James Clerk Maxwell (1831–79), the great nineteenth-century physicist (Chapter 17), imagined the parts as 'a mountain' and 'the rest of the Earth',[48] and determined that if the Earth attracted the mountain more than the mountain attracted the Earth, then the ensuing perpetual acceleration would be tiny, barely detectable. Therefore, claimed Maxwell, the Third Law was not to be justified on experimental grounds but stood, rather, as a reaffirmation of the First Law. But from today's perspective, it seems safe to say that Newton put forward his Third Law on both theoretical *and* experimental grounds—it was a law that had to apply (he thought) and a law that was borne out in practice. We shall find, however, that there *are* instances where the Third Law fails.

(Mach points out that the Third Law really amounts to a definition of mass; it comes out as 'reciprocal-acceleratability'—again. As this has already been shown to follow from the Second Law, Mach says that the Third Law is adding nothing new. But . . . economy of description isn't everything: Mach's position fails to bring out the important symmetry in the action and reaction forces. Also, acceleration is not so easy and primitive a concept as Mach supposed; it has become so only through familiarity—smooth roads, vehicles with accelerator pedals, and so on.)

Newton goes on (in the Scholium)[49] to give examples of his Third Law as applied to the simple machines (the balance, pulley, wheel, screw, and wedge) and sums up their purpose as being 'to move a given weight with a given power'. This was seized on by Thomson and Tait in their textbook *Treatise on Natural Philosophy*,[50] in 1883, as proof that Newton had, after all, discovered the concepts of work and mechanical energy. (Tait shows up, again, in a reprehensible case of chauvinistic prejudice against Mayer; Chapter 14.) However, it is clear that, while Newton did consider perpetual motion to be an absurdity, he attached no especial significance to quantities such as 'mv^2' or '$\int F\,ds$' (later linked to the concepts of kinetic energy and work, respectively), even though these quantities did crop up in his calculations.

Now, when Newton uses the Third Law to analyse examples of simple machines,[51] he strangely applies the action and reaction forces to the *same* body rather than to different bodies. As Professor Home says,[52] this would today be considered a schoolboy howler, but it seems unlikely that Newton is applying his own law incorrectly! Home suggests that the answer lies in a less restricted use of the terms 'action' and 'reaction' by Newton than that used today. For example, in the *Opticks*,[53] Queries 5, 11, 30, and 31 (my italics), Newton asks whether:

the Sun and fix'd Stars are great Earths vehemently hot, whose heat is conserved by the greatness of the Bodies, and the mutual *Action and Reaction* between them, and the Light which they emit [Query 11]

Bodies and Light *act* mutually upon one another [Query 5]

active principles, . . . are the cause of gravity, . . . and the cause of fermentation. [Query 31]

In Query 30 Newton speculates further:

Are not gross bodies and light *convertible into one another?*

In other words, Newton is giving 'action' and 'reaction' a broader meaning beyond just the accelerative force implied in his Second Law of Motion. This broader conception of 'active principles', and the suggested interconvertibility between light and bodies, looks forwards in some respects to our modern idea of energy.

However, Newton's analysis of inelastic collisions was one occasion where his force view led to a radically different result from the energy view to come. In Query 31 of the *Opticks*,[54] Newton considers a collision in a vacuum between two absolutely hard bodies of equal mass, and finds that:

They will by the laws of motion stop where they meet, and lose all their motion, and remain in rest.[55]

The total momentum is conserved, and the action and reaction forces are equal and opposite, but there is a net loss of 'motion' (we would say 'kinetic energy') of the bodies. Thomson and Tait make no comment on this. Newton goes on to say that if such a loss of motion is *not* to be allowed, then the impenetrability of the bodies will have to be sacrificed instead. This is discussed further in Chapter 7, Part I (but note, already, that there is no illogicality in Newton's supposition).

What of machines and engines? Newton refers to them briefly in the Scholium (as mentioned above) but he is strangely uninterested, and in the entire corpus of his work, the *Principia* and elsewhere, there is no other treatment of them.[56] This is in contrast to many of his contemporaries (for example, Huygens proposed a gunpowder combustion engine).

Newtonian mechanics occurs in an idealized frictionless world, mostly of celestial objects. When friction *is* treated quantitatively, it is not the loss in the efficiency of a machine, but the resistance felt by a body moving through a liquid:[57] all the various microscopic (frictional) effects are lumped together as one net force ('the resistance') and its effects on one given body are calculated.

In the celestial domain, friction, and also concepts such as 'work' and 'kinetic energy' aren't especially relevant. (Who, today, calculates the kinetic energy of Mars?) However, Newton did appreciate that even for these heavenly bodies some motion would be lost to air resistance. He estimated that at a height of 228 miles above the surface of the Earth, the air would be '1,000,000,000,000,000,000 times rarer'[58] but that motion would be lost eventually ('motion is much more apt to be lost than got and is always upon the decay'[59]), and the planets would cease orbiting if there was no 'active principle' to bring about a 'reformation'[60] of the heavenly order. Once again, Leibniz leaped into the fray, claiming to be outraged that Newton's clockwork universe needed God to wind it up occasionally.

We have seen that the force-based view and the energy view to come have many parallels. Newton's conception of force was more than something just to explain dodgem-car kinematics, but to explain everything—attractions due to gravity across vast distances, and much stronger attractions (he even estimated how much stronger) occurring at very short ranges and accounting for chemical effects, the cohesion of bodies, and the crystal structure of solids. Thus 'force' was given a cosmic role—a similar sort of cosmic role that 'energy' would acquire. Considering also Newton's suggestion of the interconvertibility of light and matter takes Newton even closer to the modern idea of energy.[61] In the case of hard-body collisions, however, Newton was prepared to accept an absolute loss of 'motion'. Newton was the greatest physicist who has ever lived—or maybe that was Einstein; such comparisons are 'incommensurable'—so does it really matter if Newton let this one, energy, go by?

Leibniz

The term 'polymath' always comes up in connection with Gottfried Leibniz (1646–1716)—the Germans refer to him as a *Universalgenie*. Born in Leipzig, he advanced philosophy, logic (his work was lost until uncovered by Bertrand Russell in the twentieth century), law, theology, linguistics, mathematics (he discovered the infinitesimal calculus later than, but independently of, Newton), physics, and other areas. He was satirized mercilessly as Dr Pangloss in Voltaire's *Candide*, chiefly for maintaining that the world we live in is the best of all possible worlds. He was clever at courting favour with influential and titled people, and at charming the ladies; for example, the Duchess of Orleans said of him 'It's so rare for intellectuals to be smartly dressed, and not to smell, and to understand jokes'.[62]

Leibniz, like Descartes, is not well known for his contributions to physics. This is strange, as Leibniz spotted what Newton had missed in the '*crucis*' experiment of inelastic impact:

[I maintain] that active forces* are preserved in the world. The objection is that two soft or non-elastic bodies meeting together lose some of their force. I answer that this is not so. It is true that the wholes lose it in relation to their total movement, but the parts receive it, being internally[†] agitated by the force of the meeting or shock . . . There is here no loss of forces, but the same thing happens as takes place when *big money is turned into small change*.[63]

The equivalence of this 'active force' to our concept of energy is striking, especially when we realize that Leibniz's concept of force is not Newton's, but is given by the quantity mv^2. Leibniz called it '*vis viva*', or 'living force' (emphatically not the same thing as a 'life force'). Apart from the factor of ½, this is identical to our modern expression for kinetic energy. But we must now go back and ask how Leibniz arrived at these ideas, and what he meant by 'active force'.

Leibniz was a rationalist, like Descartes, and built up a total picture of the world from considerations of how it ought to be rather than from just looking and seeing how it is. How 'it ought to be' meant being governed by overriding general principles, such as the Principle of Sufficient Reason, the Principle of Continuity, the Principle of Cause Equals Effect, and the Identity of the Indiscernables.

---

* Not the same as Newton's force.
  † To be fair to Newton, he was talking of a collision between absolutely hard bodies; that is, bodies having no internal parts.

From the Principle of Continuity, there could be no finite hard parts—no atoms—and matter was continually divisible.[64] Ultimately, every body had a non-material point-like origin or 'soul'—the monad. There was an infinity of such monads making up the universe, and each was a centre of activity—a brain, if you like. Each monad followed its predetermined programme of action. Each also monitored its surroundings, and kept pace with all the other monads, like perfect clocks all started together, and so keeping perfectly in time with one another. In this way the monads maintained the pre-established harmony set up by God. This had the strange consequence that the monads didn't need to (and in fact didn't) interact with each other:

no created substance exerts a metaphysical action or influence upon another ... All the future states of each thing follow from its own concept. What we call causes are in metaphysical rigour only concomitant requisites.[65]

This is strange indeed considering that the most influential precept of Leibniz's dynamics was to be that the total cause equals the total effect.

How did the monads lead to the observable features of the world? Descartes' gravity was caused by vortical pressures in the aether, and Leibniz explained the observable properties of matter (hardness, elasticity, etc.) by an infinite regress of such vortices. Unlike Descartes, however, extension was not a fundamental property of matter (as, wearing his logician's hat, Leibniz insisted that 'extension' is a predicate that can only be ascribed to a 'class', and no substance is a 'class'). 'Motion', also, was an apparent quality, and not fundamental. The underlying reality was the infinite sea of souls ... But what bearing does this have on physics, and how did it lead to the opening quote on inelastic collisions, so rooted in the observable world and in 'common sense'?

Leibniz's answer was that the monads were 'centres of activity' and so they were anything but passive—and, consequentially, all substance was also not passive:

Substance is a being capable of action ... [and] ... to be is to act.[66]

(This, of course, was anathema to the Cartesian mechanical philosophy.) In short, 'activity' was the primitive agent, and the cause of all 'effect' in the universe. Furthermore, the 'total activity' had always to be conserved—but what was its *measure*?

The measure (formula) of 'activity' was arrived at by the following steps—and, once again, the testing ground was the case of impact between bodies. First, Leibniz protested, the formula could not be

Descartes' *mv*, as this had been shown not to be conserved in all cases. Second, whereas the *directed* 'quantity of motion' (our momentum) was known to be conserved in collisions, Leibniz felt that the true absolute measure had to be a quantity without direction (a scalar), as the Principle of Sufficient Reason would demand a reason for one particular direction to have been singled out. Also, he argued, something so fundamental as 'activity' had always to be a positive quantity. Searching for a suitable scalar, Leibniz at last came upon it in the work of Huygens. (Leibniz frequently had discussions with his friend and contemporary, Huygens, at the Académie Royale des Sciences in Paris). In Huygens' work on elastic impacts, a certain quantity, total mv^2, was found to come out as a constant. This quantity, mv^2, had all the right specifications for Leibniz: it was conserved (in elastic collisions), it was a scalar, it stemmed from Galileo's results for freely falling bodies, and, ultimately, it stemmed from the impossibility of perpetual motion—in other words, from the requirement that the cause should be equal to the effect. Leibniz had no hesitation in recognizing mv^2 as the absolute measure of 'active force' that he had been looking for.

Now Leibniz, like Huygens and others, was inspired to correct Descartes. He gleefully sent off a paper entitled 'A brief demonstration of a famous error of Descartes and other learned men, concerning the claimed natural law according to which God always preserves the same quantity of motion; a law which they use incorrectly, even in mechanics'[67]—the title was anything but brief. First, Leibniz managed to *derive* his mv^2 formula. (For a body of mass, m, falling from rest through a height, h, and reaching a final speed, v, Descartes had found that $mh =$ constant, and Galileo had found that $v^2 \propto h$. Merging these two results, Leibniz obtained $mv^2 =$ constant.) Then, Leibniz went on to show that Descartes' 'quantity of motion', mv, led to error. For example, the following combinations ($m = 1$, $v = 4$) and ($m = 4$, $v = 1$) (arbitrary units) had the same 'quantity of motion', but in the first case the mass could ascend to a height of 16 units, whereas in the second case the mass could ascend to a height of only one unit. These heights were different, and so Descartes was opening the door to 'perpetual motion' and a denial of 'cause equals effect'.

Mind you, it was somewhat disingenuous of Leibniz to trumpet the fact that the cause didn't equal the effect, as if Descartes would have been unconcerned by this. Also, Leibniz said that the truth was 'all in the mathematics', as if to imply that Descartes wasn't mathematician enough to follow the reasoning. Nevertheless, Leibniz had made a valid point concerning the contradiction between the two measures of 'force'. Descartes was dead by this stage, but his followers (including Papin of the 'digester'; see Chapter 4) tried to defend his position. They claimed that the time of fall should be taken into account. Leibniz countered that time wasn't relevant—the total effect would be the same whatever the time of fall or route taken:

For a given force can produce a certain limited effect which it can never exceed whatever time be allowed to it. Let a spring uncoil itself [either] suddenly or gradually, it will not raise a greater weight to the same height, nor the same weight to a greater height.[68]

However, time *was* very relevant in Leibniz's mechanics—it was the medium through which the 'effect' was developed. In a static situation (a balanced lever, dead weight, etc.), Leibniz posited that there was a propensity to motion, or *conatus* (Leibniz took the word from Hobbes). This propensity to motion was due to a 'force' that Leibniz referred to as the 'dead force', or '*vis mortua*', and which he claimed had the same measure, *mv*, as Descartes' disputed 'quantity of motion':

I call the infinitely small efforts or *conatus*, by which the body is so to speak solicited or invited to motion, solicitations.[69]

After the motion has started, the infinitesimal *conatuses* are compounded together through time, in a gradual and continuous way (Leibniz was able to use his own integral calculus to accomplish this):

what I call dead force . . . has the same ratio in respect to living force (which is in the motion itself) as a point to a line.[70]

The *vis mortua* does bear some resemblance to our modern concept, potential energy, but its measure, *mv*, is wrong. According to Westfall,[71] Leibniz rigged the formula so that *vis viva* would come out right.[72] Also, Leibniz showed no inkling of the continual interplay between the *vis mortua* and the *vis viva*, whereby an increase in one is accompanied by a decrease in the other. In the case of free fall, for example, $v^2 \propto h$ was understood by Leibniz to apply only *at the end* of the motion— when the 'force' was totally exhausted in ascent, or fully developed in descent. More importantly, whereas our measure for potential energy is derived from an integration over space, Leibniz's was compounded over time. So, first, Galileo had battled for years trying to find a relationship involving *distance*; and then Leibniz saw the 'live force' as acting through *time*.

It was nonetheless a truly impressive achievement of Leibniz's to recognize the important role of an 'active principle', and to identify this as the quantity mv^2. What to Huygens had been nothing more than a number that happened to come out constant, took on, in Leibniz's hands, a truly cosmic significance. It was to serve as the measure of all activity, all actual motion, whatever the source—falling due to gravity, the rebound after impact, the restoring force of a spring, and so on.

However, the evidence in favour of mv^2 as a measure of 'active force' was not so overwhelming as it now appears to us. For one thing, mv^2 was not absolutely conserved, but was still only a constant for a given viewpoint (frame of reference). To this extent, it was no more useful a measure than mv (see further discussion in Chapter 7, Part I). For another thing, mv^2 was derived by reference to free-fall experiments on Earth. Why should it also apply to other examples of activity; for example, horizontal motion, or celestial motions that were not Earth-bound? Johann Bernoulli[73] was to make just this objection (Chapter 7), saying that God could have made gravity other than he did at the Earth's surface. Finally, in the vast majority of cases *in practice*, mv^2 was *not* conserved.

But it was in just these real-life cases of inelastic collisions where Leibniz's physical intuition really came to the fore, and he was the only one of his contemporaries who could truly isolate what was going on from the obscuring details of real life. Frictional effects, Leibniz saw, were *always* present, and so perpetual motion was 'without doubt, contrary to the order of things'.[74] But, at the same time, the universe must be self-sustaining (Newton's solution of a 'reformation' was not to be tolerated). Once again, the principle of cause equals effect came to Leibniz's aid: whereas a perpetually acting process could not be allowed (else the effect would be greater than the cause), a perpetual 'inertial' motion had to apply (else the effect would be less than the cause):

A perfectly free pendulum must rise to exactly the height from which it fell, otherwise the whole effect would not be equal to its total cause. But, as it is impossible to exclude all accidental hindrances, that movement soon ceases in practice, otherwise that would be a mechanical perpetual motion [perpetually acting process].[75]

Apart from the fact that Descartes' rules for impact disagreed with experiment, Leibniz had philosophical objections as well. Descartes had impact between unequal bodies as a special case, but this case was not reached continuously from the case of impacts between identical bodies. This went against Leibniz's Principle of Continuity ('nature does not take leaps'[76]). Huygens' analysis, also, was flawed, as it required instantaneous changes of direction after collisions. But, in Leibniz's philosophy, there were no perfectly hard bodies, and everything happened *gradually*:

if bodies A and B collide . . . they are . . . gradually compressed like two inflated balls, and approach each other more and more as the pressure is continually increased; but . . . the motion is weakened by this very fact . . . so that they then come entirely to rest. Then, as the elasticity of the bodies restores itself, they rebound from each other in a retrograde motion, beginning from rest and increasing continuously, at last regaining the same velocity with which they had approached each other, but in the opposite direction.[77]

This is a completely modern outlook. More than just replacing hard bodies by deformable ones, Leibniz replaced Huygens' forceless kinematics by a new dynamic analysis. It was, in fact, Leibniz who coined the word 'dynamics'.

Overview

Galileo brought in 'time', the relativity of motion, and the relationship $v^2 \propto h$; Descartes brought in the measures 'quantity of motion' (mv), and 'work of a machine' (weight × height); Huygens discovered that mv^2 was a constant; his appeal to symmetries and conservation laws was to pay dividends in a later century. Newton brought 'force' into an anaemic forceless world and founded Newtonian mechanics. Finally, Leibniz seized upon mv^2 as 'live force', the cause of all effect in the universe—the first 'block' of energy. His new dynamic analysis was modern in all respects except for one—he didn't recognize the 'motion of the small parts' as *heat*.

4

Heat in the Seventeenth Century

The Nature of Heat

From Plato, in around 400 BC, through to Bacon, Descartes, Boyle, Hooke, and Newton in the seventeenth century, heat was understood as being due to the motion of the tiniest constituents of matter—the motion of the 'small parts'. Leibniz, as we have seen, saw the motion of the 'small parts' arising from a completely different source—from inelastic collisions. But the 'small parts' were conjecture, and their motion invisible; and no one made any link between the two agencies of such motion, between inelastic collisions and *heat*.

There were other problems that would need to be resolved before heat could be recognized as one of the blocks of energy. For example: What is the quantitative measure of heat? Is heat a substance? What is the connection between heat and temperature? As energy is a difficult and abstract concept, so each of its blocks will be difficult and abstract.

The first 'modern' to see heat as a form of motion was Francis Bacon (1561–1626). Bacon put forward the view that:

heat itself, its essence and quiddity, is motion and nothing else.[1]

He went on to observe that 'pepper, mustard and wine are hot'. Bacon became famous for introducing the method of induction in science, and for stressing the need for experimentation. However, he never carried out any experiments himself, except for the one where he is reputed to have tested the preservative effect of stuffing a chicken with snow. He caught a chill and died.

Descartes attributed heat to the motion of the small parts, yes, but not directly (see Chapter 3). The small parts were set in motion by the subtle ether, his first element. This 'element of fire' may have been the very start of the heat-is-a-fluid theory.

Descartes' contemporary and fellow adherent of the mechanical philosophy was the French priest Pierre Gassendi (1592–1655). Gassendi had a material theory of cold (his 'frigorific' particles) as well as of heat. He wasn't very consistent, and went on to write: 'Anaximenes and Plutarch thought that neither hotness nor coldness should be thought of as occupying a substance, but, rather, the atoms themselves, to which must be attributed all motion and hence all action'.[2]

Gassendi's main legacy was in his promotion of the ancient Greek idea of atomism, put forward by Democritus and Epicurus, and preserved in the poem 'On the nature of things', by the Roman poet Lucretius. This magnificent poem had lain dormant for around 1700 years until Gassendi rediscovered it. Atomism, and its need for a void (between atoms), was abjured by the Scholastics, and Gassendi had to recast it in a form that didn't bring on a charge of heresy. Epicureanism was so maligned at the time that it has never fully recovered, and even today is associated merely with having a gourmet palate.

Robert Boyle (1627–91) is most famous to physicists for Boyle's Law—that the pressure of a gas is inversely proportional to its volume. We'll come to this later. Boyle's main passion was chemistry, and his aim was to free it from alchemical sympathies and antipathies, and from Aristotelian forms and qualities. He was scion to a fabulously wealthy Anglo-Irish family, and had his own laboratory and assistants, and no pressure to earn a living. A contemporary wit referred to him as the father of chemistry and son of the Earl of Cork.

Boyle was a champion of the mechanical philosophy in which all effects were due to 'those two most catholic principles of matter and motion'.[3] Heat, considered as the *motion* of the small parts, was therefore a perfect example of an effect to be explained within the mechanical philosophy. But what actually was the evidence for these miniscule, invisible motions? As Maxwell was to put it, some 200 years later:

The evidence for a state of motion, the velocity of which must far surpass that of a railway train, existing in bodies which we can place under the strongest microscope, and in which we can detect nothing but the most perfect repose, must be of a very cogent nature before we can admit that heat is essentially motion.[4]

Boyle, alone amongst his contemporaries, appreciated that the question needed answering, and demonstrated a genius for seeing explanations in the mundane:

in a heated iron, the vehement agitation of the parts may be inferred from the motion and hissing noise it imparts to drops of water or spittle that fall upon

it . . . And . . . fire, which is the hottest body we know, consists of parts so vehemently agitated, that they perpetually and swiftly fly abroad in swarms, and
dissipate or shatter all the combustible bodies they meet with in their way.

He summed up the requirements for the motion to count as heat: it had
to be 'very vehement'; its 'determination [direction] . . . very various';
and it applied to particles 'so minute as to be singly insensible'. He followed this up with some perceptive observations of how this internal
motion could be engendered from the *outside* by friction, percussion,
or attrition:

The vulgar experiment of striking fire with a flint and steel, sufficiently declares what a heat, in a trice, may be produced in cold bodies by percussion or
collision.

Most interesting of all, Boyle made a distinction between heat and bulk
motion:

though air and water be moved never so vehemently as in high winds and cata-
racts; yet we are not to expect that they should be manifestly hot, because the
vehemency belongs to the progressive motion of the whole body

and

if a somewhat large nail be driven by a hammer into a plank, or piece of wood,
it will receive divers strokes on the head before it grow hot; but when it is driven
to the head so that it can go no further, [then only] a few strokes will suffice to
give it a considerable heat; for whilst . . . the nail enters further and further into
the wood, the motion that is produced is chiefly progressive, and is of the whole
nail tending one way; whereas, when that motion is stopped, then the impulse
given by the stroke . . . must be spent in making a various vehement and intes-
tine commotion of the parts among themselves.

This is the only precursor that I can find to Leibniz's conversion of 'large
money into small change' (Chapter 3).

However, Boyle did on occasion admit the possibility of Descartes'
subtle element of fire. For example, he attributed the gain in weight
of a metal upon burning (calcining) to the absorption of 'fire-atoms';
and, when adding water to salt of tartar, he says: '[the agitation was]
vehement enough to produce a sensible heat; especially if we admit the
ingress and action of some subtle and ethereal matter, from which alone
Monsieur Des Cartes ingeniously attempts to derive the incalescence.' It
is interesting to note that Boyle only ever considered a heat-substance in
chemical processes. In the eighteenth and nineteenth centuries, we shall

find that the caloric (material or substance) theory of heat was mostly adopted by chemists, while a dynamic theory of heat was mostly taken up by physicists (the terms 'physicist' and 'chemist' weren't introduced until the nineteenth century).

Finally, Boyle carried out many 'Experiments touching Cold etc'.[5] He noted the very large forces resulting from freezing ('congelation') and he speculated on the possibility of 'frigorific particles', but could find no change in weight in a block of ice as it thawed in the balance pan.

Robert Hooke (1635–1703) was Boyle's laboratory assistant in Oxford, and subsequently a natural philosopher in his own right. From 1662 he was employed as curator of experiments at the Royal Society in London. He is known to physicists for 'Hooke's Law'—possibly the least of his scientific accomplishments—rather than for his masterful work, *Micrographia*, a book with many detailed drawings of the world revealed by his microscope. Hooke was one who had no doubts about the redundancy of 'frigorific particles' and 'fire-atoms'. In sarcastic vein, he wrote:

we need not trouble ourselves to examine by what Prometheus the element of fire came to be fetched down from above the regions of air, in what cells or boxes it is kept, and what Epimetheus lets it go; nor to consider what it is that causes so great a conflux of the atomical particles of fire, which are said to fly to a flaming body like vultures or eagles to a putrefying carcase, and there to make a very great pudder.[6]

Thermometry

How should heat be quantified? And what is the relation of heat to temperature? Does a teaspoon of boiling water have less heat in it than a whole bath of tepid water? Does a given temperature interval correspond to a given change in heat, and is this true anywhere within the temperature scale and for any thermometric substance (e.g. alcohol, mercury, or water) or any thermometric property (expansion, the colour of material upon heating, etc.)? Is there a common zero of heat and temperature when all motion of the 'small parts' ceases? What is the connection between temperature, heat, and the *sensation* of heat?

Galileo made the point that the sensation of heat didn't exist by itself; it was only there if someone was around to do the sensing: 'there's no tickle at the end of the feather'.[7] But this was a truism. To answer the last question of the previous paragraph first, temperature *is* a quantification of the sensation of hotness, but it is a not a very accurate method, and it only applies in the range of temperatures in which humans can do any sensing. Now it is a curious fact that there is no one physical property that fits this task of mapping out the temperature. This is to be contrasted with the measure of pressure, which is of one kind and has definite units (of force divided by area). Temperature only achieves this status when an absolute scale of temperature is defined—by Kelvin in the nineteenth century (Chapter 16).

Sensitivity of the skin, as we have mentioned, is an obvious choice of thermometric property, but it is not at all objective: for centuries it was thought that ponds and caves were colder in summer than in winter, presumably as fingers were colder in winter than in summer. Volume change is the physical property that most readily presents itself as a suitable measure of temperature—but this still leaves open the question of which substance to use.

Galileo chose the property of expansion, and the substance of air, for his thermometer (strictly speaking, a 'thermoscope'—an instrument giving only a qualitative indication of temperature). A student of Galileo's tells of the following lecture demonstration:

Galileo took a glass vessel about the size of a hen's egg, fitted to a tube the width of a straw and about two spans long; he heated the glass bulb in his hands, and turned the glass upside down so that the tube dipped in water contained in another vessel; as soon as the bulb cooled down the water rose in the tube to the height of a span above the level in the vessel. This instrument he used to investigate degrees of heat and cold.[8]

(In the complete works of Galileo, there is no mention of a device comprising coloured glass 'teardrops' in a cylinder of fluid, now sold as 'Galileo's thermometer'.)

Rey, a French country doctor, was the first to use the expansion of a *liquid* (around 1630). He employed the same glassware as in Galileo's thermoscope, but with the water and the air swapped over. In a letter to Mersenne, he explains:

[my thermoscope is] nothing more than a little round phial with a very long and slender neck. To use it I put it in the sun, and sometimes in the hand of a feverish patient, after first filling all but the neck with water.[9]

(Father Mersenne (1588–1648) was a Minim Friar, and acted as a sort of distribution network for scientific information in the mid-seventeenth century.) Rey's was the first liquid-in-glass thermoscope, but all such instruments open to the air were unreliable (the 'atmosphere' was soon to be discovered—see the next section). The first *sealed* thermometer was invented by the Grand Duke of Tuscany, Ferdinand II, of the famous Medici family. (Galileo had named the moons of Jupiter the Medicean stars, after Ferdinand's father, Cosimo de Medici.)

Ferdinand was one of the founders of the scientific society, the Accademia del Cimento, in Florence in 1657, and his thermometer came to be known as the Florentine thermometer. It was widely distributed to other newly formed scientific societies and royal families across Europe. Boyle first came across one in 1661: 'a small seal'd weather-glass, newly brought by an ingenious traveller from Florence, where it seems some of the eminent virtuosi that enobled that fair city had got the start of us in reducing seal'd glasses into a convenient shape for thermoscopes'.[10] The Duke found that ponds and caves were, after all, colder in winter: and Edmé Mariotte (*c.*1620–84), of the French Royal Academy of Sciences, found that the cellars of the Paris Observatory varied by no more than 0.3 °C (modern units) between winter and summer. (Monitoring the temperature of these cellars became something of an obsession for French scientists right up to the French Revolution; for example, the Royal Astronomer Jean-Domenique Cassini (1748–1845) went up and down the 210 steps almost daily in 1788 and 1789 to take readings.)

Although water and mercury were investigated, 'spirit-of-wine' (alcohol) was the preferred thermometric liquid, as its freezing point was lower. The Florentine academicians found that water had its minimum volume just *above* the freezing point (Hooke flatly refused to believe this).[11] Degrees were marked off with beads of glass or enamel fused on to the stem, but there were, as yet, no fixed points. Even so, the Florentine thermometers were of unsurpassed accuracy for the times (some fifty thermometers found in a trunk in Florence in 1829 were still all in agreement).

Fixed points were needed, and snow, melting butter, blood-heat, boiling water, and, of course, deep cellars were variously suggested. How should degrees be defined? Hooke and Huygens both tried defining a standard volume of substance (at a given fixed point) and then a degree represented a given fractional increase in this volume. The more reliable method of two fixed points was suggested by two Italians, Fabri in 1669 and Bartolo in 1672.

The question still remained: what was the thermometer measuring? In 1680, Eschinardi, another Italian,[12] had the idea of having the heating effect of one standard candle to measure out 'one degree', and the heating effect of two standard candles to measure out 'two degrees', and so on. Here, at last, was a quantitative link between heat and temperature. Carlo Renaldini (1615–98), professor at Padua, and former 'Accademician', had a similar idea. He had two reservoirs of water, one boiling and the other ice-cold, and a one-ounce dipper. A mixture of one ounce of boiling water and 11 ounces of cold water was defined as having a temperature of 'one degree', two ounces of boiling water and ten ounces of cold water as having a temperature of 'two degrees', and so on.

Renaldini's method made the implicit assumption of a *linear* relation between degrees of temperature and quantity of heat. However, there were many results that didn't tie in with this. For example, the Accademia del Cimento found that, at one temperature, the same volumes of different fluids caused different amounts of ice to melt. It wasn't until Joseph Black, 100 years later, that such problems were sorted out, and a real understanding and robust measure of heat began to emerge (Chapter 6).

The Vacuum, the Atmosphere, and the Concept of Pressure

Boyle made the first sealed 'weather glasses' in England and wrote:

At the beginning I had difficulty to bring men to believe there would be rarefaction and condensation of a liquor hermetically seal'd up, because of the school [Scholastic] doctrine touching the impossibility of a vacuum.[13]

The Scholastics were churchmen who rigidly followed the teachings of Aristotle. They didn't believe in the vacuum: 'nature abhors a vacuum', they said. However, the mechanical philosophy required a vacuum between the atoms[14] and, furthermore, avoided emotive agents such as 'abhorrence'.

One bit of phenomenology that needed explaining was the long-known fact that syphons couldn't lift water over hills greater than around 34 feet high, and pumps couldn't raise water to more than this height. Galileo argued that the strength of materials was due to the 'force-of-the-void' binding atoms together, and that the 'breaking strength'[15] of water was equivalent to a column of water 34 feet high. (So, even the great Galileo had a blind spot.[16])

Evangelista Torricelli (1608–47), a brilliant young admirer of Galileo, joined the debate. He argued that it was the *weight* of the air rather than Galileo's vacuum-force that held up the column of water: therefore, for liquids other than water, the height of the column would depend on the liquid's density. Torricelli set about checking up on this supposition, and experimented with seawater, honey, and finally mercury (this was mined in Tuscany). He specially commissioned a glass tube, sealed at one end, and with extra-thick walls. He filled this to the brim with mercury, closed off the open end with his finger, and inverted the tube into an open dish of mercury—the first barometer (for a schematic representation of barometers, see Fig. 4.1). The level of mercury in the tube fell until its height was 29 inches or, in other words, exactly 1/14th of the height of the column of water. As mercury was 14 times as dense, this result supported Torricelli's explanation.

However, it was at least twenty years before Torricelli's explanation was accepted. Some of the objections he had to answer included the following:

(1) If the dish of mercury is covered over with a lid, why isn't the weight of the air cut off?

(2) Weight acts downwards, so how does the weight of the air go round corners and push the column of mercury upwards?

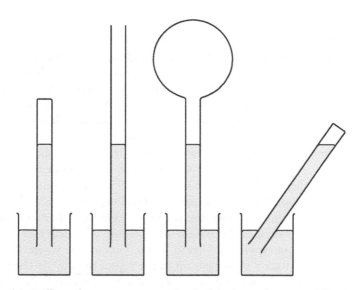

Fig. 4.1 Different barometers, variously tilted (redrawn from Westfall, *The Construction of Modern Science*, Cambridge University Press, 1977).

Torricelli had devised and played with hydrostatic toys—such as the so-called 'Cartesian diver'—and had built up an intuitive understanding of hydrostatic *pressure*. He not only answered all the objections satisfactorily, but went on to realize that the concept of pressure could apply to the air as well as to water. He was therefore the first to appreciate a new idea, captured in his evocative proclamation: 'we live submerged at the bottom of an ocean of air'.[17]

There was still the pertinent question of what was in the space above the mercury in the barometer. (Torricelli, by the way, barely mentioned his barometer again—he had visited Galileo during the latter's house arrest, and presumably wanted to avoid a similar fate.) The Scholastics argued, variously, that the space contained a bubble of distended air, ethereal vapours emitted by the mercury, or ethereal vapours seeping in from the outside through fine pores in the glass—but why did just so much ether go through and no more? Torricelli arranged for some glass tubes to be blown with an extra-large spherical shape at the sealed end. He found that whatever the shape and size of the space, and whatever the inclination of the tube, the mercury column remained at exactly 29 inches. What was more, by inclining the tube the space could be made to disappear altogether. In other words, the amount and shape of the 'ethereal vapours' made no difference—they just weren't there.

Blaise Pascal (1623–62), of Triangle fame, carried out a famous demonstration in the 1660s. He repeated the Torricellian 'balance of weights' experiment but on a grand scale, using two glass tubes approximately 50 feet in length, and supported against the mast of a ship in the harbour at Rouen (the glass industry in Rouen must have been very advanced). He filled one tube with wine, and the other with water—one can only wonder at the practicalities of doing this. The crowd was asked to predict the result: the established view said that the wine, being more spirituous, should stand at a lower height than the water (the spirits pushed the surface down); the 'vacuists' argued that the wine, being less dense, should stand at a taller height than the water. We all know the outcome: the water stood lower than the wine, as Pascal had expected.

Pascal, however, was a logician: perhaps nature *did* abhor a vacuum, but just to the extent of the 'weight' of a column of mercury 29 inches high. After more experiments, such as the famous one where Pascal's brother-in-law took a barometer up to the top of the Puy de Dome in 1648, and the brilliantly ingenious vacuum-in-a-vacuum,[18] Pascal did eventually concede that the vacuum exists. (The vacuum-in-a-vacuum experiment clinched the irrelevance of spirituous ethers.)

Otto von Guericke (1602–86), from Magdeburg in Germany, constructed a tightly sealed spherical container made from two hemispherical halves of strong material (copper), and then used a water pump to pump the air out (he had, in effect, constructed the first air pump). He then demonstrated the 'force' of the vacuum by the number of horses required to separate the halves. (Apparently, two teams of eight horses were required—so either the horses were exceptionally feeble or von Guericke was exaggerating the strength of his vacuum.)

Boyle joined in the debate (about the existence and properties of the vacuum) and in his private laboratory in Oxford he asked his assistant, Hooke, to improve on von Guericke's air pump. Hooke did this, and also made a 'receiver'—a glass bell jar resting on a plate, making a completely sealed chamber around the exit-tube from the pump, and having a special stopper that allowed specimens and apparatus to be introduced (without destroying the vacuum). In the 1650s, Boyle and Hooke carried out many such 'pneumatics' experiments, with the receiver variously evacuated, pumped with extra air, or with equal air inside and outside.

From experiments such as introducing a barometer, or a carp's bladder, into the receiver, Boyle reaffirmed that air had weight, and that the vacuum exists. He also found out something new—the air had 'spring';[19] it tried to expand into any gaps, and pushed against the walls of a container. For example, the carp's bladder slowly expanded when placed in a vacuum, presumably because of the expansion of residual air left within the bladder. Boyle concluded that air was an 'elastic fluid', which always expanded to fill any container.

The Scholastics, in the person of the Jesuit Father Linus, thought they now had Boyle cornered: if air was supposedly 'elastic' and 'expansible', it should also be 'compressible'—plainly absurd, thought Linus. Boyle countered by carrying out the experiments for which he is known to posterity. A glass tube was curved into the shape of the letter 'J', with the shorter end being sealed. Some air was trapped in this shorter end by a column of mercury poured into the longer end. By increasing the amount of mercury, the 'spring' (we now say 'pressure') on the trapped air could be increased and the volume change noted. The air was indeed compressed; and Boyle determined that the volume of the air for successive increments in the column of mercury 'supported the hypothesis that supposes the pressures and expansions to be in a reciprocal proportion'[20] (Boyle's Law). (In France, this is known as Mariotte's Law, as Mariotte carried out similar experiments—but after Boyle.)

Thus, the vacuum had arrived, as had the atmosphere and the pressure of an elastic fluid. Pressure would turn out to be a concept almost—but not quite—as intimately linked to heat, and to energy, as was temperature.

Pneumatics and Further Studies on Heat

The air pump ushered in a host of 'pneumatic' experiments. The vacuum was found not to support life. Philosophers and invited guests watched with morbid fascination as various animals (a bird, a mouse, an eel) were put in the receiver and seen to die as the air was pumped out.[21] Also, the flame of a candle went out in the vacuum and the sound of a bell became very faint, but magnetic and electrical effects persisted. Von Guericke, Boyle, and Johann Bernoulli noted an occasional luminous glow in an evacuated receiver, but could not account for it. Heat was understood to be required for water to boil, but 'ebullition' was observed in a dish of merely tepid water if the air was pumped out.

Light was transmitted through the vacuum, enabling all these experiments to be witnessed (according to the Scholastics, if a vacuum could be made to exist at all, then it should be opaque). A 'burning glass' (lens) outside the receiver could be used to focus light on to something insider the receiver, and perhaps set that something alight. In a celebrated demonstration, Mariotte caused gunpowder to explode using an 'ice lens'. In this and other experiments, it was shown that 'heat-rays' could be focused, gunpowder needed no air in order to burn, and 'heat-rays' as well as light could be transmitted through the vacuum. (It was hard to explain this last result with the motion theory of heat—in a vacuum, there were no 'small parts' to be set moving.)

Mariotte, in his 1679 'Essay du chaud et du froid',[22] describes other experiments showing the properties of radiant heat (as we would now call it). Dark colours and rough surfaces absorbed heat more readily, while light-coloured shiny surfaces reflected the heat. Mariotte put a concave metal mirror near a fire, and found that he could not long endure the heat when his hand was at the focus. However, putting a glass plate between the fire and the mirror, then the *heat* could no longer be felt but the *light* of the fire was undiminished. This was the first time that the light and the heat from a given source had been separated. However, when the Sun was the source, then the light and heat could *not* be separated.

We can begin to appreciate the enormous difficulties in identifying and separating out the various phenomena all associated with the one concept, heat. Before we moderns become too complacent—we do, after all, know how the story will turn out—consider the puzzling case of the reflection of cold. In the *Saggi di naturali esperienze* of the Accademia del Cimento, there is a description of an

> experiment whether a concave mirror exposed to a mass of 500 lbs of ice would make any perceptible reverberations of cold in a very sensitive 400-degree thermometer located at its focus. In truth this at once began to descend, but because the ice was nearby it remained doubtful whether direct cold, or that which was reflected, cooled it most. The latter was taken away by covering the mirror, and (whatever the cause may have been) it is certain that the spirit immediately began to rise again.[23]

With the hindsight of the Second Law of Thermodynamics (Chapter 16), we can argue that the reflection of cold is impossible—well, not *absolutely* impossible; just extremely unlikely (Chapters 17 and 18).

Edmond Halley (1656–1742) understood that the heating effect of the Sun's rays was proportional to the sine of the angle of incidence. He further understood that this led to a differential heating of the Earth's surface, which then led to the trade winds. (Lister, on the other hand, thought that the trade winds were due to the diurnal respiration of the Sargasso seaweed.) While Halley explained the winds as being due to hot airs being 'rarefied'[24] and therefore rising, he could not understand the process of evaporation: how could a heavy substance like water rise through a lighter substance like air? (Halley suggested that the 'water-atoms' became distended and filled with a 'finer air'). Halley, Mariotte, and Perrault all made estimates of the total rainfall, the rate of flow of rivers, and the evaporation to be expected from lakes and seas. In this way, it began to be appreciated that there was a closed cycle of water—and so occult processes, involving subterranean caverns, or eternal springs, were no longer needed.

While the vacuum did not support life or flame, increasing the air pressure within the receiver appeared to make animals more vigorous, plants grow more strongly, and burning coals glow more brightly. Rey found that gently heating lead and tin in air led to an increase in their weight: he proposed that some of the air had 'condensed'. Mersenne objected that, on the contrary, air was 'rarefied' upon heating.

Another doctor, John Mayow (1640–79), performed experiments that brought out common aspects of respiration and combustion: when a mouse in an evacuated receiver died, then, also, a candle could no longer be ignited. However, the idea that some component of the air was being used up did not readily suggest itself to contemporary investigators. For one thing, the reduction in the volume of air was very small: when Mayow placed a bell jar over water, then a burning candle caused the water level to rise by only 1/14th. To Mayow, this small reduction indicated a loss in the 'elasticity' of the air. (In modern terms, about as much CO_2 is put out as O_2 is taken up, so the initial change in gas volume is slight. However, the CO_2 is more soluble, and it slowly dissolves in the water, leading to an eventual reduction in the volume of gas.)

Mayow (and also Hooke) postulated that both respiration and burning required a nitrous spirit, and that this could be found in the air. This 'nitro-aerial' spirit was supposedly present in a concentrated form in gunpowder, and explained why the latter could burn without air. In fact, the nitro-aerial spirit was thought to be responsible in a more general sense for animal heat, animal motion, and plant growth: a 'vital or life force', not to be confused—yet!—with Leibniz's 'live force'.

Heat-engines

The study of pneumatics and heat opened the way for a new technological advance—the heat-engine. The main incentive for the new 'fire-engine' (as it was usually called) was the need to pump water out of increasingly deep mines.

The Marquis of Worcester (we have heard of him being saved from the Tower of London so that he could perfect his perpetually rotating wheel; Chapter 2) invented his 'water-commanding engine' around 1663 (see Fig. 4.2). Water was heated in a boiler and the steam collected in a chamber. When the heat supply was withdrawn, the steam condensed and water was sucked up from below. This water was then forced up a vertical pipe by the pressure of the next round of steam. The engine was erected at Vauxhall in London, and raised water to a height of 40 feet. The Marquis obtained the rights to 99 years' use by Act of Parliament, but the engine was never used commercially (or at all?).

Of more success was the engine of Thomas Savery (1650–1715), in 1698. This engine worked on the same principal as the 'water-commanding' engine except that externally applied cold water was used to speed up the condensation of the steam. A partial vacuum was created and then atmospheric pressure forced water into the engine chamber, and fresh steam drove this water up the up-pipe. In this way, mines could be 'drayned'—or so it was claimed (see Fig. 4.3).

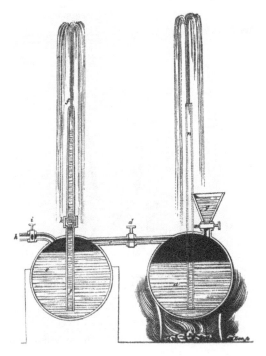

Fig. 4.2 The Marquis of Worcester's 'water-commanding engine', described in his book *The Century of Inventions* (1655), as depicted in Robert Stuart's *A Descriptive History of the Steam Engine* (1824).

Steam pressure was used directly by the French inventor Denis Papin (1647–1712) in his 'digester'—basically, a pressure cooker. He describes how beef bones could be softened by cooking in this manner:

I took beef bones that had never been boiled but kept dry a long time, and of the hardest part of the leg; . . . Having pressed the fire until the drop of water would dry away in 3 seconds and ten pressures, I took off the fire, and the vessels being cooled, I found very good jelly in both my pots . . . having seasoned it with sugar and juice of lemon, I did eat it with as much pleasure, and found it as stomachical, as if it had been jelly of hartshorn.[25]

An important development was the transfer of the piston-in-a-cylinder technology of the air pump into the heat-engine. This happened via the following route. First, von Guericke suggested the transmission of power by connecting a piston in a long narrow cylinder to one in a short wide

Fig. 4.3 Thomas Savery's Steam Engine, 1698 (from Stuart, as in Fig. 4.2).

cylinder (the same principle of operation as used in today's air brakes). Second, Huygens imagined a piston in a cylindrical chamber for his hypothetical gunpowder-engine. Third, Papin conceived a method of transmitting power over long distances, by using compressed air travelling through a tube (something like the system of canisters in pipes employed in the department stores of 50 years ago). Finally, in 1690, Papin proposed an engine based on Savery's design, but with steam condensing in a cylinder and drawing down a piston. (He was by this time a Huguenot exile in London, employed as curator of experiments at the Royal Society, at the invitation of Huygens.) Papin was thus the very first proposer of the cylinder steam engine. It is probable that Newcomen, of the famous Newcomen engine, had never heard of Papin's proposal when he came to invent his own cylinder steam engine in 1712 (Chapter 5).

In the very last year of the seventeenth century, Amontons put forward a completely different sort of engine. Guillaume Amontons (1663–1705), son of a lawyer from Paris, had a childhood illness that left him almost completely deaf. He devoted himself to mathematics and physical science, despite some opposition from his family. Perhaps his deafness was the inspiration behind his idea of a system of hilltop signal-ling using spyglasses and separate signals for each letter of the alphabet. It's not certain whether this was ever carried out (there was reputedly a demonstration in front of Louis XIV), but it would have been a first for telegraphy. A definite first was Amontons' invention of a heat-engine, using the expansion of air to provide the motive force.

Amontons had already been impressed by the large expansivity of air, and had exploited this in his air-thermometer. This was like a barometer except that it was sealed at *both* ends. A pocket of air at one end was warmed, and its temperature increased: its pressure also increased, and this raised a column of mercury. It wasn't possible to ensure that the volume of the air remained exactly constant, but Amontons nevertheless managed to establish a rough proportionality between temperature and pressure, for air between the freezing and boiling points of water (this later became known as Charles' Law). Amontons also made careful in-vestigations of the amount, and rate, of expansion of air on heating, and surmised that if the pressure could be kept constant, the volume would increase in proportion with the temperature.

These investigations sowed the seed of an idea in Amontons: if ex-panding air could move a column of mercury, could it not also be made to raise water for a machine? This led him on to a new enterprise, a rad-ical departure—his proposal for a 'fire-engine', or 'fire-mill', as he called it. The design consisted of a wheel with a hollow rim and hollow spokes, partitioned into cells with interconnecting valves. The wheel was to be filled with water on one side and air on the other side. A fire would be lit on the air side, and then the heated air would expand and force the water through the cells. This would cause the wheel to overbalance and then start to rotate. (The design possibly drew from Amontons' adoles-cent interest in perpetual motion machines.)

Amontons made a detailed plan (Fig. 4.4) and a small-scale model, but didn't manage to fulfil his grand ambition of constructing a fire-mill of some 30 feet in diameter.

Even so, Amontons went far further than any of his contemporaries: he considered the performance criteria of a generalized heat-engine. He commented on the tremendous power of heat, such as in the 'violent action' of the cannon:

Nobody doubts that fire gives rise to violent action . . . But could it not also be used . . . to move machinery where one usually employs limited animate forces, such as men or horses?[26]

Fig. 4.4 Amontons' 'fire mill', 1699 (from Stuart, as in Fig. 4.2).

More than this, Amontons made an attempt at quantification. From his experiments on the heating and cooling rates of air and water, he estimated that his wheel would rotate once in 36 seconds, and that it would exert a force sufficient to replace 39 horses. By observing glass-polishers and their rate of working, Amontons estimated that the power of a man is 'about a sixth part of the labour of a horse';[27] and, by 'labour of a horse', he meant the 'force exerted by the horse, multiplied by the speed of the horse at its point of application'.[28] This is essentially the same as our modern definition of power, or the *rate of doing work*.

Amontons also estimated what would later be called the 'duty' of the engine—the work done for a given amount of fuel—and claimed that his fire-mill would compare favourably with the pre-existing sources of power; namely, wind, water, animal, and man.

Finally, Amontons tackled for the first time the problem of losses in duty because of friction. He carried out experiments and found that:

(1) Friction did not depend on the area of the surfaces in contact, but only on the weight on those surfaces.

(2) For lubricated surfaces, friction was roughly independent of the materials.

(3) For frictional losses in a machine with rotating parts, the important thing to calculate was the 'couple' (weight times distance) of the frictional forces at the bearings. He estimated that in hauling up a weight over a pulley, the power lost would be around 1/30th of the total power. For his own fire-mill, Amontons estimated that frictional losses would be of the order of 1/8th.

These thoughtful experiments and insights were prescient, and surely earn Amontons the title of 'grandfather of thermodynamics'.

Overview

The change in outlook in European science in the seventeenth century is sometimes referred to as the Scientific Revolution. We have already seen (Chapter 3) the birth of mechanics, culminating in Newton's great work, the *Principia*. The air pump, telescope, and microscope were invented during this century, and all opened up new worlds. Also, the new clock, thermometer, and barometer helped to chart the parameters of this revolution. In particular, the last two instruments introduced the new 'pneumatic variables' of pressure, P, volume, V, and temperature, T, which described the properties of an 'elastic fluid'. The relations $P \propto 1/V$ (Boyle's or Mariotte's Law), $P \propto T$ (Amontons, and later Gay-Lussac), and $V \propto T$ (Amontons, and later Dalton and Gay-Lussac) were found to hold, more or less.*

The phenomenology of heat had increased enormously. It was understood to have a role in expansion, evaporation, melting, and boiling; and was an agent in chemical decomposition, as in combustion, fermentation, and putrefaction. Heat also had links with light and with life. In fact, 'heat' had so many facets that it was impossible to embrace or even recognize all the phenomena within one concept, and one body of theory. But all were agreed that heat was the 'motion of the small parts'. Nevertheless, this consensus would still leave room for two competing theories to emerge in the eighteenth century, a competition that would influence, and even delay, the understanding of energy.

---

* The symbol '$\propto$' means 'proportional to'.

While the motion of the small parts was taken for granted, it was still not agreed what the fundamental measure of this motion was—*mv* or *mv*²? (This controversy is covered in Chapter 7, Part I.) Also, the question of whether such macroscopic measures could apply to microscopic constituents was never asked (except by Newton: in his *Opticks*, Query 31, he speculated that his Laws of Motion might be applicable to microscopic effects).

In the very last year of the century, there were three designs of heat-engine—using atmospheric pressure, steam pressure, and the expansion of air. While none was in the least bit practicable (except perhaps for Papin's digester), there was an understanding by a few visionaries that engines were the way of the future. Amontons, in particular, envisaged a generalized heat-engine that could do 'work' and had given fuel requirements and a given efficiency. No one, of course, had any idea of the huge role that the heat-engine would have in the understanding of energy.

5

Heat in the Eighteenth Century

Hot Air

In the eighteenth century, heat and gas studies were inextricably bound together. The two main theories of heat hinged on different conceptions of what a gas was: the dynamical theory saw heat as a process—the motion of the constituent particles—and its first quantitative working out was in the gaseous case; the substance theory had heat as a thing, a subtle 'elastic fluid' or ether—in other words, a gas of sorts.

The phenomenological links between heat and gas were becoming more and more evident: combustion, and also 'calcination' (the gentle heating of metals), were known to occur only when certain 'airs' were present; various 'airs' were released on heating, fermentation, and rotting; respiration was necessary to life and was somehow connected with animal heat; and heat was connected to the processes of evaporation and boiling. Perhaps the most telling feature linking heat and gas was the enormous expansivity of water as it turned into steam, and of air when heated—both exploited in various sorts of heat-engine. Lastly, the fact that 'fire' and 'air', two of Aristotle's four elements, shared this property of 'elasticity' was thought to be highly significant by most philosophers of the day.

Amontons (Chapter 4) had shown that air increases its volume, on heating, by around one third between 0 °C and 100 °C (modern units) and, therefore, assuming linearity, by around 1/300th per degree. It seemed as if, at last, the perfect thermometric substance—better even than mercury—had been found. (We now understand the reason why gas has this special link with heat: the forces between gas molecules are very small (at usual densities) and so we see the unadulterated effects of heat—naked heat, you might say.)

Our modern view of a gas is roughly as follows: a bucket of gas at room temperature and pressure contains around 10^{23} molecules, weighing around 10^{-26} kg each, moving in random directions, with speeds of a few thousand kilometres per hour. The eighteenth-century conception of a gas was utterly different. The particles of the 'aeriform fluid', as it was often called, were essentially *static*. Each particle occupied a definite location, and had a tiny jiggling motion about this mean position. (I call this the 'passion fruit pip jelly' model.) But then what could account for the 'spring' of the air?

Boyle had shown (Chapter 4) that the air had 'spring'—it expanded to fill its container. He said:

[I conceive] the air near the earth to be such a heap of little bodies . . . as may be resembled to a fleece of wool. For this . . . consists of many slender and flexible hairs; each of which may indeed, like a little spring, be easily bent or rolled up, . . . still endeavouring to stretch itself out again.[1]

But Newton realized that Boyle's springy particles would only account for the air's resistance to compression, not for its tendency to expand without limit. Now Newton had conjectured that all physical phenomena in the world were due to just two kinds of forces—forces of attraction and repulsion (Chapter 3). Perhaps the expansive property of air was due to a repulsive force acting between air particles and (Newton calculated) if this repulsive force acted only between nearest neighbours, and varied in inverse proportion to the distance separating these neighbours, then the air (elastic fluid) would satisfy Boyle's Law. However, Newton was careful to add that he didn't know if the air really *did* behave like this:

But whether elastic fluids do really consist of particles so repelling each other, is a physical question . . .[2]

Newton also toyed with the idea of ethers;[3] for example, an ether that could transmit gravity (thus eliminating the need for action-at-a-distance). Such was the authority of Newton that, despite his circumspection, his followers embraced his theories and forgot their speculative origins: and so Newton's model for a gas, and his 'aether', gave legitimacy to a subtle-fluid theory for heat. The expansive property of heat could then be explained by the repulsive force acting between neighbouring 'fire particles'. Newton, all the while, still preferred a motion theory of heat.

Newton's insights often sound so crisp and modern (especially compared with Boyle's prolixity), such as when he touches on black bodies in the *Opticks*. He suggests that they heat up more than bodies of lighter colour because the incident light is 'stifled and lost'[4] within them. His most famous contribution to heat studies was 'Newton's Law of Cooling'[5]—even though he published it anonymously (he had a reputation to uphold, and was by nature guarded and secretive[6]). The law stated that the rate of cooling of a body was proportional to its excess temperature over the surroundings. (Newton's purpose was to use this law to extend the temperature scale above the maximum for mercury-in-glass thermometers: the times taken for various molten metals to cool down and solidify gave fixed points on a line, and this line could be extended to higher temperatures.)

In the year of Newton's death (1727), Hales published his *Vegetable Staticks*;[7] he was the first investigator to give priority to the study of gases. (One might say that in going from Thales to Hales, one goes from 'water is all' to 'gases are all'.) Hales always used the term 'air', or 'airs', even though Van Helmont, in the sixteenth century, had already coined the word 'gas', taking it from the Greek root, *kaos*.[8]

Stephen Hales (1677–1761) was a physician and clergyman from Teddington, Middlesex. The poet Alexander Pope, from neighbouring Twickenham, makes mention of 'plain parson Hales'.[9]

Hales heated many substances, including hog's blood, amber, oyster shells, beeswax, wheat, tobacco, gallstones, and urinary calculi, and collected the yields of 'air'. He invented the 'pneumatic trough' for this purpose (Fig. 5.1). The trough used water, and this resulted in some soluble gases being lost: however, Hales had no idea that there were *different* gases, with different properties, and he saw the purpose of the water merely as a method of washing the air clean of impurities. (Later in the century, Cavendish and Priestley improved on the trough, by collecting gases over mercury.)

Hales made the startling discovery that prodigious quantities of gas were released when things were heated. For example, an apple yielded 48 times its volume of 'air'. Hales commented that if the air had to be forced back inside, presumably requiring a pressure of some 48 atmospheres, then the apple would explode. He concluded that there was a repulsive force in 'elastick air', but this was converted into an attractive force when the gas became 'fixed' within the solid—air was 'amphibious',[10] he said.

In the early eighteenth century, air was not considered elemental, and was thought to be variable in quality, like soil. So it is not surprising that Hales made no attempt to try to identify the different airs. (Water, on the other hand, *was* taken to be elemental.)

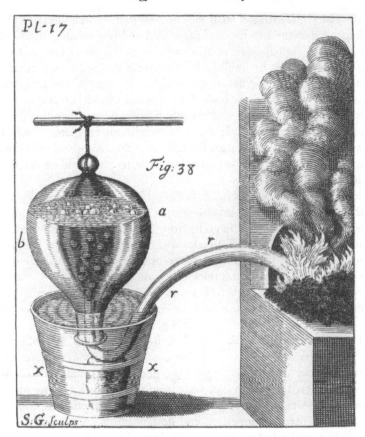

Fig. 5.1 Hales' pneumatic trough, adapted from *Vegetable Staticks*, 1727.

The Phlogiston Theory

Combustion was another arena where heat and gas studies overlapped, and where the concept of an ether was again invoked. It was noticed by Boyle, Hooke, Mayow, and others that combustion could only occur in air (apart from special cases such as the combustion of gunpowder; see Chapter 4). It was also noticed that the volume of the air was slightly reduced after combustion had taken place. It may seem obvious to us now that a component of the air is being used up, but this was not the natural inference to make in the eighteenth century (for one thing, oxygen had still to be discovered). The contemporary explanation was a physical rather than a chemical one—that the 'elasticity' of the air had lessened.

It was nevertheless thought that a vital ingredient was required for combustion (Mayow called this the 'nitro-aerial' spirit): this extra, subtle, ingredient was supposedly present in air (and also in gunpowder), and was responsible for the air's elasticity. An alternative view developed in Germany, whereby the combustible essence was taken to be in the burning body rather than in the air. In fact, as we shall see, everything about this alternative combustion theory was an inversion of the former approach.

First Becher, in 1667, put forward his theory that combustible bodies contain a substance ('oily earth') that gets used up and released during combustion, leaving a stony 'vitreous earth' residue. Then Stahl, in 1703, revived Becher's theory, and called this combustible essence 'phlogiston', after the Greek prefix *phlog*, or *phlox*, for flame. Charcoal was thought to be an almost pure embodiment of phlogiston. Part of the evidence was that when charcoal burns, only a few ashes are left behind—the leftovers after the phlogiston had been released.

When a metal is heated in air, a 'calx' is formed (what we now know to be the metal oxide), and when the calx is heated in the presence of charcoal, the phlogiston from the charcoal is absorbed by the calx and combined with the metal. The phlogistonists saw the fact that the metal was shiny, lustrous, and malleable as evidence that it contained more phlogiston ('fiery essence') than the dry, brittle calx. One problem presented itself straight away: the calxes weighed *more* than their respective metals (Rey and Boyle had noticed this), even though the metals were now combined *with* phlogiston. The solution?—the phlogiston was given the property of having *negative* weight, or 'levity'.

The Substance Theory of Heat

One person who did not concur with the phlogiston theory was Hermann Boerhaave (1688–1738), of Leyden in Holland—but he was the initiator of yet another subtle-fluid theory, the substance theory of heat. Boerhaave's book, *Elementiae chemicae*[11] (1735), was translated into French, German, and English, and was widely distributed and read. In it, Boerhaave established the morphology of heat studies for the next 30 years.

Boerhaave was a remarkable and influential person in many other ways as well. In medicine, he was the founder of a great teaching tradition at Leyden, and was the first to instigate a systematic approach to clinical practice (he introduced the use of the Hippocratic Oath, the post mortem, and Fahrenheit's thermometers). His capacity for hard work was phenomenal. He held three professorships simultaneously at the University of Leyden—in medicine, chemistry, and botany—and showed almost fanatical zeal in his experiments on mercury, undertaken to show that the transmutation of mercury into gold was not possible. (The mercury was purified by straining it through leather, washing in seawater, and then being shaken in a fulling mill for 8½ months. In another experiment, a gold–mercury amalgam was distilled 877 times, and in yet another, some mercury was heated continuously for 15½ years—no changes in its weight or purity were detected.)

In the *Elementiae chemicae*, Boerhaave made no reference to his contemporary, Stahl, or to Stahl's phlogiston theory, and he utterly rejected Stahl's ideology of animism. On the other hand, Boerhaave very much admired the English philosophers, Boyle and Newton, and their adherence to the mechanical philosophy. But Boerhaave still didn't go along with their motion theories of heat. In this regard, he looked even further back, to Descartes' 'element of fire', and can therefore be regarded as the founder of the substance theory, or subtle-fluid theory, of heat.

Boerhaave attributed heat—or 'fire' as he called it—with truly cosmic significance:

Fire is the universal cause of all changes in nature . . . Thus, were a man entirely destitute of heat, he would immediately freeze into a statue.[12]

The 'fire particles' were too minute to be visible, and were weightless. Boerhaave checked up on this by weighing hot iron continuously as it cooled, and as it was re-heated: there was no detectable change in weight.

For Boerhaave, it was axiomatic that 'fire' cannot be created or destroyed:

The quantity of fire in the universe is fixed and immutable.

He therefore did not agree with Boyle that 'fire' could be created anew by percussion, friction, or attrition: 'Fire . . . is a body *sui generis* not creatable or producible *de novo*'. But Boerhaave could not deny the common experience of heating in such cases, and he gave the rather lame (or, at least, odd) explanation that attrition is very swift, and only fire—the swiftest thing in nature—can keep up.

A controversial aspect of Boerhaave's views was his 'distribution theory' of fire—that fire is to be found equally everywhere:

Pure or elementary fire is equally present in all places; nor is there any point of space, or any body, wherein there is more found than in any other.

Even the coldest regions would therefore contain fire:

For if, even in Iceland, in the middle of winter, and at midnight, a steel and flint be but struck against each other . . . a spark of fire will fly off.

He agreed that an absolute zero of temperature would imply a total absence of heat, but thought that this could never happen in practice. Another thing that Boerhaave deemed to be against nature was that different substances should attract heat to differing extents:

Among all the Bodies of the universe that have hitherto been discovered and examined, there never was yet found any one that had spontaneously, and of its own nature, a greater degree of Heat than any other.

And this gives us occasion to adore the infinite wisdom of the great creator. For had there been anything that attracted fire, it would have become a perpetual furnace: and how unhappy must the condition of man have been under such circumstances?

Higher and lower temperatures could still occur in forced cases (where things are especially heated or cooled, for example, by using a burner, or dousing with cold water), and they implied, respectively, higher or lower densities of the heat-fluid.

Boerhaave understood the difficulty of defining the quantity of heat and noted that 'although we are able to discuss the Force of Fire by its sensible effects; yet we cannot from its Force make certain judgements of its quantity'. Nevertheless, he did attempt to measure these 'sensible' effects, the chief of which was the expansion or contraction of the given thermometric substance. He asked Fahrenheit to perform the experiments.

Daniel Fahrenheit (1686–1736), famous for his temperature scale, did not have a very auspicious start to his career. His parents died of mushroom poisoning, and his legal guardians arranged for him to train as a book-keeper; but Daniel wasn't interested, stole some money, and ended up with a warrant being issued for his arrest. His guardians attempted to pack him off to the East Indies, but Daniel managed to evade both arrest and the East Indies, and subsequently led a peripatetic existence between Germany, Sweden, and Denmark, making thermometers all the while. He eventually settled in Amsterdam (around 1717), the warrant having been cancelled by this time.

Fahrenheit carried out various experiments for Boerhaave, using—on some occasions—the drying rooms of the local sugar refinery as a source of heat. For example, a dog, cat, and sparrow were observed as they died of heat exhaustion. On a less gruesome note, Fahrenheit examined the 'method of mixtures', which had been initiated by Renaldini in Padua (see Chapter 4), and independently investigated by Brook Taylor (1685–1731), in London. Taylor found that the temperature of a mixture of x volumes of ice-water and y volumes of boiling water was the same as the arithmetic mean temperature: $\{(x \text{ vols}) T_0 + (y \text{ vols}) T_{100}\}/(x + y)$vols. (Brook Taylor was a mathematician and gentleman-scientist of private means. He is most famous as the inventor of the 'Taylor series' in mathematics, an idea that was sparked by discussions in a London coffee house.[13])

Fahrenheit tried mixing mercury and water together but the final temperature was nowhere near the mean; in fact, it was always closer to the water temperature. By trial and error, Fahrenheit eventually found out what proportions, by volume, were required in order to make the final temperature equal to the mean. He says in a letter to Boerhaave:

I changed the proportions, taking instead [of] equal volumes, three volumes of quicksilver and two volumes of water; and then I obtained roughly the mean temperature . . . A most remarkable aspect of this experiment is that although the quantity of matter in the quicksilver in relation to the water is about 20 to 1, its effect is no greater than if water were mixed in the proportion of 1 to 1. I leave the explanation of this to you, Sir, and would only add that it may be possible to deduce here from the reason why quicksilver thermometers are more sensitive than alcohol thermometers.[14]

In other words, despite the fact that the quicksilver was some 14 times denser (implying around 20 times the mass), a temperature halfway between the two initial temperatures was only obtained when water and mercury were mixed in the volume ratio of 2 to 3.

Boerhaave conceded that 'the matter of fire' was clearly not distributed in proportion to density, although this had been his first guess. However he took the volume ratio of 3 to 2 as sufficiently close to a volume ratio of 1 to 1 that he was prepared to conclude that 'fire' is distributed in direct proportion to volume—his 'distribution theory of heat'.

George Martine (1700–41), a young Scottish physician from St Andrews, was not convinced by Boerhaave's conclusion. He knew Boerhaave well, as he had obtained his doctoral degree from the University of

Leyden. However, Martine's approach was less influenced by metaphysical speculations and, for example, he always used the pragmatic term 'heat' rather than the emotive term 'fire'. Back in St Andrews, between 1738 and 1739 (the last few years of his short life), Martine carried out experiments on rates of heating and cooling, and found that:

contrary to all our fine theory, Quicksilver, the most dense ordinary fluid in the world, excepting only melted Gold, is, however, the most ticklish next to Air; it heats and cools sooner than Water, Oil, or even rectified Spirit of Wine itself.[15]

While Boerhaave's controversial 'distribution theory of heat' was not adopted by many investigators, Boerhaave's overall conception of heat as a subtle fluid *was* taken up, and was the dominant view by the middle of the eighteenth century. These subtle fluids were proliferating—there was one for combustion (phlogiston) and others for electricity, magnetism, and gravity. All were subtle (invisible), and some were imponderable (weightless) and elastic (self-repelling). Eventually, the subtlety, levity, and imponderability of these elastic fluids became too much for the scientific community to ponder. However, 'eventually' was a long time in coming (we shall return to the cases of caloric and phlogiston later) and, in the case of electricity, it could be argued that the subtle-fluid theory never went away.

The Kinetic Theory

We now turn our attention back to the mechanical, or motion, theory of heat. The first attempt to describe mathematically the connection between heat and the 'motion of the small parts' was made by Jakob Hermannn (1678–1733), from Basle in Switzerland, in his book *Phoromania*,[16] published in 1716. (The word '*phoromania*' comes from the Greek root *phoros* meaning 'motion'. It seems to have been a fashion in the eighteenth century to use words from antiquity.)

Hermann considered the case of elastic fluids (gases), and put forward the following prescient proposition:

Other things being equal, heat is proportional both to the density of a hot body, and to the square of the agitation of its particles.[17]

'Other things being equal', apparently, meant that this applied to all 'bodies of a similar texture'. Hermann also went on to explain that 'The agitation is the *mean* of the individual speeds' (my italics). This insight (regarding 'mean', or 'average', speeds) was not taken up again until Maxwell some 150 years later (Chapter 17).

The next investigation into the motion theory of heat, as applied to a gas, was by Daniel Bernoulli, son of Johann. Daniel published his great work, *Hydrodynamica*[18] (see Chapter 7), in 1738 in Strasbourg, although the work itself was carried out while Daniel was at the Russian Academy of Sciences at St Petersburg—he went there at the invitation of Peter the Great. Hermann also spent some years in St Petersburg, and Daniel and Jakob Hermann would certainly have had many useful discussions together. This was in contrast to the help that Daniel could expect from his father: Johann Bernoulli not only went out of his way to block his son's chances of getting a professorship in mathematics at Basle, but published his own work, *Hydraulica*, and predated it so that it looked as if the work was completed before Daniel's masterpiece (Chapter 7).

Despite Johann's efforts, *Hydrodynamica* sealed Daniel Bernoulli's reputation across the whole of Europe. (It includes the effect for which Daniel Bernoulli is famous today—the different rates of fluid flow for different routes, and which can result in aerodynamic 'lift'.)

In Chapter X of *Hydrodynamica*, Daniel gives his kinetic theory of gases. He considers the 'elastic fluid' (air) to be contained within a cylinder with a weighted piston at the top (Fig. 5.2). The fluid contains 'very minute corpuscles, which are driven hither and thither with a very rapid motion'.[19] (The expression 'hither and thither' is the only hint that the speeds and directions are *randomly* distributed.) The pressure of the air is due to the impacts made by these corpuscles with the walls and piston of the cylinder (this explanation of pressure is generic to all subsequent kinetic theories of gases).

Now, when the volume of the air is reduced by pushing down on the piston, 'it appears that a greater effort [pressure] is made by the fluid for two reasons: first, because the number of particles is now greater in the ratio of the space in which they are contained, and second, because each particle repeats its impacts more often'. According to Bernoulli, if the volume, V, is reduced to V/s, then the number of particles 'contiguous' to the piston will be increased by $s^{2/3}$, and the mean separation of the particles will be decreased by $s^{1/3}$. This will result in a pressure increase from P to Ps. Thus the product of the new pressure, Ps, and the new volume, V/s, will equal PV. In other words, the product of pressure and volume will remain unaltered. This is Boyle's, or Mariotte's, Law, a law

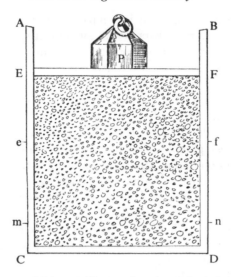

Fig. 5.2 Daniel Bernoulli's kinetic theory of an elastic fluid, from *Hydrodynamica*, 1738 (by permission of Dover Publications, Inc.).

that is confirmed by 'manifold experience'. Bernoulli is admirably cautious and adds 'whether it [Boyle's Law] holds for considerably denser air I have not sufficiently investigated'. Also impressive, Bernoulli's attention to important experimental detail is impeccable, and he notes that 'the temperature of the air while it is being compressed must be carefully kept constant'.

He goes on to note that heating the air will increase the internal speeds, v, of the particles, and will lead to an increase in the pressure, P, on the piston: 'it is not difficult to see that the weight P should be in the duplicate ratio of this velocity [$P \propto v^2$] because, when the velocity increases, not only the number of impacts, but also the intensity of each of them increases equally'. In other words, Amontons' experimental finding—that pressure is proportional to temperature—is given a theoretical underpinning.

Daniel Bernoulli's novel, and essentially correct, description of a gas was almost totally ignored for over a hundred years. (I say 'almost' because his work *was* picked up by the Swiss school of natural philosophy, and was an important influence on Prévost—see Chapter 6.) Nowadays, Bernoulli's description of a gas as particles in rapid random motion feels completely intuitive, especially in comparison with a static 'passion fruit pip jelly'—but intuition changes with the centuries.

The modern kinetic theory incorporates some aspects that were totally foreign to the eighteenth-century mind.

First, Bernoulli's analysis did not consider a spread of particle speeds (at a given temperature). Even though Bernoulli was one of the founders of probability theory, and he advised Catherine the Great on smallpox inoculation, the idea of a *distribution* of speeds was too advanced. (One can, nevertheless, glimpse the concept of randomized motions in Bernoulli's description of the corpuscles as moving 'hither and thither'.)

Second, the passage of sound was actually *easier* to understand in a jelly, whereas in the kinetic theory one needs to understand the tricky (statistical) concept of 'mean free path', and that this path must be much less than the wavelength of sound. (Curiously, the erroneous substance theory of heat, in the hands of Laplace, was able to account for sound quantitatively, and correctly.)

Third, the kinetic theory found it harder to explain the transmission of heat from the Sun to the Earth across vast tracts of empty space. (The substance theory also struggled with this, and Boerhaave admits that 'it appears very extraordinary that the sun, after a continual emission of corpuscles of fire upwards of 5,000 years, should not yet be exhausted'.[20])

Finally, and most importantly, a conservation principle for heat did not follow automatically from the kinetic theory, but seemed quite natural in the substance theory (if the 'substance of heat' was conserved, then heat would be conserved). True, the heat-fluid was 'subtle' (invisible) and 'imponderable' (weightless), but then the particles of a gas were also subtle and individually weightless as far as the technological capabilities of this era were concerned.

That the wrong theory should prevail is not so unusual, but what is remarkable is that there were so few tell-tale clashes with phenomenology: after all, a gas does *not* approximate to a 'passion fruit pip jelly'. While pausing (until Chapters 8–12) to consider what these clashes may have been, it is interesting to observe that there have been other famous theories that were 'wrong' and yet were not in great conflict with experiment: the Ptolemaic system for the motion of the planets (superseded by the Copernican system); and Newton's Law of Universal Gravitation (superseded by Einstein's Theory of Gravitation).

One thing that Daniel Bernoulli did achieve, albeit without due recognition, was the introduction of the conceptual model that would dominate heat and gas studies right up to the present day. This was the scenario of a gas trapped in a cylinder, enclosed by a piston. It was no accident that Bernoulli had incorporated this scenario into his kinetic theory—it was a feature of the new 'fire engine' technology (as in Papin's suggested engine at the end of the seventeenth century; Chapter 4). Bernoulli was particularly interested in these advances in 'power technology' (Chapter 7, Part III); and had heard rumours of a revolutionary new engine in England, most probably Newcomen's engine. This engine had everything to do with heat and gas, as we shall see.

The Newcomen Engine

Not much is known about the life of Thomas Newcomen (1663–1729), inventor of the first true steam engine. He was an ironmonger and blacksmith from Dartmouth in Devon, a member of the Calvinist sect of Anabaptists, and he died in London, probably of the plague.

Newcomen's first acquaintance with the steam engine came when he was called upon to advise on problems with a Savery engine that was being used to pump water out of the mines in neighbouring Cornwall. Newcomen and his assistant, the glazier John Cawley, carried out a series of tests on Captain Savery's engine. These tests led Newcomen to make some improvements to the engine that, taken individually, were impressive enough but, taken altogether, were momentous, and resulted in the invention of the first steam engine that actually *worked.*[21]

The most noticeable difference between Newcomen's 'fire engine' and Savery's 'water-commanding engine' was that the steam chamber in Newcomen's engine was now a cylinder with a piston (Fig. 5.3). The piston was connected by a chain to one end of a great beam oscillating about a central pivot; the other end of the beam was connected to a counterweight—the mine-pumping gear itself. A boiler, directly under the cylinder, filled the cylinder with steam and, at the same time, the descending counterweight pulled the piston up and allowed the steam to expand (note that the steam pressure alone was not enough to drive up the piston.)

A second difference between the Newcomen and Savery engines was the introduction of various valve mechanisms that led to a more automated engine. Thus, when the piston reached the top of the cylinder, the great beam caused two valves to trip: one to cut off the supply of steam and the other to switch on an internal spray of cold water (the latter causing the steam to condense). The steam condensed, and then both

Fig. 5.3 Newcomen's 'fire-engine', Oxclose, Tyne & Wear, 1717 (Science and Society Picture Library, Science Museum, London).

the warm condensate and the condensing water were collected through an 'eduction' pipe (another novelty) and returned to the boiler. Now, as the steam condensed, a partial vacuum was created under the piston. Atmospheric pressure then drove the piston down, and the counterweight was brought back up: this constituted the power-stroke of the engine. When the piston reached the bottom of the cylinder, the two valves were tripped again—one switching off the spray of cold water, the other turning on the supply of steam—and the cycle recommenced.

Timing was everything: the great lumbering beam could move only slowly, yet the valves had to be tripped with snap-action precision. As Cardwell says, Newcomen's valve mechanism represented 'a display of inventive ability amounting to genius'.[22] Prior to Newcomen, all engines required the continual presence of operators to start and stop the supplies of steam, cold water, and so on. For the first time, a steam engine could now be truly self-acting.

The final modification introduced by Newcomen combatted the problem of dissolved air. This air ended up in the cylinder, and no amount of cold water would make it condense (the Savery engine regularly stalled on this account). Newcomen overcame this by his invention of the 'snifting valve', which allowed the engine to 'sniff' away the unwanted air.

Newcomen knew nothing of the various theories of heat or gases, but through his superb engineering intuition he had invented an engine that was of as much historic importance as the Gutenberg press or the weight-driven clock.[23] The influence of the engine was unstoppable, and it spread rapidly to all the mining areas of England (first used in Dudley in 1712), and then to the continental mines of Dannemora (Sweden), Schemnitz (now Slovakia), Passy (France), and on to North America. While the physics of the engine was not well understood (for example, it was still not known that all matter exists in one of three phases—solid, liquid, or gas), it was appreciated, for the first time, that *heat* ('fire') was the prime mover of the engine. The success of the heat engine ('fire engine') could not be questioned—the physics would just have to catch up.

6

The Discovery of Latent and Specific Heats

Throughout the whole of the eighteenth century, Paris was the undisputed centre of science in Europe, even during the French Revolution (in 1789). There were other lesser centres, such as Basle and Geneva in Switzerland, St Petersburg in Russia, and Stockholm in Sweden. After Boerhaave, Dutch science waned and German science was at its lowest ebb.

In Britain, Newton was a hard act to follow. The Royal Society in London degenerated into something akin to a gentleman's club, and the Universities of Oxford and Cambridge were moribund as far as natural philosophy was concerned. The action moved north to Scotland (Scotland had five universities to England's two), and then on to Birmingham and Manchester.

Much has been written about the rise of the Scottish school of science, and about the Scottish Enlightenment. However, in the case of the discovery of latent and specific heats, climate was probably as influential as any other factor: the two discoverers, Black and Wilcke, came from the cold of Scotland and Sweden, respectively, and both men attributed observations of snow and ice to their discoveries.

Black

Joseph Black (1728–99) advanced our understanding of heat more than any other philosopher in the eighteenth century. He was born of Scottish parents in the French province of Bordeaux (his father was a wine merchant) and educated at a private school in Belfast, Aged only 16, he went to study medicine at the University of Glasgow in Scotland. He soon found himself more interested in chemistry than in medicine. The turning point came when he started to attend the chemistry lectures of

William Cullen (1710–90). Cullen was a doctor, and also founder of the science of nosology. He was quite a character 'known everywhere for his huge peruke, bigger hat, big coat-flaps sticking out, and huge sand-glass to measure patient's pulses'.[1] Quite quickly, the relationship between Cullen and Black developed from master and pupil to professor and valued assistant, and then on to a lifelong friendship. In their Edinburgh days, they were to meet at the Oyster club, a weekly dining club started off by Black and two other friends, the economist Adam Smith and the geologist James Hutton—all figures in what became known as the Scottish Enlightenment. They also met at the Poker Club where, over a sherry, a shilling-dinner, and claret, members were 'roasted' for their views.[2]

Black's dissertation for his MD at Glasgow University was nominally an examination of the treatment for urinary calculi ('stones'), but it led Black to a major discovery: he was the first person to reform the generic mixture, 'air', into chemically distinct components—by identifying one of these as 'fixed air' (what we now call carbon dioxide). He did this by perfecting the chemical balance, and using it more systematically than any chemist had ever done before.[3] He was thus able to attribute the decrease in weight of *magnesia alba* (magnesium carbonate) to the expulsion of 'fixed air', rather than to the gain of a causticizing agent, the negatively weighted phlogiston.

Black's other major discoveries were in the field of heat, but all shared this feature of respectful attention to a *quantitative* analysis. This, Black said, was more inclined to lead to knowledge and understanding than 'vague reasoning'.[4] The influence of the philosopher David Hume is evident. (Hume was a leading figure in the Scottish Enlightenment—and also a member of the Poker club, and a friend and patient of Black's.)

General Properties of Heat

Black recognized the crucial distinction between heat and temperature:

Heat may be considered, either in respect of its quantity, or of its intensity. Thus two lbs. of water, equally heated, must contain double the quantity [of heat] that one of them does, though the thermometer applied to them separately, or together, stands at precisely the same point.

This still didn't reveal how much heat there was in an absolute sense (but if one could count the individual particles making up the heat-fluid—if there *was* a heat-fluid—then, presumably, one would have an absolute measure of heat). Black, however, was content to adopt a purely operational procedure, and in his later measurements of latent and specific heats, the heats were all relative to temperature changes in a given mass of water.

Black read more significance than anyone else before him into the common observation that different bodies, left to themselves, will all acquire the same temperature:

if we take one thousand, or more, different kinds of matter, such as metals, stones, salts, woods, corks, feathers, wool, water and a variety of other fluids, although they be all at first of different heats, let them be placed together in the same room without a fire, and into which the sun does not shine, [then] the heat will be communicated from the hotter of these bodies to the colder, during some hours perhaps, . . ., at the end of which time, if we apply a thermometer to them all in succession, it will point precisely to the same degree.

He elevated this observation to the status of a law:

We must adopt, as one of the most general laws of heat, the principle that all bodies communicating freely with each other, and exposed to no inequality of external action, acquire the same temperature as indicated by a thermometer

and realized that this was really quite a remarkable fact, not suggested by any properties of the individual bodies:

No previous acquaintance with the peculiar relation of each body to heat could have assured us of this, and we owe the discovery entirely to the thermometer.

He further understood that this equality of temperature was not the same thing as an equality of heat, but represented an '*equilibrium* of heat' between the various bodies.

This idea, of bodies arriving at a common temperature, is one we are very familiar with—so familiar, in fact, that it is hard to see that it is not an obvious result. However, bodies are generally *not* found to be all at the same temperature: people and animals maintain their own specific temperatures, marble may feel cold, metals may feel hot or cold to the touch, the ground is warmer than the air temperature at night, lakes and ponds are often warmer or colder than their surroundings, and so on. To recognize that all bodies are striving to reach the *same* temperature was a major achievement. However, Black did miss one important and 'obvious' result—that bodies are generally striving to cool down rather than heat up.

Latent Heat

As we have seen, Black made the crucial distinction between temperature and quantity of heat. He then carried out a series of ingenious and simple mixing experiments in order to be assured that each degree of the thermometer represented an equal quantity of heat. (This 'method of mixtures' had already been used by Brook Taylor in 1723; see Chapter 5.) It was this understanding—of the difference between heat and temperature—that led to Black's first discovery in the science of heat: the discovery of the concept of latent heat. Once again, Cullen was the starting point.

Cullen had noticed a surprising and inexplicable phenomenon—the intense cold produced when 'spirituous' liquids, such as ether, are allowed to evaporate, especially when this occurs in an evacuated receiver. He communicated these findings to Black. At the same time, Black was perusing Boerhaave's *Elementa Chemiae* (see Chapter 5) and was intrigued by a description of one of Fahrenheit's experiments. Fahrenheit had managed to cool a sample of water some eight degrees* *below* the freezing point (32 °F). The water remained liquid (we would now say it was supercooled) provided that the vessel was small and was left completely undisturbed. The slightest agitation, even air currents on the surface of the water, resulted in the formation of feathery ice crystals spreading rapidly in a network. Most intriguing of all, as the water froze its temperature jumped discontinuously from 24 °F to 32 °F. As Black put it, the 'heat unnecessary to ice' was liberated from the water.

Black mused on these findings, and also on his observations of the snow around Glasgow. When the weather grew milder and the temperature rose above freezing point, Black noted that this did not result in an immediate thaw of all the snow:

were the ice and snow to melt . . . suddenly, [then] the torrents and inundations would be incomparably more irresistible and dreadful. They would tear up and sweep away everything, and that so suddenly, that mankind should have great difficulty to escape from their ravages. This sudden liquefaction does not actually happen; the masses of ice or snow melt with a very slow progress . . . [which] enables us to preserve it easily during the summer, in the structures called Ice-houses.

---

* Almost all the temperatures in this chapter are on the Fahrenheit scale.

Black saw that the very slow melting of snow, the cooling when nitrous ether evaporates, and the heat liberated in Fahrenheit's experiments all pointed to the fact that substantial amounts of heat were involved. These quantities of heat were specifically for the purpose of 'rendering the ice [into a] fluid' or converting liquid into vapour—for changes of state, in other words. What's more, the heat that enabled a change of state did not cause any change of temperature. Amontons had already shown that water, while boiling, stays at a constant temperature; but though the temperature remained constant, a transfer of heat *was* taking place, as evidenced by the continuous supply of heat being drawn from the boiler. Black called the heat required to enable a change of state, but not showing up as a change in temperature, the hidden, or 'latent', heat. (Note that Black's concept of latent heat presupposed a commitment to the *conservation* of heat: rather than saying that heat disappears during a change of state, he held that heat *is* conserved but becomes latent.)

But it was not sufficient to argue just from 'vague reasoning'—experimental confirmation was required. Black carried out the experiments in the large hall adjoining his college rooms, as it was isolated and sufficiently warm (47 °F was evidently considered warm in eighteenth-century Scotland[5]). A sample of ice and an equal mass of ice-cold water were observed as the former thawed and the latter heated up:

> I then began to observe the ascent of this thermometer, at proper intervals, in order to learn with what celerity the water received heat, stirring the water gently with the end of a feather about a minute before each observation.[6]

The time taken for the ice to just melt away completely was measured (it took 10½ hours), and this was compared to the time taken for the water to warm up from 33 °F to 40 °F (half an hour). Assuming that the room's temperature, and other factors, had remained constant for the whole 10½ hours, Black took it that the ice had received 21 times as much heat as the water had received.

Black also attempted the reverse experiment; namely, a measurement of the heat given up when a certain amount of water freezes. However, it proved much harder to find a source of cold that was reliably constant—there was no inkling at this stage in history that there was any asymmetry between the processes of cooling and heating.

Black soon extended his ideas to the case of the vaporization of water. As before, he argued that the change of state couldn't happen instantaneously, or 'water would explode with the force of gunpowder'. He attempted to obtain supercooled steam but this proved too tricky. He then fell back on the usual methods; in other words, he measured the time required to just boil away a given mass of water, and measured the temperature increase in another equal mass of water during the same time interval. Also, he checked with a local distiller about the constancy of his furnace. The distiller assured Black that he could tell, to a pint, the quantity of liquor he would get in an hour. Nevertheless, Black carried out his own experiments, using his laboratory furnace to see how long it took to boil off small amounts of water. Combining all these careful measurements, Black found that the heat absorbed to convert water into vapour was equal to the heat that would have raised the temperature of an equal mass of water by 810 °F—if the water hadn't all boiled away first (this is about 17% lower than the modern figure). He stopped these experiments for a couple of years but then, in 1764, came a great stimulus to further enquiry; this was the year in which James Watt discovered a new, improved version of the Newcomen steam engine. We shall take up the threads again in the section on Watt in Chapter 8.

Specific Heat

As with Black's discovery of latent heat, his discovery of specific heat (he called it 'heat capacity') was started off by his reading of an experiment of Fahrenheit's, again in Boerhaave's *Elements of Chemistry*. Fahrenheit had mixed water and mercury together, and found that in order to obtain a final temperature equal to the mean, he needed to mix three volumes of mercury to every two volumes of water (Chapter 5). We saw how this had led Boerhaave to conclude that heat is distributed 'in proportion to [a body's] extension'. This was something that Black could not agree with, especially as Fahrenheit's experiment appeared to contradict it (there was 50% more volume of mercury than water). Furthermore, Black was trying to mesh Fahrenheit's findings with those of George Martine from 1740. Martine had found that:

spirit of wine both heats and cools faster than water, and that in a much greater proportion than the inverse ratio of their Specific weight does require, as we observed likewise of oil. But still quicksilver, however dense, is more ticklish and easier affected by heat and cold than any of these fluids. Common brandy, upon trial, I did not find to differ sensibly from water in this respect.[7]

All of a sudden, it was clear to Black how to explain both Fahrenheit's results and Martine's results in one theory—by supposing that mercury has a *smaller store of heat* than an equal mass of water. A greater mass of mercury than water would be required in Fahrenheit's mixing experiment; while, in Martine's experiment, a smaller quantity of heat would be implicated for a given mass of mercury than for the same mass of water; and therefore a given temperature change would occur more quickly for mercury than for water.

Black went on to assert that all other substances also had their own characteristic capacity for heat—however, it must be admitted that he couldn't fully justify this assertion from his researches. Take the experiment of placing a cube of iron and a cube of wood in the same oven for an equal length of time. Upon removal from the oven, the iron felt much hotter, and so Black argued that its capacity for heat was the greater. In fact, *wood* has a greater heat capacity but its conductivity (e.g. to fingers) is much lower. Black didn't take account of factors such as conductivity, emissivity, and so on. In fact, he thought that the actual speed of heat entering a body was the same for all materials—except, apparently, for those of a spongy consistency.

In summary, Black's researches and his intuition had ushered in a new understanding of heat. However, apart from a few desultory measurements, he never carried out a systematic survey of specific heats, and he never wrote up any of his ideas. It was left to his students to publish his lecture notes and also to defend his priority. By all accounts, Black was a brilliant lecturer, mostly *ex tempore*, and with many experimental demonstrations to help develop the argument. A student from Black's Edinburgh years described him as follows:

He wore black speckless clothes, silk stockings, silver buckles, and either a slim green silk umbrella, or a genteel brown cane . . . . The wildest boy respected Black. No lad could be irreverent towards a man so pale, so gentle, so elegant, and so illustrious. So he glided, like a spirit, through our rather mischievous sportiveness, unharmed.[8]

Wilcke

While studying the shapes of snowflakes and ice crystals in Sweden in 1769, Johan Wilcke (1732–96) made the observation that water that was cooled below 32 °F became warm on freezing. In the winter of 1772 (so the story goes) Wilcke was trying to sweep up snow from a small courtyard and, tiring of the broom, he tried melting the snow with hot water. He was surprised at how ineffective the hot water was—a large amount only melted a small quantity of snow.[9]

In this way, around ten years after Black, and quite independently of him, Wilke came to exactly the same conclusions from almost exactly the same evidence, and formulated the concepts of latent and specific heat all over again. As, unlike Black, Wilcke did publish his work, he got much of the credit.

(The Finnish scientist Johan Gadolin (1760–1852) continued Wilcke's work on specific heats. The element gadolinium was named after him.)

Irvinism

Black's work on specific heats was extended by one of his students and a future professor at Edinburgh, William Irvine (1743–87). Irvine had the following ingenious idea—all bodies contain a certain absolute quantity of heat, dependent only on their temperature and their capacity for heat. Just as a tank may hold a certain volume of water dependent only on its cross-section and its height, so a body may hold a certain volume of heat, 'poured in' up to a certain level. The fluid theory of heat is virtually written into Irvine's model.

Irvine argued that any change in heat capacity would lead to a change in temperature and vice versa. In other words, Irvine did not believe in latent heat. He had it that heat used up or given out during a change of state or chemical reaction was due solely to changes in the heat capacities of the reactants or the products. As specific heats could be measured, this laid Irvine's theory open to experimental test. Irvine's theory threw up another interesting possibility—what if the tank was empty? This would represent a state of zero heat, and would occur at the absolute zero of temperature. Predictions of this absolute zero of temperature provided another quantitative check of Irvine's theory.

Lavoisier and Laplace, in their work on heat in 1783 (see Chapter 8), found that Irvine's theory was not borne out. Both the data on specific heats and the predictions for absolute zero (the latter varying wildly from −600 to −14,000 °C) contradicted Irvine.[10]

Irvine, like Black, did not like to put pen to paper, and it was left to his son to tidy up his father's papers, and publish Irvine's work posthumously.

The Specific Heats of Gases

The last of the Scottish school to continue Black's work on specific heat was Adair Crawford (1748–95). He was a surgeon at St Thomas' Hospital in London, and visited Scotland to learn about the latest theories on heat, and to attend some of Irvine's lectures. Crawford had an ulterior goal—he wanted to explain the perplexing problem of animal heat.

In Glasgow, in the summer of 1777, he measured the heat capacities of a variety of 'animal, vegetable and mineral substances'[11] and—a first—of 'atmospherical and fixed air'. In other words, he was the first scientist to measure the specific heat of a *gas*. He used a thin hog's bladder, washed it with soap and water, dried it, and forced any residual air out by rolling packing thread over it. After the given 'air' had been introduced into the bladder, it was heated in a Dutch oven, wrapped in flannel to keep it warm, and then pressed out of this bladder using weights. The extracted warmed gas was then forced through a 'worm' (a crooked glass tube) that contained half an ounce of cold iron filings, and the temperature rise of these was measured.

Needless to say, the experiments were not very accurate. They were repeated at a later date, but this time using a brass cylinder to contain the gas and a water-bath for the heat to be transferred to. However, the results in both sets of experiments showed what Crawford wanted them to show—that the heat capacity of 'fixed air' (the 'air' that is breathed out by animals) was considerably less than the heat capacity of 'atmospherical air' (the 'air' that is breathed in). This, Crawford presumed, was the source of animal heat.

Unwittingly, Crawford had maintained the test-gas at a roughly constant pressure when using the hog's bladder, and at a constant volume when using the brass cylinder. This distinction was to be important in the future study of the specific heats of gases (Chapter 10), but Crawford was unaware of it.

Radiant Heat

An obscure apothecary in Sweden discovered that there were two kinds of heat, never mind two kinds of heat theory. This was Carl Wilhelm Scheele (1742–86), from Stockholm, Uppsala, and, finally, Koping (on his deathbed in Koping, Scheele quickly married the pharmacist's daughter to enable her to inherit the pharmacy).

Scheele was an outstanding chemist, who discovered oxygen (before Joseph Priestley), chlorine (before Humphry Davy), and many other chemicals besides. Recently (1982), there has been speculation that the pigment known as Scheele's Green (copper arsenite) may have been implicated in Napoleon's death (the wallpaper in Napoleon's room on St Helena was green).

By careful observations and clear thinking about everyday phenomena, Scheele identified 'radiant heat' and distinguished it from 'familiar heat' and from light.[12] The observations required no special equipment and could, in principle, have been made centuries earlier—but they weren't.

(The German autodidact and polymath Johann Heinrich Lambert (1728–77) was another to discover radiant heat—'obscure heat',[13] as he called it. In the eighteenth century, it was still possible for a very gifted and hard-working individual to encompass the whole sweep of human knowledge. Lambert was such a one, and he certainly was hard-working: he worked from 5 a.m. to midnight with only a two-hour break at noon.[14] He is famous for his photometer, for his proof that π is irrational, and many other things.)

Scheele noted that warmed air rose upwards from a hot body, such as an oven, and you could see the air shimmering where convection currents (modern expression) were occurring. However, he noticed that a completely different sort of heat travelled in rays, in all directions (not just upwards), and didn't involve the air. In fact, Scheele found that the air could be disturbed, as in a draught, and the 'heat-rays' were unaffected. Also, the flame and ascending smoke of a candle burning were not deflected when crossed by such rays; and the shadow of a window frame didn't quiver, but the shadow of a hot iron or stone did quiver.

This new heat (Scheele coined the term *radiant* heat) was similar to light: it travelled in straight lines, and followed the optical laws of reflection. Indeed, heat sources were usually also light sources, as in the prime example of the Sun. But Scheele saw that radiant heat was not only different from convective heat; it was different from light as well. Light was reflected from a polished metal surface, and was transmitted through and reflected by glass, while radiant heat was reflected by the metal but *absorbed* by the glass—'remarkable' commented Scheele. Scheele also noted that the greatest radiant heat was obtained not from a brightly burning fire, but from the glowing charcoal left behind. Also, a fire could be seen from far away, but its warmth felt only from nearby ('at about three ells' distance'). (An ell is the distance from shoulder to fingertips.)

Simple experiments convinced Scheele that a hot body was just 'hot', whatever kind of heat was used to warm it. Also, the different kinds of heat could be converted into, and separated from, each other. For example:

at two ells' distance in front of the stove, by means of a small metallic concave mirror, a focal point can be produced which kindles sulphur. Such a mirror can be held for a very long time in this position without its becoming warm; but if it is coated with some soot over a burning candle, [then] it cannot be held for four minutes in the [same] former position before the stove, without burning the fingers upon it . . .
and
the metallic concave mirror and the plate of metal rapidly become hot [by conduction] as soon as they touch a hot body, although they do not become the least warm from the [radiant] heat proceeding from the stove.

Scheele's work on radiant heat was continued by the Swiss school at Geneva, especially de Saussure (1740–99), Pictet (1752–1825), and Pierre Prévost (1751–1839). These scientists were part of a new breed—mountaineers. They explored the rugged Alpine landscapes for fun (a new Romantic idea, and a far cry from Thomas de Quincy, who lowered the blinds of his coach to shut out the Lake District views[15]). De Saussure, for example, was one of the first to climb Mont Blanc. The mountains were nature's laboratory, and intelligent observations led to advances in geology, meteorology, and associated physics.

Pictet was the first philosopher to attempt to measure the speed of radiant heat, but could only conclude that it travelled 'perhaps as rapidly as sound or even light'.[16]

De Saussure and Pictet showed that a hot but not glowing bullet at the focus of one concave metal mirror produced a rise in the thermometer placed at the focus of another mirror. In order to prove that heat rather than light was involved, Pictet repeated the experiment with a non-luminous source—a glass vessel of boiling water. Once again, the 'obscure heat' was detected, and brought to a focus at a secondary mirror.

Pictet went on to repeat the arrangement yet again, but this time placing *ice* at the focus. Even so, Pictet was sceptical, saying that 'cold was only privation of heat' and 'a negative could not be reflected'.[17] He found, however, a positive result—the temperature at the second mirror *was* lowered—but this is now explained away by numerous experimental difficulties (see Evans and Popp:[18] these authors also show how the experiment may be carried out at home). Pictet's apparent radiation of cold caused a stir amongst natural philosophers. Its chief benefit was to provoke the physicist and literary figure (he had translated Euripides) Pierre Prévost into examining it.

According to Prévost, the heat-fluid (by now, 1791, called 'caloric') was made up of discrete heat-particles moving rapidly in all directions. (Thus, while Prévost adopted the mainstream caloric theory, it's clear that he was also drawing from the motion theory, and from the work of his countryman, Daniel Bernoulli, published half a century earlier.) The crucial aspects of Prévost's theory were as follows: the heat-particles are tiny and almost infinite in number; they move very fast in all directions; a body at a higher temperature has faster heat-particles and therefore emits caloric at a higher rate; and all bodies are, at the same time, continually emitting and absorbing heat-particles to and from their surroundings.

As a hot body emits caloric at a faster rate then, in any given time interval, the hot body will emit more caloric than it receives, and will cool down; likewise, the cooler body will emit less caloric than it receives, and will heat up. When the rate of emission and absorption are equal, then the body stays at a constant temperature. This was Prévost's 'Theory of Exchange'[19] or concept of 'mobile equilibrium'.

Prévost's theory explained how bodies reached an equilibrium of temperature—but what did it have to say about Pictet's experiment, about the transmission of cold? In fact, Prévost had shown that the transmission (and also the reflection) of 'cold' is impossible.[20] This asymmetry between 'hot' and 'cold' passed without a murmur, but was to have big repercussions 50 years later: it would lead to a new law of thermodynamics (see Chapters 12 and 16).

Pictet remarked that Prévost's theory might be applicable to the 'electrostatic machine', as this simultaneously acquired and discharged electricity. Another link between heat and electricity was made by the Dutch scientist Ingen-Housz (1739–99). He found that good conductors of electricity were also good conductors of heat, and devised an ingenious way of measuring specific thermal conductivities: first, wires were made up (from the materials to be tested), then these wires were coated with a standard thickness of wax, and then they were heated at one end. A ring of wax melted when the heat reached it, and Ingen-Housz could measure the speed with which this wax 'melting point' travelled along the wire. Finally, Ingen-Housz and other researchers noted yet one more link between heat, light, and electricity—they could all be generated by friction.

Overview

Black had established a clear distinction between heat and temperature, defined 'quantity of heat', shown that bodies are constantly striving to achieve a common *temperature* rather than a common heat, and defined specific heat and latent heat. As the end of the century approached, the substance theory of heat was in the ascendant. (Lavoisier made heat one of his chemical 'elements', and called it *'calorique'* (Chapter 8).) Black professed to be undecided on the question, but privately he held to the heat-substance theory. (Although it's nowhere in his lectures, Black must surely have made some connection between the containing of his 'fixed air' (carbon dioxide) within a solid, and the 'fixing' of the heat-substance into latent heat.)

Radiant heat was discovered by Scheele, and there were more and more parallels being noticed between ordinary heat, radiant heat, light, and electricity.

Prévost's theory of 'mobile equilibrium' was a fundamental advance. It supplied a mechanism by which Black's common temperature could be achieved, but much more than this, it introduced a totally new idea—the idea of a *dynamic* equilibrium. As so often, the new idea seems intuitive to us now, but was strange and unfamiliar in the beginning. The heat-fluid in Irvine's 'tank' found its level, but this was a once-and-for-all process. Prévost's equilibrium was quite different: it was maintained by continual small adjustments. Count Rumford—we shall meet him in Chapter 9—thought this was ludicrous; how could a body be both emitting and absorbing something at the same time? Maxwell, on the other hand, saw Prévost's theory as underpinning his own kinetic theory of gases (Chapter 17). Lastly, Prévost's dynamic equilibrium, having its roots in both the substance and the motion theories of heat, showed that these theories were not as antithetical as is sometimes assumed.

The eighteenth century had a preoccupation with the 'heat-of-a-body', and this was taken to its logical extreme in 'Irvinism'. We now don't consider the 'heat-of-a-body' to be a particularly telling parameter—it is not of universal importance for the very reason that it is too body-specific. In the next chapter we shall find that the 'force-of-a-body-in-motion' was also a misleading concept.

7

A Hundred and One Years of Mechanics: Newton to Lagrange via Daniel Bernoulli

Part I: The 'Force of a Body in Motion'

We will put the study of heat to one side and consider some other 'blocks' of energy that started to be recognized in the eighteenth century—the 'blocks' of mechanical energy. Mechanical energy emerged from mechanics, the study of bodies in motion. The bodies could be as small as the particles or atoms from which, some said, everything is made, or as large as a celestial body, or any size in between.

The cause of prime motion was uncertain, and generally taken to be God-given. After this, the cause of different motions was due only to the mutual interactions of the bodies. (While Newton had introduced forces, he still thought of these as ultimately due to a *body* and not an 'abstract mathematical point'.) Although it was of course known that a body's motion could be altered by the intervention of mind (as when I decide to pick up an apple), the mechanics was not so ambitious as to attempt an explanation of this sort of thing. In fact, such effects were rigorously excluded. According to Leibniz, even God didn't interfere after He had set everything up.

Although there were important precursors (see Chapter 3), mechanics as we know it today was essentially started off by Newton. We shall remind ourselves of Newton's three Laws of Motion, as set out in the *Principia* (see Chapter 3):[1]

LAW I: Every body perseveres in its state of rest, or of uniform motion in a right line, unless it is compelled to change that state by forces impressed thereon.

LAW II: The alteration of motion is ever proportional to the motive force impressed; and is made in the direction of the right line in which that force is impressed.

LAW III: To every action there is always opposed an equal reaction: or, the mutual actions of two bodies upon each other are always equal, and directed to contrary parts.

What Newton had done in stating his laws of motion was amazing: from the plethora of phenomena—a horse pulling a cart, a fly walking on water, magnets attracting each other, a drop of water being drawn up a narrow glass tube, the motions of the planets and of the Moon, the spinning of the Earth, the tides, a boy playing with pebbles, and so on—to wrest out the concepts of 'nothing is happening', 'something is happening', and 'force'. Force explains all the cases where something is happening in just one way—the motion is changed from uniform to accelerated. This was all the more amazing when we remember that, apart from free fall, there were no smooth accelerations around in those days—no cars with accelerator pedals, only carriages jolting along rutted roads. Newton wasn't sure, but he suspected that *all* the phenomena in nature, including chemistry and the action of light, would be explainable in this way.

We might think that all the natural philosophers would retire to their firesides, light their clay pipes, and start writing their memoirs—but this didn't happen. Newton was lauded for his law of Universal Gravitation (except by the followers of Descartes, who deplored the implied 'action-at-a-distance'), for explaining the motion of the Moon, the figure (shape) of the planets, and the path of Halley's comet—but his Laws of Motion went almost unnoticed. Even in Cambridge—Newton's university—the mechanics textbook used was that of the Cartesian Rohault, albeit with Newtonian annotations.

There was, of course, the time required for the dissemination of the work. Newton's *Principia* was published in 1687, but it was only after 1727 that his work was first promoted in France, by the French *philosophe* Voltaire (1694–1778). Voltaire had judiciously absented himself from the environs of the French court, after two stays in the Bastille. He visited London, where he was greatly impressed by Newton and Locke, and all things English. Upon returning to France, Voltaire won over his mistress, the Marquise du Châtelet (1706–49), to Newton. She helped Voltaire with his *Elements de la Philosophie de Newton*, and subsequently she translated the *Principia* into French—until recently, it was the only French version. The Italian philosopher Algarotti (1712–64) wrote *Newtonianism for Ladies, Dialogue on Light and Colour*, but this concentrated on optics. It was in Holland where Newton was most admired and where 's Gravesande published his textbook, *Mathematical Elements of Physics*, which was an exposition of Newton's physics.

The *Principia* is possibly the most influential book ever written (by a single author) even while it is one of the least read. However, Newton was secretive, defensive, and combative. Not surprisingly, there followed an era of conflict and debate as the world struggled to absorb the new world-views.

The chief opponent was Leibniz, who had to defend himself against Newton's charges of plagiarism regarding the discovery of the calculus. As well as the calculus, the conflict was about fundamental differences in the metaphysical position of the Newtonians and the Leibnizians. It included such questions as: Were space and time absolute and universal, and was space a void? Were there atoms in this void, and were they hard and impenetrable? Could forces act—and act instantaneously—across this void? Did Newton's clockwork universe need God to wind it up from time to time (to use Leibniz's metaphor)? All these issues were interrelated but we shall concentrate on just one—what is the best measure of 'force'? This is because the debate evolved into two distinct outlooks: the Leibnizian 'energy view' and the Newtonian 'force view'.

The controversy was initially between Leibniz and Descartes (although Descartes had in fact died 36 years earlier), and was sparked by a paper published by Leibniz in 1686, the year before the *Principia* appeared. In this paper (we have covered it already in Chapter 3), Leibniz trumpeted the fact that Descartes' quantity of motion, mv, was *not* universally conserved, despite Descartes' claims, and, worse still, could lead to a perpetual motion. Leibniz claimed that his own measure for 'force' (*vis viva*, given by mv^2) *was* always conserved and did not lead to a perpetual motion. Leibniz therefore started off the whole debate, and established the two competing measures of 'force': the Cartesian mv versus the Leibnizian mv^2 (m is the mass, and v is the speed, of the given body).

The Newtonians, ever ready for a fight, especially against Leibniz, were only too willing to join in the controversy. When some Newtonians, Johann Bernoulli and Willem 's Gravesande, 'deserted'[2] to the *vis viva* camp, the Reverend Dr Samuel Clarke was outraged, and wrote (in 1728) that they were attempting to 'besmirch the name of the great Sir Isaac Newton'.[3] Johann Bernoulli (1667–1748) was a member of the illustrious Bernoulli family of mathematicians from Basle in Switzerland. He was not only a 'deserter' but became Leibniz's most ardent supporter and hostile to anything Newtonian. For example, Johann backed

Leibniz in the priority dispute over the calculus, and promoted Leibniz's version of the calculus on the continent.

As regards the other natural philosophers, the two camps were almost entirely split along national lines. This prompted the Scottish philosopher, Thomas Reid, to suggest the following absurd compromise:

I would humbly propose an amicable solution upon the following terms:

(1) In all books, writings or discourses made within Great Britain and the dominions thereto belonging . . . the word force shall be understood to mean a measure of motion proportional to the quantity of matter and the velocity of the body on which that force is imprest; unless where the contrary is expressly declared.

(2) In all other places the word force shall be understood to mean a measure of motion proportional to the quantity of matter and the square of the velocity.

(3) All hostilities between mathematicians on both sides shall cease from the time above specified.[4]

While the controversy between Leibniz and the Cartesians was thus extended to include Newton, the Newtonians still held to the Cartesian measure of force, mv, known today as momentum, albeit with the crucial modification that it now had direction as well as magnitude. This may seem puzzling to the modern reader—why wasn't the force given by $F = ma$, as defined by Newton himself in his Second Law of Motion (where m is the mass and a is the acceleration)? There are a number of answers to this.

First, the quantity 'mass times velocity' had a long history, stretching back to antiquity. It expressed the condition of equilibrium in the lever of Archimedes (when the lever was balanced, the masses at each end were in inverse proportion to their velocities). It was extended by Jordanus de Nemore in the thirteenth century, Stevin in the sixteenth century, and Galileo in the seventeenth century to be the condition of equilibrium for all the simple machines (inclined plane, pulley, screw, wedge, etc.) before being taken up by Descartes.

Second, Newton's own definition of force, implicit in his Second Law, was not in fact the familiar $F = ma$. He defined force as the change in a body's momentum, yes, but with no mention of the rate or time interval involved (see Chapter 3). In fact, it wasn't until 1750 that Newton's force was properly defined (by Euler; see Part II) and even then, not given in the modern vectorial form, $\mathbf{F} = m\mathbf{a}$.

Surprising as it may seem, another reason (I believe) why force was not given as $F = ma$ is that there were few easy-to-interpret confirmations of this relationship, apart from the single example of freely falling bodies. Newton had managed to determine the path of the Moon falling to Earth under gravity ('the only problem which gave me a headache',[5] Newton said) but the Moon's orbital motion did not suggest '$F = ma$'. A more straightforward example was the mutual attraction between floating cargoes of lodestone and iron, but the motion would soon be damped by the water. There was Hooke's Law for springs, and also the motion of vibrating strings—but the motion was oscillatory, and again did not suggest '$F = ma$'. None of the above cases was such as to bring out the simplicity of $F = ma$; I push something and it goes faster. All this is yet further testimony to Newton's genius—that he could found a relationship for which there was no pressing *experimental* need or evidence.

So let's see how mv^2 and mv shape up. The testing ground was initially collisions between bodies, as these were at the heart of the mechanical philosophy. The findings were as follows: in the case of elastic collisions, both mv^2 and mv were conserved, whereas for inelastic collisions, only mv was conserved. 'Elastic' was a word that started to be used in the late seventeenth century: with reference to a fluid, it meant 'expanding to fill its container' (from the French word *elater*); while, with reference to a spring or a body, it meant 'able to take up its original shape after compression'; and 'elastic' as applied to a collision was taken to mean that the participating bodies were, confusingly, either absolutely hard or absolutely deformable (springy).

Jumping ahead to the modern definition of an 'elastic collision', we are plunged into tautology: the collision is elastic if the total kinetic energy is conserved; total kinetic energy is only conserved if the collision is elastic. The definitions—both then and now—are only saved from circularity when it is realized that there are *other* ways in which elastic and inelastic collisions can be differentiated, apart from the constancy of mv^2. In inelastic collisions, the bodies may have their shapes permanently deformed, end up being joined together, or be shattered into smaller pieces (or be warmer . . .).

These distinctions may seem obvious to us now, and some of the philosophers of the day (eighteenth century) thought so as well—too obvious, they thought, for it to be necessary to carry out actual experiments. Some, such as d'Alembert (Part IV), even thought that mechanics was a branch of mathematics, and was therefore not 'contingent' on the results of experiments. But the differences between elastic and inelastic collisions *are* contingent, and further enlightenment would only come when experimental as well as mathematical investigations were carried out.

However, checking up experimentally on collisions was no trivial matter—how were constant initial velocities to be achieved, and how were final velocities to be measured? How were friction and air resistance to be sufficiently minimized, and how were deformations to be gauged? There were no air-tables or electronic timers, and even an accurately flat surface was beyond the casting and planing techniques of the day.

Willem 's Gravesande (1688–1742) was one natural philosopher who did go to great pains to carry out the experiments. He was famous, at Leyden, for his lecture demonstrations of various physical principles. (We have met him before, in Chapter 2—it was he who went to visit the Landgrave of Hesse, to check up on Orffyreus' perpetual motion machine.) 's Gravesande's allegiance was initially to *mv*, and to the Newtonian school. However, after carrying out various experiments, he broadened his view to include mv^2 as a useful measure (much to the disgust of the Reverend Dr Clarke). His crucial experiment consisted in letting various masses fall into a trough of clay (Fig. 7.1). He found that the depth of the impression in the clay depended on mv^2 rather than on *mv*, and he reputedly said: 'Ah, c'est moi qui me suis trompé'[6] ('Aha, it is I that is mistaken').

But 's Gravesande didn't leave it at that. He wanted to determine if mv^2 was an important measure in collisions as well as in free fall. So he rotated his area of enquiry into the horizontal plane—by letting spherical bodies suspended from a pendulum collide. Once again, the motion was brought to a halt by clay or wax, and once again the depth of the impression was found to be proportional to mv^2. 's Gravesande also measured the *mv* before and after collision, and found that the total *mv* was conserved, and that this was true whether the collision was elastic or inelastic (he used an arrangement of ratchets and springs to measure this, in itself begging a lot of questions). All in all, he concluded that *both mv* and mv^2 were important measures, but which came into play depended on the effects being investigated: *mv* was important when it came to a determination of the velocity after a collision; mv^2 was important when the 'force' was totally 'consumed'—for example, at the end of a fall. 's Gravesande was on the right track, but it still wasn't evident to him that mv^2 and *mv together* are required for a full determination of the velocities after a collision. While some 75 years earlier, Huygens had found that mv^2 was conserved in certain collisions (see Chapter 3), it was still hard to be sure what the significance of this was—was it just a number that

The trough was filled with clay; the test body was released from the top of the stand; its depth of penetration into the clay was measured.

Fig. 7.1 Poleni's apparatus (similar to 's Gravesande's) for free fall into clay (photograph from the Boerhaave Museum, Leiden). (The trough would be filled with clay, and the ball dropped from the top.)

happened to come out constant, or was there some new physical entity involved?

It appeared that mv had the edge over mv^2 as it was conserved in elastic *and* inelastic collisions, and conserved *at all instants* throughout an elastic collision. Despite Leibniz's vaunting of the cosmic conservation of mv^2, in fact it wasn't conserved except at the very start and end of the collision.

So where was the contest? It seemed as if the Newtonians were winning hands down (mv was conserved at all instants, and in all types of collision), but the difficulty for Newton arose in a parallel metaphysical problem, the problem of hard-body collisions. In Query number 31 in the *Opticks* (we have met it already in Chapter 3), Newton proposed that for two identical, absolutely hard bodies approaching each other with equal speeds, as the bodies were impenetrable, their motion simply stopped.

I believe that Newton's conclusion would have been just as startling and alien to the eighteenth-century philosophers as it is to us today. The trouble is that it simply never happens like that. You may say, well, those were ideal, unobservable 'atoms'—what about real, *composite* bodies? But real bodies can be pretty close to being hard and impenetrable, and yet the outcome of a head-on collision never even approximates to an abrupt cessation of motion. Quite the reverse—the harder the bodies, the more strongly they rebound. (Mariotte (1620–84)—see Chapter 4—tried collisions using balls of ivory, and of glass, as close to perfectly hard as was possible in those days.) You may try to help Newton again and say, well, what about collisions between two absolutely soft bodies—say, two lumps of putty? Yes, the motion does stop, but then we don't have the same, tricky starting conditions, which assume non-deformable, impenetrable bodies. (For cases of intermediate hardness, Newton had it that motion *was* lost, just to the extent that the body was deformed.) Newton gave no explanation as to where the motion had gone. He simply said that if this absolute loss of motion is 'not to be allowed',[7] then we shall have to admit that bodies can never be truly impenetrable.[8]

Johann Bernoulli didn't like the discontinuities that occurred in any collision, let alone in Newton's query: while the *total* momentum was conserved, the momentum of an *individual* body could change abruptly, in magnitude and/or direction. 'Nature does not make jumps',[9] he complained. Also, Bernoulli and his son, Daniel (1700–82), did not like the implied loss of *cosmic* 'motion' that resulted from Newton's query. Leibniz had said that the total effect must neither exceed nor be less than the total cause, and both Bernoullis agreed with Leibniz (and, remarkably, with each other) on this.

Why wasn't Newton more bothered by the counterintuitiveness of his thought experiment? Newton, like Leibniz and all the other natural philosophers, didn't sanction a net increase in motion coming from no-where. His Third Law of Motion was just such as to prevent the 'ab-surdity' of such a perpetual motion (see Chapter 3). However, Newton's thought experiment does not generate an increase but, rather, a net *de-crease* in motion. This is something that *is* observed on the cosmic scale. As Newton observed, 'motion is much more apt to be lost than got, and is always upon the decay'.[10] Perhaps this is why Newton allowed his thought experiment to remain.

The resolution came with compromises, eventually. Real bodies might be impenetrable, but they were compressible to a certain extent, and approached absolute hardness only in the limit. (We can begin to ap-preciate the symbiotic relationship between concepts, experiment, and new mathematical ideas, such as 'in the limit'.) Johann Bernoulli helped physical intuition along by using a good analogy—an air-filled bal-loon. As the balloon was pumped up to a greater and greater pressure, it became harder and harder, and also more and more elastic. At the limit of absolute hardness, the balloon became absolutely elastic instead of Newton's absolutely inelastic.[11] D'Alembert (1717–83) made a compari-son between the mathematical concept of a point and the abstract con-cept of hardness—neither was achievable in practice, but both could be defined and then became usable concepts. Also, by stretching an instant of time to a small but finite interval,[12] instantaneous changes (implying infinite forces) could be smoothed away.

So, what has happened to the contest? It seems as if mv has managed to recover, after allowing for hardness to be an abstract idea, achievable only in the limit, and allowing for impact to occur over a short but finite time. The expression mv^2 also manages to survive, provided that we understand 'conserved at all times' to really mean 'recoverable' at the end of the collision. We still need to explain what happens in the case of colliding *soft* bodies (putty, for example), and ask why we're not upset at the loss of 'motion' in this case. Leibniz had the answer (and we have given it already in Chapter 3). He said that while the 'bulk motion' of the bodies did cease, no overall 'motion' (*vis viva*) was lost, as it was pre-served in the 'motion of the small parts'.

This consideration brings out a real difference between mv and mv^2. Leibniz had been referring to mv^2, but what about the momentum, mv—can it also be dispersed in the same way, by a transfer to the 'small parts'? The trouble, we now understand, is that bulk momentum has an overall direction,[13] and therefore it can't all be transferred to random microscopic motions (these are in *all* directions). Leibniz suspected this trouble, as he claimed that only a quantity with no direction could be of universal significance.

We now argue as follows. Consider the case of a soft body, a ball of putty, colliding with a wall: where has the initial momentum of the putty gone? We can answer in two ways. First, we say that the wall is not truly immoveable but does, in fact, move a tiny bit, in accordance with Newton's Third Law of Motion. As it has an almost infinite mass by comparison with the putty, the wall moves very slowly, just enough to ensure that total momentum *is* conserved (but not enough to jeopardize the conservation of kinetic energy[14]). Alternatively, we treat everything microscopically: the ball is deformed and heats up slightly, the wall also heats up, and there may be shock waves through it, and sound waves through the air (the thud of the ball on impact). All in all, the bulk motion of the putty is 'randomized' away.[15]

A lot of this feels like arguing after the fact. It's no wonder that the philosophers of the day struggled with the concepts of *vis viva*, force, and hardness: the ideas were difficult, the mathematics was new, and the phenomena were detailed and various. Physical understanding and enlightenment (this was the Enlightenment, after all) would only come when experiment, mathematics, and ideas all moved forward together, and when many different scenarios were examined.

So far, we have only looked at the case of collisions. Despite the high aims of the mechanical philosophy, this was perhaps not the best test case for a resolution of the controversy. In a collision, the 'happening' is over so quickly that there is no time to witness the accelerations and decelerations (later—in Chapter 17—we shall find that the collisions in a gas serve no other 'purpose' than to randomize the directions of the molecules[16]). Besides which, Leibniz made truly cosmic claims for the applicability of *vis viva*. It was therefore necessary to examine as many different settings as possible, in addition to the case of collisions.

Johann Bernoulli, despite being an ardent supporter of *vis viva*, made a pertinent criticism of Leibniz's use of mv^2. He said that gravity at the Earth's surface was just one example of 'activity'—what justification did Leibniz have in extending the validity of the formula, mv^2, derived from the specific case of free fall on Earth, to *all* phenomena? Leibniz had no good answer to this.

Turning our attention now to mathematics, we find a surprising thing—'geometers' (applied mathematicians) had been using Newton's definition of force (as in $F = ma$) for years. (The geometers included Varignon, the Bernoullis, Hermann, Euler, d'Alembert, and Clairaut—we shall meet them again in Parts II and IV.) More surprising, no one had attributed this relationship to Newton—least of all Newton himself (he attributed it to Galileo). More surprising still, both mv and mv^2 were regularly coming out of the mathematics. Clearly, the geometers and the natural philosophers weren't talking to each other, even when these roles were combined into one person.

Newton had kept strangely quiet during the *vis viva* controversy (mind you, he was an old man at this time—he died in 1727, at the age of 84). However, he did pipe up at one point when he thought he had *vis viva* in trouble. When the force is constant (as in free fall), the acceleration is uniform, and so equal amounts of mv are generated in equal intervals of time. However, in the case of *vis viva*, more is generated when the body's speed increases from 10 to 15 m s^{-1}, say, than when its speed increases from 5 to 10 m s^{-1}. Newton argued that this could only occur (in the case of free fall) if the body's weight increased with time. This was clearly absurd, and so *vis viva* was evidently not the crucial determinant of 'motion'. This seemingly paradoxical result occurred because Newton was looking at the effects following on from a compounding of force (Newton's F) through time. By looking instead at the compounding of force through *space*, a dependence on mv^2 rather than on mv appeared. Newton himself had discovered this some 30 years earlier: he had found that the area under the force versus distance curve was proportional to the square of the speed.[17] However, the quantity mv^2 that arose out of his calculations held little significance for him—so little, it seems, that he promptly forgot all about it.

Eventually, by the middle of the eighteenth century, came the realization that a summation (strictly, an integration) of the force (Newton's F) for successive intervals of *time* generated a given change in mv, while a summation of the force for successive intervals of *distance* (ds) generated a given change in mv^2:

$$\int F \mathrm{d}t = mv \tag{7.1}$$

$$\int F \mathrm{d}s = 1/2\, mv^2 \tag{7.2}$$

At last we can see the factor of a ½ appearing before the quantity mv^2, turning it from *vis viva* into the familiar expression for kinetic energy, ½mv^2. (In the mid-eighteenth century, units and constants hadn't yet settled down to common convention: a mere factor of ½ was not going to rumble Leibniz's new concept of *vis viva*.)

Once again, we might think that all the natural philosophers would pick up their clay pipes and declare the controversy settled—clearly, *both* mv and ½*mv*² were important measures. Some philosophers ('s Grave-sande, d'Alembert, and others) did say just that, but most carried on just as partisan as ever. There were even some staunch Newtonians—such as Atwood (1745–1807)—who adhered strictly to Newton's Laws of Motion, and rejected *both* mv and ½*mv*². Just as happens today, it was one thing to do the mathematics, and another thing to take on board the full meaning and implications of that mathematics.

Part of the difficulty lay in the fact that it was hard for the contemporary philosophers to accept the physical significance of an integration of force through space. Ironically, 100 years earlier, Galileo had struggled with the idea of introducing '*time*'. Now, in the mid-eighteenth century, no one wanted to consider anything but time. We have already seen how Newton tried to derail the *vis viva* concept by noting that it didn't increase uniformly with time. For Leibniz as well (a rare point of agreement between them), time was the crucial variable—in fact, Leibniz's whole philosophy demanded it. Everything from monads through to macroscopic bodies followed its destiny or 'world-line' through time. That any evolutionary change could come about by a mere change in position was ridiculous. But, as we shall find out in Part II, this is exactly what can and does happen in nature.

In the meantime, what about the controversy? Which concept had won, *mv* or ½*mv*²? The curious thing is that the whole controversy, which had raged for over half a century, simply went off the boil. There were a number of reasons for this.

One is that the whole concept, the 'force of a body in motion', was fatally flawed. Galileo, with his Principle of Relativity, had shown that the motion of an isolated body can't be determined absolutely; and Newton showed in his First Law of Motion that a body in uniform motion needs no force to keep it moving (so, for example, the *Starship Enterprise* needs no fuel just to keep going). From our vantage point of over 200 years into the future, we can see that the controversy was really about energy—were Newton's concepts of force and momentum enough to describe the phenomena, or was some other ingredient necessary? Atwood insisted that ½*mv*² was merely a shorthand—Newton's Laws were complete and sufficient in themselves. But the quantity ½*mv*² was cropping up in too many places to be dismissed in this way.

There are two ways in which motion can stop being relative and start being absolute. The motion can be altered and then the *change* relative to the former motion can be determined absolutely; or, the motion can be defined relative to other bodies *within a system*. These two ways are epitomized in the Newtonian 'force view' and the emerging 'energy view'. In the Newtonian view, an isolated body is the centre of attention, and we track its twists and turns, its accelerations and decelerations, as it moves along its path and encounters the slings and arrows (the external forces) of fortune. In the energy view to come, a system is examined in its entirety, and we look at the interplay between the various bodies within this system, and within a certain time interval or cycle.

Paradoxically, the controversy waned just as the energy view was in the ascendant. Mechanics began to take a new turn, away from an individual body and towards the system. The new systems view demanded its own new calculus (the 'variational mechanics' of Euler and Lagrange; Part IV), but this was so mathematical, abstract, and divorced from physical concepts that most geometers, engineers, and natural philosophers couldn't keep up.

The final reason why *vis viva* went off the boil is that it simply didn't answer to all that Leibniz had asked of it. Leibniz had shown great daring and intuition in inventing the concept—he had, in effect, discovered kinetic energy. But he wanted *vis viva* to play an even larger role—to account for all 'activity', whether cosmic or local, and to explain his metaphysic that the total cause is equal to the total effect, whatever those causes and effects might be. But it turned out that the expression, $\frac{1}{2}mv^2$, wasn't conserved except in the rather special case of elastic collisions. It had still to be realized that, yes, kinetic energy (*vis viva*) was a 'block' of energy (when scaled by ½), but it was conserved, as mechanical energy, when combined with another 'block'—potential energy.

Part II: Potential Energy

Introduction

It is potential energy that is the missing partner to kinetic energy. In a closed mechanical system (defined, it must be admitted, in a begging-the-question way, as one from which there are no losses), it is only the combination of potential and kinetic energy, *taken together*, that is conserved: when kinetic energy increases, it is at the expense of potential energy; when kinetic energy is consumed, potential energy is increased in equal measure. But what is potential energy? Whereas kinetic energy had a specific formulation, and a discoverer (Leibniz), we shall find that the formulation and understanding of potential energy just sort of slipped into use—a gradual process that took 100 years or so.

There were three routes along which ideas about potential energy began to develop. The first was the intuitive understanding that the live force doesn't just disappear, or appear from nowhere—there has to be some quantity that represents live-force-in-waiting; but this intuition didn't immediately yield a formula for potential energy. The next route was the solution of special problems by 'geometers': the orbits of planets, the motion of vibrating strings, the shape of the Earth, and so on. Finally, the third route was via the engineers—they defined a certain quantity that was useful when comparing the performance of different machines or engines. This was the quantity called 'work'.

Work

We have emphasized how new concepts in physics are only forged when experiment, mathematics, and the ideas themselves all move forward together. Now we see yet another factor that is important—technology. We shall examine this last case first.

For the most part, the conservation of momentum was a strangely unhelpful rule for the engineer (although in ballistics, the conservation of total *mv was* demonstrated, in the Robins' ballistic pendulum). Even stranger, Leibniz's *vis viva* also had only a limited utility. A much more useful measure began to appear—the quantity called 'work'. This, the engineers took to be the force (tacitly, Newton's *F*) multiplied by the distance through which the force acts. We have already seen (see Chapter 4) how Amontons, as early as 1699, had said that the 'labour of a horse' was the force exerted by the horse multiplied by the speed of the horse at its point of application (strictly speaking, this is the *rate* of doing work).

The engineer's incentive was the need to make comparisons between a growing variety of machines: the water-wheel and the windmill, the Archimedean screw and the hand pump, the horse-drawn plough, the labourer with wheelbarrow or shovel, and the increasing number of steam-driven water pumps. The variety was important. That the concept of work, initially derived for the engineer's expediency, could be used in all these cases meant that here was something fundamental and universal. Mathematically, it can be seen that 'work' is none other than the geometers' integration of force over distance (Equation 7.2). As this is equal to live force, the *work done is equal to live force*. But the geometers and the engineers moved in different worlds, and so this connection took time to be noticed.

However, in the special case of gravity, the connection between 'work' and live force was axiomatic (weight times height is proportional to mv^2; see Chapter 3, 'Leibniz'). Furthermore, the weight-raising machine maintained a special hold on the engineer's attention during the eighteenth century, as the most ubiquitous machine was the water pump for draining mines (around 20 times more water, by volume, came out of English mines than ore[18]).

The lever was the quintessential weight-raising machine, but by the sixteenth century, Stevin had shown that all the simple machines (the pulley, wedge, screw, and inclined plane) were really equivalent to the lever. It was then a gradual but inevitable step to generalize the definition of work, from the lever to all simple machines, and then to *all* machines, simple or not, and weight-raising or not. Amontons, again, was prescient in applying the concept of work to his 'fire-engine'—a link that was years ahead of its time.

One other philosopher who was years ahead of his time was the French engineer–scientist Parent (1666–1716). In 1704, he formulated his 'Theory of the Greatest Possible Perfection of Machines'.[19] He considered water-wheels, and made the crucial connection between the motion (*vis viva*) of a stream and the 'work' ('travail') that it could carry out. Also, the 'maximum power' ('effet') was equal to the mass of water times the maximum height reached when the stream was directed upwards, like a fountain. Moreover, he realized that all water-wheels worked in essentially the same way, whether exploiting a fall of water (as in the overshot wheel) or a moving stream on the flat (undershot wheel). Impressively, he used the new differential calculus, but he did make some unwarranted assumptions, and determined that the maximum efficiency of any water-wheel was 4/27. This was soon found to be a serious underestimate.

Another French engineer, Déparcieux, used another original line of argument (in 1752). He said that the maximum efficiency of an overshot wheel had to be 100%, as one overshot wheel would be able to drive a second one *in reverse* (using, for example, scooped buckets). (This assumed 'mechanical perfection'—no friction or loss of water due to splashing and so on.) This idea, of 'reversibility', would be used again, with great consequence, by the Carnots, both father and son (see Chapters 8 and 12).

By experimentation rather than argument, the English engineer Smeaton, in 1759, found that overshot wheels could do around twice as much work as undershot wheels. The Newtonian 'force-mechanics' could give no explanation, as total momentum was conserved in either case. But the continental 'live force mechanics' *could* explain it: the undershot wheel was situated on a fast-flowing stream, and there was considerable 'shock' to the wheel-blades, turbulence, and spray; whereas the overshot wheel was fed with slow-moving water from a millpond, and there was therefore less *waste* of live force (and less waste of water).

Thus a new concern, an *engineer's* concern—the efficiency of machines—led to new insights.

There were some natural philosophers who weren't that impressed by all this arguing from technology. D'Alembert, as we have noted, was one who tried to axiomatize mechanics from pure reason. Even Newton, while a gifted experimenter, was curiously silent on the new 'power technology' of water-wheels and 'fire-engines'.

But the links between work and *vis viva* slowly became more and more compelling, although they were not put into a theoretical framework until the work of Lazare Carnot in the second half of the eighteenth century (see Chapter 8). This new concept, work, justified its existence by its utility to the engineers, and by its universality, being applicable whatever the force and whatever the machine.

Stored Live Force

Let's now turn back to the first route to potential energy—the philosophers' intuitive understanding of potential energy, negatively, as a shortfall in kinetic energy. They understood potential energy as both a reservoir and a sink of kinetic energy.

Leibniz had the germ of an idea of potential energy. He called it *vis mortua* (dead force) and considered that it existed in cases of equilibrium, where motion had yet to begin—it was a 'solicitation to motion'.[20] Once new motion had been generated, then 'the force is living and arises from an infinite number of continuous impressions of dead force'.[21] But Leibniz did not consider that there was anything symmetrical about these two kinds of 'force'; there was no double-act, no constant exchange or interplay between potential and kinetic energy. In fact, *vis mortua* and *vis viva* were, for Leibniz, two very different kinds of entity:

living force is related to bare solicitation as the infinite to the finite, or as lines to their elements in my differential calculus.[22]

Johann Bernoulli, who, as we've seen in Part I, was a vociferous supporter of Leibniz's *vis viva*, recognized many examples of stored 'live force'; for example, in a stretched spring and a raised weight. Johann also moved a step closer to perceiving an equivalence between the dead and the live forces. He stated that *vis mortua* was consumed whenever *vis viva* was generated, and that the reverse was also true. This *had* to be so in order to guarantee that the cause was equal to the effect, and vice versa. Nevertheless, Johann still agreed with Leibniz that the relation of *vis mortua* to *vis viva* was like 'a line to a surface' or 'a surface to a volume'.

In modern terms, we agree with Johann Bernoulli that 'live force' may be stored, as in a longbow, a pendulum, a spring, a boulder, and so on. But we now appreciate that there's more to potential energy than mere potential. The common feature of all these examples is that they all exhibit a dependence on *positions* (the bowstring must be pulled back a few inches, the bob raised to a certain height—and also moved laterally— the spring squashed in or stretched out, and the boulder moved up a mountain from ground level). This last example brings out another telling feature—it is only the *relative* position that is important. The same is, in fact, true for all the other examples—the dependence is always on relative rather than absolute position.

The concept of potential energy did not win a quick or an easy acceptance. There was a tacit use of it for years before it was brought out into the open—and named (by Daniel Bernoulli; Part III). Something 'actual', like motion, is easier to understand than something latent. Also, *vis viva* and force are more intuitively obvious concepts, as they both relate to an individual body. Potential energy relates instead to the relative positions of bodies, or parts of a body, *within a system*—it should more properly be called the energy of configuration. As the geometers began to deal with special applications, involving more and more complicated scenarios (systems), the concept of potential energy began to emerge. The solutions to these problems would forge the new discipline of mechanics that would in retrospect be called Newtonian mechanics (although the practitioners, at least those outside Britain, did not feel that they were contributing to a Newtonian legacy). We now describe some of these special applications.

Special Problems

Clairaut

A problem that led to one of the most entertaining episodes in natural philosophy was that concerned with the 'figure' (shape) of the Earth.

Newton predicted that it was fatter towards the equator, whereas the French Royal astronomer Cassini (1677–1756) thought it was 'fatter' at the poles. The French Royal Academy of Sciences arranged and funded two expeditions to test these competing predictions. The philosophers had to measure the length, over the Earth's surface, of a one-degree change in latitude, along a given line of longitude. One team would visit Quito on the equator; the other team would go to Lapland in the polar region. It is with this second group that our story continues.

The philosopher and mathematician Maupertuis (see Part IV), the youthful Clairaut, the Swedish Celsers (of the Celsius scale), and others left Paris for Lapland in 1736. They had many adventures, including being welcomed by the King of Sweden, being ice-bound, having their signalling fires removed for firewood, enduring plagues of gnats, and so on. The exigencies of this undertaking can hardly be exaggerated (the expedition took almost two years, while the team in South America took over ten years, with some never returning). The eventual results supported Newton's theory. Voltaire was delighted, and described Maupertuis as the 'flattener of the poles and the Cassinis'.[23] (He also lampooned Maupertuis for bringing back to Paris two attractive Lapland women.)

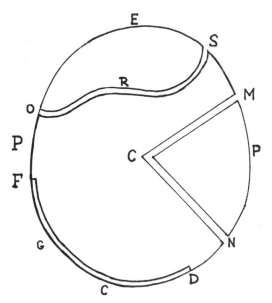

Fig. 7.2 Clairaut's 'canals', in his *Théorie de la figure de la terre*, 1743.

Now comes the bit of relevance to energy. Clairaut (1713–65) determined the equilibrium shape of a rotating, self-gravitating, fluid mass (the Earth) in his book *Théorie de la figure de la terre*,[24] in 1743. With a totally original argument, he imagined hypothetical canals within the body of the Earth (Fig. 7.2), and a hypothetical small test mass that could travel along these canals. He saw that equilibrium would only be maintained so long as no work was done by the test mass in going from one end of a canal to the other. Mathematically, the work would indeed be zero if the canal end-points had the same value of 'some function'—the potential energy function. This was the very start of potential function theory, continued by Laplace at the end of the eighteenth century (see below) and formulated rigorously by Green, in 1828 (see Chapter 13).

Johann Bernoulli

Johann, as we have seen earlier, understood the idea of potential energy as stored 'live force'. He used the ancient 'rule of accelerative forces', $F = ma$ (of course, this had nothing to do with Newton as far as Johann was concerned), to solve various problems. For example, in 1727, he considered the vibrations of a string (such as a violin string) fixed at both ends. (The mathematician and musician, Brook Taylor—see Chapter 5—worked independently on the same problem.) Johann implicitly identified the product of the tension and the extension of the string with potential energy, and found that the maximum potential energy was equal to the maximum *vis viva* of the string. He also carried out experiments, and found that Hooke's Law (Mariotte's Law in France)—that the extension was proportional to the force—didn't always apply (he was using gut, which is anomalous in this regard).

Johann also had the idea of 'work'. However, he made no link between the engineer's concept of work and his own. This was because he used it to solve problems in statics rather than in power generation. In fact, he used the idea of work to solve the case of static equilibrium in a novel way, and in a way that was to have enormous ramifications for mechanics. This was Johann's Principle of Virtual Work, and it was to be a founding principle of Lagrange's mechanics (see Part IV). It was never published by Johann, but was explained in a throw-away letter to Varignon in 1717 (fortunately, Varignon didn't throw it away). We shall discuss it further in Part IV.

Daniel Bernoulli

Like his father, Johann, Daniel attached huge importance to the concept of live force and also understood potential energy as a store of live force. In fact, Daniel was the philosopher who first introduced the term 'potential'. But Daniel went further than his father, as he appreciated the *equal* status of potential and actual live force. For example, for Daniel, the fact that the centre of gravity of an isolated system maintains a constant height was explained by the '*equality between the actual descent and potential ascent*'.[25] This was preferred to those 'Paternal words', the '*conservation of live forces*'.[26] There are countless other references to 'actual' and 'potential' live force in Daniel's book *Hydrodynamica*,[27] published in 1738.

Daniel's later work in celestial mechanics showed that he made the connection between the geometer's equation, $\int Fds = \frac{1}{2}mv^2$ (Equation 7.2), and the equality between potential and actual live force. This led Daniel on to making the association between $\int Fds$ and potential 'energy'—a totally modern outlook.

Daniel was one of the first (*the* first?) to intuit the important link between the stability of certain mechanical systems and the fact that the 'potential energy' was at a minimum in these cases. He passed this intuition on to Euler, and it helped the latter take the first steps towards the variational mechanics (see Parts III and IV).

Finally, Daniel Bernoulli was the only one of his contemporaries who saw kinetic and potential energy in a wider context, outside the confines of mechanics, and in the real world of engines and machines. One can sometimes substitute the modern word 'energy' for the old term 'live force' in his work without disturbing the meaning. It is for this reason that we consider him to be a hero of energy in the eighteenth century. His life and work is reviewed in Part III.

Euler

The leading applied mathematician of the eighteenth century—and, perhaps, of all time[28]—Euler (1707–83) certainly was the most prolific, publishing major books, and an average of a paper a week, for the whole of his working life. This was in the midst of the hubbub of family life (he had 13 children—eight of whom died in infancy—and 30 grandchildren). Overwork led him to lose the sight of one eye in 1735. In later life, his other eye deteriorated and he was forced to do his calculations in large characters on a slate. Finally, in 1766, he lost his sight altogether, but his work continued unabated (he carried out the calculations in his head). His publications also continued unabated, as he dictated them to his pupils and his children—in one case, even his tailor was enlisted (the tailor wrote down Euler's *Elements of Algebra*).

In 1736, Euler published his *Mechanica, sive motus scientia analytice exposita*.[29] This was a landmark in mechanics, flanked by Newton's *Principia*, in 1687, and Lagrange's *Analytique mécanique*, in 1788. In it, Euler put mechanics on rigorous analytical foundations: formulating Newton's Second Law as $F = ma$; defining angular momentum, and introducing its conservation as a separate axiom in addition to the conservation of linear momentum; defining the point-mass, the rigid body, and the continuum; identifying the rotation of a rigid body about three independent axes; and much more. In later papers, he recognized kinetic and potential energy for what they are (in today's terms), and saw that energy was conserved in time-independent cases. Also, it is largely due to Euler that our understanding of kinetic energy was extended from translations (linear motions) to motion along a curve. The rotational kinetic energy was still '½ × mass-term × speed²', but the 'speed' was an angular speed, and the mass distribution around the axis of rotation had to be determined in each case.[30]

Why, then, do we not consider Euler, also, to be a hero of energy? The answer is that Euler did not base a whole physics on energy and, indeed, hardly ever used the term 'live force'. He was hostile to Leibnizian philosophy (as espoused by Leibniz's follower, Wolff), especially the 'monadology' that implied an active force within each monad. As Euler explained in his *Letters to a German Princess*,[31] the force had always to be *external* to the body in question. Euler's views were therefore more aligned with Newton than with the Leibnizians. He may also have had religious objections to the latter: he was a staunch Lutheran and believed in a God by revelation rather than in a God who had rational requirements, such as the principle of sufficient reason and so on. (Wolff retaliated by saying that 'Euler is a baby in everything but the integral calculus'.)[32]

Boscovich

A Jesuit priest from Ragusa (Dubrovnik), in Dalmatia, Roger Boscovich (1711–87) had some far-reaching and iconoclastic views. In order to defend the principle of continuity, and to solve the problems of hard bodies in collision, Boscovich developed a new theory of matter in which atoms were 'points of force'[33]—indivisible and without extension. For small separations between these mass-points, there was a strong repulsive force that increased to infinity as the points approached each other. This prohibited the points from ever touching. As the separation was increased, the force alternated between repulsion and attraction, and eventually followed Newton's inverse square law of gravitational attraction. Boscovich illustrated his theory by his famous curve of force against distance. He recognized that the area under the curve was a constant, but he didn't make any links with 'energy'. In fact, Boscovich was antagonistic to the idea of *vis viva*, perhaps for the same reasons as Euler. In defining force as something that acts at an abstract point in space, Boscovich's work was a precursor to modern field theory; it is known that Faraday (see Chapter 15) studied Boscovich with care.[34]

Boscovich went on to write poetry, and to annotate Stay's *Principia* in verse form. (This was a pastime of many natural philosophers in the eighteenth century. For example, Erasmus Darwin, grandfather of Charles, put forward a proto-theory of evolution in the poems 'The Temple of Nature' and 'The Loves of the Plants'.[35])

Laplace

Some 30 years later, in 1785, the French mathematician Laplace (1749–1827) calculated the gravitational attraction between masses. This was no mere repetition of Newton's formula ($F = mMG/r^2$), because Laplace considered an extreme case: one mass was huge and extended (like the Earth) and the other one was vanishingly small, an abstract test-point (like a *jeu de paume* ball of unit mass). Also, instead of examining just one position for the test mass, Laplace, in effect, seeded the whole of space surrounding the large mass with possible positions (x,y,z) for the test mass. What emerged was new and interesting: the strength of the attraction at every (exterior) point could be determined by knowing a certain function, V, which depended *only* on position (e.g. V did not depend on the speed of the test particle). Also, the distribution of mass within the large body made no difference (so, for example, the *jeu de paume* ball would feel exactly the same effect, even if the mass of the Earth were concentrated at an off-centre point, like a huge, biased bowling ball). The function V took on a life of its own. It was the *energy*, in fact, the gravitational *potential energy* exterior to the large mass (once the mass of the test particle had been fed into the problem). The strength of the attraction at any point in space could be determined by seeing how V varied with respect to close *neighbouring* positions. This was determined in the famous Laplace's equation, which had the remarkable property that it represented the *smoothest* possible variation of V in space. (Laplace's equation was also remarkable in another way—it turned out to be applicable to a whole range of problems in physics.) The potential function, V, was a far cry from the simple, one-dimensional formulation, $mg\Delta h$ (the change in potential energy of a mass, m, for a change in height, Δh, near the surface of the Earth); however, it wasn't until well into the nineteenth century that V could be incorporated into the rest of physics, and seen as a bridge between Clairaut's and Green's potential functions (see Chapter 13).

Summary

By the second half of the eighteenth century, the ideas of potential and kinetic energy were beginning to emerge (but still not with these names). Kinetic energy was the energy of motion, now generalized to include rotations as well as translations. Potential energy was more than just kinetic energy-in-waiting, it was the energy of configuration.

Potential energy and work were eventually seen to be one and the same (an integration of force over distance). However, the terms 'work' and 'potential energy' grew to have a subtly different usage: 'work' was energy used up or generated by a force acting along a specific *path* in space, whereas 'potential energy' could be defined at a *point* (albeit relative to some given reference location). The concept of potential energy thus carries over into the modern field theory, where various properties (such as the height and direction of a blade of grass!)[36] can be defined at every point within the field. On the other hand, the fact that work is defined *along a path* is a feature that will have especial relevance to the next advance in mechanics, the 'variational mechanics' (Part IV).

The curious thing is that while mv^2 and force, F, were discovered by specific people, Leibniz and Newton, respectively, and arrived with much fanfare and controversy, potential energy just sort of slipped in uninvited, like the thirteenth fairy at the party.

Part III: Daniel Bernoulli, Unsung Hero of Energy

We have said how the controversy over *vis viva* went off the boil, but if there was one philosopher who kept it at least simmering, it was Daniel Bernoulli (we have met him before, in Chapter 5). In his great work *Hydrodynamica*, he stated the conservation of live force as his most important principle:

The primary one [principle] is the conservation of living forces, or, as I say, the equality between *actual descent* and *potential ascent*.[37]

In one fell swoop, Daniel Bernoulli gave primacy to energy conservation, and introduced, for the first time, the term 'potential'. This was a progression from his father's views, and an important step towards seeing kinetic and potential energy in equal terms.

Daniel, like his father, had started his studies in medicine rather than in physics, and at the end of his career he held professorships in both physiology and physics. This gave him a wider stage on which to see the applicability of the concept of live force. His doctorate on respiration, and an early paper on the effect of motion on the muscles, utilized the concept, and in a talk on the work-rate of the heart he made explicit reference to the conservation of 'living force'.

At the same time, Daniel had no woolly ideas about some vague, vital spirit permeating everything. Always, his use of the principle was strictly quantitative. We have seen, from Feynman's parable of Dennis' blocks, and from Galileo's exhortation to describe the world mathematically, that new concepts *must* be allied to formulae. Daniel Bernoulli certainly upheld this requirement. Time and time again, in *Hydrodynamica*, he uses wordy constructions in order to ensure that all physical concepts such as force, live force, and potential live force are quantitatively defined. For example:

the force of a fluid dashing against a perpendicular plane at a given velocity is equal to the weight of the cylinder of fluid erected above that plane, of which the altitude is such that from it something moveable, by falling freely from rest, would acquire the velocity of the fluid.

In two other respects (as well as his advocacy of the energy principle) Daniel was different from his contemporaries: he believed in checking his theories out against experiment, and he was interested in machines— the limits to their perfectibility and their utility to mankind. Bernoulli was, in other words, more of a *physicist*, whereas the others[38] were akin to applied mathematicians. For example, in Chapter IX of *Hydrodynamica*, Bernoulli introduced the word 'potential' again, this time in the context of a hydraulic machine, and defined the *absolute potential* as 'a weight, an activated pressure, or other so-called dead force . . . multiplied by the distance which the same moves through'. He then went on to make quantitative comparisons between the *absolute potential* of the machine and the 'work endured by day labourers elevating water'. It is a bit startling to read that the work 'is not to be interpreted in a physiological, but in a moral sense', but then he continues: 'morally I estimate neither more nor less the work of a man who exerts, at some velocity, a double effort than that of one who, in the same effort, doubles the velocity, because certainly either one achieves the same effect, although it may happen that the work of the one, despite being no less strong than the other, is very much greater in a physiological sense'. Also, he carried on, by a thread, the tradition that examines the *efficiency* of machines—from Amontons and Parent before him to the Carnots, father and son, around the turn of the nineteenth century (see Chapters 8 and 12). Bernoulli concluded (with regard to a hydraulic machine) that:

there is some definite termination of perfection beyond which one may not be able to proceed.

and

With the same absolute potential existing, I say that all machines which suffer no friction, and generate no motions useless to the proposed end maintain the same effect, and that one [design] is therefore not to be preferred to the other.

Such observations were to have revolutionary repercussions for thermodynamics in the nineteenth century.

In the last chapter of *Hydrodynamica*, Bernoulli put forward an innovative scheme for driving ships (what we would now refer to as jet propulsion):

I do not see what would hinder very large ships from being moved without sails and oars by this [the following] method: the water is elevated continually to a height, and then flows out through orifices in the lowest part of the ship, it being arranged so that the direction of the water flowing out faces towards the stern. But lest someone at the very outset laugh at this opinion as being too absurd, it will be to our purpose to investigate this argument more accurately, and to submit it to calculation.

The most original part of the book from the perspective of the developing concept of energy occurs in the chapter on elastic fluids (gases). We have already seen (in our Chapter 5) how Bernoulli proposed a kinetic theory of gases. His vision of a gas as made up of an almost infinite swarm of minute particles in ceaseless motion foreshadows our modern view (and contrasts sharply with the contemporary passion-fruit-pip-jelly model). He accounts for the pressure of a gas as being due to collisions of the gas particles with the container walls, and predicts that Boyle's (Mariotte's) Law will be followed, and that the pressure will be proportional to the squared speed of these gas particles. An interesting speculation of Bernoulli's is that a temperature increase would lead to an increase in the particle speed, and this in turn would lead to an increase in pressure. These ideas are almost identical to our modern ones.[39]

Bernoulli continued with a further amazing feat: he managed to derive a *quantitative expression* for the live force contained within the gas. How he did this was as follows. He considered the gas cylinder again, but this time the piston was imagined loaded with an extra weight that drives it down. He then applied Newton's Second Law to the falling piston, and so found its speed and its kinetic energy ('actual live force'). He then equated this to the 'potential live force' of the piston (its weight times the fall-height). But the two were not equal, and Daniel drew the correct conclusion—the discrepancy was due to the extra live force needed to compress the gas (what we now refer to as 'the work done by the piston on the gas'):

I say, therefore, that air occupying the space A cannot be condensed into the space $(A - h)$ unless a live force is applied which is generated by the descent of the weight, p, through the height, $A \ln(A/(A - h))$, however that compression may have been achieved; but it can be done in an infinite number of ways.

Crucially, Bernoulli now had an expression for this live force: p times $A \ln(A/(A - h))$. Granted that $A \ln(A/(A - h))$ is proportional to a change in volume, dV, and p is proportional to the gas pressure, P, then we end up with the formula PdV for the work done by the piston on the gas (or, as Daniel Bernoulli would have it, the 'live force contained within an elastic fluid'). The expression PdV was to become an iconic formula in classical thermodynamics—it represented another 'block' of energy. However, this part of Bernoulli's work seems to have been by-passed, and the expression was lost and then rediscovered over 50 years later (by 'Giddy' Gilbert in 1792), after Watt's invention of the steam engine (see Chapter 8).

Of course, Daniel Bernoulli was not able to see into the future, but he did appreciate the importance of his formula, as the following excerpts show:

This is an argument worthy of attention [the amount of live force in an elastic fluid], since to this are reduced the measures of the forces for driving machines by air or fire . . .

Since it happens in many ways that air is compressed . . . [naturally] by nature . . . [then] there is certainly hope that . . . great advances can be devised for driving machines, just as Mr. Amontons once showed a method of driving machines by means of fire . . .

I am convinced that if all the live force which is latent in a cubic foot of coal, and is brought out of the latter by combustion, were usefully applied for driving a machine, more could thence be gained than from a day's labour of eight or ten men.

Bernoulli immediately put his new formula to the test. Hales, in his *Vegetable Staticks* (see Chapter 5), had found that 'Half a cubick inch, or 158 grains, of Newcastle coal yielded 180 cubick inches of air'. Therefore (Bernoulli calculated), a cubic foot of coal would yield 360 cubic feet of air, and this would correspond to the 'falling of a weight of 3,938,000 pounds from a height of one foot'.

Of steam, Bernoulli wrote:

water [changed] to vapour by means of fire possesses incredible force; the most ingenious machine so far, which delivers water to a whole town by this principle of motion, is in London.

This was very likely Newcomen's revolutionary new engine (see Chapter 5). Finally, Bernoulli wrote of 'the astounding effect which can be expected from gunpowder'. His calculations showed him that:

in theory a machine is given by means of which one cubic foot of gunpowder can elevate 183,913,864 pounds to a height of one foot, which work, I would believe, not even 100 very strong men can perform within one day's span, whatever machine they use.

While today we no longer adhere to Hales' or Bernoulli's view that coal and gunpowder contain 'air' compressed to a very high degree, we see that Daniel Bernoulli's analysis demonstrates a highly original interpretation of the concepts of actual and potential live forces. He brings them ever closer to our modern conception of energy. His work was picked up by only a few members of the Swiss school (for example, Prévost; see Chapter 6) and then only to the extent of the kinetic theory itself, and not these later sections of Daniel's Chapter X in *Hydrodynamica*).

The energy principle was so important to Daniel that it permeated almost all his subsequent work (after *Hydrodynamica*). For example, he considered the case of a deformed elastic band (think of a ruler flexed into various shapes). In a letter to Euler, written on 7 March 1739, he proposed that when the potential energy of the band is at a minimum, the equation of the elastica (the shape of the curve) would follow:

I have today a quantity of thoughts on elastic bands . . . I think that an elastic band which takes on of itself a certain curvature will bend in such a way that the live force will be a minimum, since otherwise the band would move of itself [carry on moving]. I plan to develop this idea further in a paper; but meanwhile I should like to know your opinion on this hypothesis.[40]

Euler replied (on 5 May 1739):

That the elastic curve must have a maximum or minimum property, I do not doubt at all . . . but what sort of expression should be a maximum was obscure to me at first; but now I see well that this must be the quantity of potential forces which lie in the bendings: but how this quantity must be determined I am eager to learn from the piece which your Worship has promised . . .

But Euler still had years to wait. In 1742, Bernoulli wrote:

My thoughts on the shapes of elastic bands, which I wrote on paper only higgledy-piggledy, and long ago at that, I have not yet been able to set in order . . .

And finally, in 1743:

I express the potential live force of the curved band by $\int ds/r^2$ . . . Since no one has perfected the isoperimetric [variational] method as much as you, you will easily solve this problem of rendering $\int ds/r^2$ a minimum.

Daniel's idea that the stable state of the band occurred for a minimum in the potential energy was important in the emerging 'variational' mechanics (to be covered in Part IV). As Euler was to write:

the method of maxima and minima [could not be carried out until the] most perspicacious Daniel Bernoulli pointed out to me . . . a certain formula, which he calls the *potential force*.[41]

(It seems as if Daniel wanted help with the maths and Euler with the physics.)

There were other cases where Daniel employed a minimum principle. For example, already over 70 years old, he analysed the motion produced in a bar that had been struck at its mid-point. By assuming that the shape adopted was such as to minimize the kinetic energy, he was able to predict the subsequent motion of the bar. (This became, in the twentieth century, the Rayleigh–Ritz Principle.)

While Daniel's father, Johann Bernoulli, was deeply antagonistic to all theories Newtonian, Daniel actively supported Newton and promoted his work on the continent (in *Hydrodynamica*, he refers to Newton as the 'crown prince' of physics). In fact, it was Daniel's open-mindedness, and his ability to merge the British Newtonian and the continental Leibniz-ian traditions, that really opened the door to the new energy physics of the nineteenth century.

In a paper in 1738, Daniel used Newton's inverse square law of gravi-tational attraction to solve the problem of the motion of the Moon in a Sun–Earth–Moon system. However, Daniel's approach was energy-based rather than force-based—completely novel in this celestial setting. He calculated the *vis viva* of the Moon and the work done by it as it was brought to its orbit from infinity. This was only slightly more out-rageous an approach in the mid-eighteenth century than it would be today. (Who now would calculate the kinetic energy of the Moon?) Eu-ler's evaluation of this paper showed its importance:

I was pleased no end about His Honourable etc [Bernoulli's] '*Dissertation de Principio virium vivarium*', (firstly) for the idea itself to apply the '*Principium Conservationis Virium Vivarum*' to this material, as well as for the usefulness and incredible advantage which one gains by it, to find out the movement of the Moon, which otherwise, and only approximately so, must be derived by the most intricate equations.[42]

Daniel planned to extend this work with a more general treatment con-sidering several force centres. This he carried out 12 years later, in 1750. He considered a system of two attracting bodies, then three, and finally the general case of n attracting bodies; and he implicitly assumed central forces (forces dependent only on the distance between the body and the force centre). In effect, he calculated the potential energy of a system of n interacting bodies, and found that the potential plus kinetic energies summed to a constant value—but he preferred to interpret the results in terms of the route-independence of changes in kinetic energy. Bernoulli had considered only two types of central force—a constant (parallel)[43] force field, and Newton's inverse-square attraction. He stated his con-cern that the applicability of inverse-square attractions in non-terrestrial regions had not yet been demonstrated by observations. However,

Bernoulli understood that his energy principle was of general validity, and would still follow, so long as the forces were central. It is noteworthy that while developing these ideas of kinetic, potential, and total mechanical energy, Bernoulli still adhered to Newton's concept of force, considering it as real, and not just a convenient shorthand.

Why did Bernoulli's work go unnoticed? Why is he unsung, even to this day?

Bernoulli was ahead of his time in many ways: the frequent use of the 'energy principle', the kinetic theory of gases, calculating the work done by a gas, his use of minimum principles, his premonition of stored live force as a fuel, and his ideas about engines. He was an original and creative thinker in other areas as well. In the foundations of mechanics, he examined the parallelogram law for adding forces—was this law empirically or necessarily true? He also generalized collisions between bodies to include the case of angular momentum transfer. In analysis, he described the vibrations of a stretched string, fixed at both ends, as a superposition of an infinite number of simple harmonic terms. He also tackled some perennially hard problems, such as the tides, and rolling and sliding friction. His contributions to experiment and to engineering are too numerous to mention here. (He is most famous today for the 'Bernoulli effect'—the continuity condition requiring that fluid flowing around a curve speeds up relative to fluid flowing via a more direct route: the faster-flowing fluid exerts less pressure to the sides, and this results in aerodynamic 'lift'; for example, for air flowing around an aircraft wing.)

Outside of mathematical physics, Daniel Bernoulli was a leader in the area of probability and statistics. For example, he calculated the change in life expectancy of a population before and after inoculation against smallpox ('variolation')—this work influenced Catherine the Great.

However, in the area of mechanics, Bernoulli was overshadowed by Euler (Parts II and IV), the greatest applied mathematician of the age. Also, he was by-passed by a move towards a more abstract mechanics, led by d'Alembert, Euler, and Lagrange (to be covered below). This wasn't Bernoulli's style: he liked to keep a grip on the physical implications of his theories, and also to carry out experimental investigations to get a physical feel for things. In short, Daniel Bernoulli was that rare and unfashionable thing at the time, a true physicist.

Of course, Daniel Bernoulli may just not have had the right sort of personality to promote his ideas. We know very little about his private life, but he seems to have been more affable than his father or uncle. These two were often bitter adversaries, whereas Daniel was very close to his brothers (also mathematicians), even though they were more favoured by his father. (This is evidenced in Daniel's letters, where we find, for example, that he was very upset by the sudden death of his older brother, Nikolaus II, in St Petersburg; also, there is no mention of rivalry, or falling out, between Daniel and his brothers, or anyone else.) He corresponded with all the leading mathematicians of the age, and was a close friend to some (Euler and Clairaut). On finding that his father had pre-dated the *Hydraulica*, and used many of Daniel's results without acknowledgement, Daniel wrote in complaint to Euler:

Of my entire *Hydrodynamica*, of which indeed I in truth need not credit one iota to my father, I am robbed all of a sudden, and therefore in one hour I lose the fruits of a work of ten years. All propositions are taken from my *Hydrodynamica*; nevertheless my father calls his writings *Hydraulica, now first discovered, anno 1732*, since my *Hydrodynamica* was printed only in 1738 . . . In the beginning it seemed almost unbearable to me; but finally I took everything with resignation; yet I also developed disgust and contempt for my previous studies, so that I would rather have learned the shoemaker's trade than mathematics.[44]

Fortunately for physics, Daniel Bernoulli's wish could not be granted, although it is true that he never again reached such a peak in creativity. His book *Hydrodynamica* was written while Daniel was at St Petersburg between 1725 and 1733 (he was invited there when Peter the Great founded the Russian Academy of Science). Daniel encouraged Euler to join him at St Petersburg (Euler's relative Hermann—see Chapter 5— was also there at this time). But the Russian climate didn't agree with Daniel, and he eventually moved back to Basle, to become professor of botany and anatomy. Daniel's father held the mathematics chair at Basle, but when Johann died in 1748, the chair (due to his father's scheming?) skipped Daniel and went to his younger brother, Johann II. However, two years later Daniel acquired the physics chair, which he held until his death in 1782. (He also managed to swap botany for physiology). It is recorded that his lectures were accompanied by entertaining experimental demonstrations. He never married, and his nephews helped him with lecturing duties in the last few years of his life.

Part IV: Least Action, a Very Principled Tale

Introduction

We come now to a radical change of approach: from conservatory principles (such as for energy and momentum) to optimization principles. In this new approach, the quantity under investigation, while it isn't necessarily conserved, it is at least always used 'economically'. But what is the link with energy, and what of Feynman's allegory, of the need for energy to be conserved?

It turns out that the quantity that is optimized (in most cases, minimized[45]) is something called 'action', and that action has dimensions of energy × time. Also, the optimization of action will show that it is energy, rather than force, that truly characterizes a mechanical system. Finally, it turns out that that mighty cornerstone of physics, the principle of the conservation of energy, is not put in jeopardy, but, on the contrary, energy is shown to be such a crucial concept that it is essential to be able to analyse systems that are not necessarily isolated, and where energy can trickle in or out in a controlled fashion.

A Brief History of Least Action

It was noticed, even in ancient times, that optimization principles were operating. For example, according to Greek mythology, Queen Dido was granted land to found a city (Carthage) on condition that the land area was limited to the size of a bull's hide. Dido shrewdly cut the hide into thin strips, joined them end to end, and with the resulting very long strip chose just the right contour to maximize the land area (a semi-circle positioned up against the coast). Also: bee hives used the minimum amount of wax to store the maximum amount of honey; light rays took the least time to travel between two points (Fermat's Principle, 1662— think how impressive it was to know that light took *any* time to travel between two points); and the time for a body to fall between two points on a curve was minimized if the curve's shape was chosen correctly (Johann Bernoulli set this problem as a challenge in 1692; Newton took up the bait, anonymously, but 'the lion was recognized by his claw').[46]

Then, in the mid-eighteenth century, Maupertuis (we have heard earlier of his trip to Lapland), head of the Berlin Academy of Sciences, put forward a new principle, the Principle of Least Action. He proclaimed that the path of a particle of mass m, speed v, and small distance increments s, was always such as to make the quantity 'mvs' (the *action*) sum to a minimum. He vaunted his new Principle, extending it to the non-physical realm (his Principle of Maximum Happiness), and claiming that it was evidence for the existence of God. But many criticized Maupertuis—saying that he claimed too much, he stole the idea from others (Leibniz), and that, anyway, the Principle was philosophically flawed (how did the particle *know* what the right path was?). However, an influential colleague at the Berlin Academy, the famous Euler (Part II), was the perfect ally: he corrected Maupertuis' mistakes and defended Maupertuis' priority, even to the extent of down-playing the fact that he had discovered the Principle himself the year before.[47]

Not only did Euler cede priority to Maupertuis, but he actively promoted Maupertuis' case in the notorious König affair. This is an interesting tale, evocative of the spirit and characters of the eighteenth century, so we shall let it divert our attention for a short while.

The König Affair

According to the *Dictionary of Scientific Biography*, de Maupertuis was a difficult person: 'spoilt, intransigent ... of unusual physical appearance— very short and always moving, with many tics, careless of his apparel'.[48] However, things had been going relatively smoothly at King Frederick's Berlin Academy of Sciences (late 1740s), and Maupertuis had invited Euler to join him there. Maupertuis also ensured the election of a former protegé, Samuel König (1712–57). Some ten years earlier, Maupertuis and König, along with Voltaire, and Voltaire's mistress, Emilie du Châtelet (Part I), had formed a little group that actively promoted Newton's ideas on the continent. They had sometimes met to discuss philosophy at Mme du Châtelet's mansion at Cirey.

When König arrived in Berlin, in 1750, he was warmly received, but relations turned sour when König started to attack the validity of the Principle of Least Action, and, worse still, to say that it was really Leibniz's idea all along, as evidenced by a letter written from Leibniz to Herman. Maupertuis was furious and demanded to see the letter; but the letter could not be produced—it seems it went via a certain Henzi, of Berne, who had been decapitated. Then Voltaire arrived on the scene (in a distressed state, as Emilie had recently died in childbirth) and matters became even worse. Voltaire, who had once flattered Maupertuis with the soubriquet the 'great Earth-flattener' (see Part II), now turned against him and sided with König. King Frederick took the part of Maupertuis, president of his Academy, and generally tried to calm things down—but whether due to past jealousies over Emilie or current jealousies over the attentions of the king, relations deteriorated further. Euler, having an antipathy towards anything Leibnizian, sided with Maupertuis.

The affair reached a peak with the publication of Voltaire's vicious satire *The Diatribe of Dr Akakia, Citizen of St Malo*. Through the fictional Dr Akakia, Voltaire accused Maupertuis of plagiarism, of having a tyrannical rule over the Academy, and other things besides. (The syllable 'kak' would have sounded just as offensive in the eighteenth century as it does today.) Maupertuis became sick and, pursued by vitriolic volleys from Voltaire's pen, left Berlin for his home town of St Malo. He never really recovered and died some years later in 1759, in Basle, while stopping off at Bernoulli's (the younger Johann) on his way back to Berlin.

The Principle of Virtual Work

The first link between optimization principles and energy was due to Johann Bernoulli. His prescient idea, the foundation of what we now call the Principle of Virtual Work, was proposed in a letter in 1717 (fortunately it didn't get thrown away; see Part II). According to his idea, a certain quantity had to sum to zero rather than be maximized or minimized—but, in fact, all are examples of 'optimization'.

The Principle applies to statics—nothing is moving—and the system is said to be in (static) equilibrium. There has always been a problem with analysing such systems—is the fact that there is no movement an indication that there are no forces present, or are all the forces exactly balanced? For example: is the boulder on the mountain stationary because its weight is exactly balanced by an equal and opposite reaction from the ground, or (pretending that we know nothing about gravity) do the boulder and the surface of the mountain merely occupy adjacent positions in space? What is the difference between a rock-shaped pile of sand and a sandstone rock? Is the tent loosely stacked together (like a house of cards) or are there strong forces in the poles and stays? Is the lever stationary because all the forces are balanced (and all the turning moments too) or because there are, in fact, no forces?

Our instinct is to disturb the system ever so slightly, and see what happens. We try to roll the boulder, pick up the rock, wobble the tent, and jiggle the lever. In the Principle of Virtual Work, the same strategy is adopted, but the disturbance is a *mathematically imagined* jiggling rather than a real, physical one. A tiny *virtual* (imaginary) displacement is applied wherever there is a mass (strictly speaking, mass-*point*).[49] If there happens to be a force acting at the mass-point, then 'virtual work' will be carried out, and this work will be either positive (the displacement is along the direction of the force) or negative (the displacement is against the direction of the force). The total virtual work is then the sum of the separate tiny dollops of virtual work from all the relevant mass-points and, to guarantee equilibrium, this total must equal zero. As Johann Bernoulli expressed it, in his letter to Varignon:

In every case of equilibrium . . . the sum of the positive energies will be equal to the sum of the negative energies, taken as positive.[50]

What immediately commands our attention is the fact that Johann coins a new term for this virtual work—he calls it the *energy*. This is the first time that the word 'energy' is used in physics (this honour is usually wrongly awarded to Thomas Young, who arrived on the scene almost a hundred years later—see Chapter 10).

In essence, what is being postulated is that the system responds to a virtual nudge by reactive virtual tremors: the system shifts and resettles itself in exactly such a way as to counteract the nudge—the virtual adjustments act *in concert*, as a *system*. Also, a stronger virtual nudge will elicit a stronger virtual response. This is reminiscent of Newton's Third Law of Motion, which is also a postulate. However, Newton's Third Law does not apply in one go, to a whole system, but applies to one pair of interacting particles, and to another pair, and so on, considering each pair *separately* from the others (it also asserts a balance of forces rather than a balance of energies). It turns out that the Third Law is not *universally* applicable—it is only guaranteed in the interactions of stationary rigid bodies.[51]

D'Alembert's Principle

D'Alembert (whose biographical details are given later on) made an enormous leap, which extended Bernoulli's Principle from statics to dynamics. Now in dynamics things *are* moving. The forces (applied, like gravity, or internal, like cohesive and constraint forces) are no longer in balance (we are not in a state of static equilibrium) and so there *is* some residual force and some consequential motion of masses.[52] In fact, this residual motion is exactly that accelerative motion predicted by Newton's Second Law. D'Alembert's great leap forward was his realization that a 'mass undergoing an acceleration', in and of itself, constitutes a force. This newly postulated force, **I** (we give it this nomenclature[53] as it was later called the 'inertial force'), has magnitude *ma*, but acts in the *opposite* direction to the acceleration, in other words, we have $I = -m\mathbf{a}$. So far, so unremarkable (anyone can rename anything as they like), but d'Alembert understood that I really was a force—it could do 'forcey' things (such as push boulders over cliffs), and so it had to be included along with the above-mentioned applied and internal forces. When *all* these forces were considered together, then the Principle of Virtual Work should apply in the usual way.

This idea, of including the inertial forces (one for every mass undergoing an acceleration) along with the applied and internal forces, and then applying the Principle of Virtual Work, is what is known as 'd'Alembert's Principle'. In effect, what d'Alembert postulated was that the actual accelerations were exactly the right motions needed to bring the system into equilibrium—an equilibrium of a dynamic rather than a static kind.

Let us follow through how this works out in the simple case of just one applied force, F, and one mass, m. We have $F = m\mathbf{a}$ as usual. But now '$-m\mathbf{a}$' itself constitutes a new force, I, and so Newton's Second Law becomes $F = -I$. Rearranging, we end up with $F + I = 0$. But then this is a 'sum of forces totalling zero', and so equilibrium is assured. Still, it is most disconcerting to find that, in effect, $F = m\mathbf{a}$ has merely been rearranged to $F - m\mathbf{a} = 0$. How can such a trite rearrangement lead to anything new? What is even more astounding is that d'Alembert's Principle turns out to be the cornerstone of the whole of mechanics, and is the bridge that takes mechanics from the classical realm on to Einstein's General Relativity and quantum mechanics.[54] Underpinning d'Alembert's Principle is Bernoulli's sum of virtual energies—and so *energy* is playing a crucial role in all of physics.

There are many ironic twists along the way. First, forces still play a large part in the Principle, especially the discredited 'force of a body in motion' (Part I), now in the guise of 'inertial motion'. Second, d'Alembert reviled the concept of force, inertial or otherwise, considering it 'obscure and metaphysical'[55]—yet his Principle relies on an equilibrium of such forces. Finally, d'Alembert's Principle uses the content of Newton's Second Law, that a force causes a body to accelerate, and yet d'Alembert could hardly bare to admit this into his mechanics, which he tried to establish on purely rational grounds. He wrote (*Traité de dynamique*)[56]:

Why have we gone back to the principle [that $F = ma$], . . . supported on that single vague and obscure axiom that the effect is proportional to its cause . . . We shall in no way examine whether this principle [$F = ma$] is a necessary truth or not. We only say that the evidence that has so far been produced on this matter is irrelevant. Neither shall we accept it, as some geometers have done [Daniel Bernoulli], as being of purely contingent truth, which would destroy the exactness of mechanics and reduce it to being no more than an experimental science [!]. We shall be content to remark that, true or false, clear or obscure, it [$F=ma$] is useless to mechanics and that, consequently, it should be abolished.

That d'Alembert should have been the architect of his own Principle is truly amazing.

Before proceeding with the Principle, we shall skim through some biographical details of this 'sinister personality'.[57] We therefore leave the exalted plane of rational mechanics and land instead at the school of hard knocks, as Jean Le Rond d'Alembert was a foundling, left, when only a few hours old, on the steps of the church of St Jean le Rond, in Paris, on 16 November 1717. The police eventually traced his parentage to the famous *salonniere*, Mme de Tencin, and a cavalry officer, Chevalier Destouches. Mme de Tencin never acknowledged her son, but d'Alembert's father arranged for him to be fostered by a glazier and the glazier's wife, paid for his education, and left him a small annuity. D'Alembert wrote his best mathematical and literary works while living with his foster mother for 48 years. He finally 'weaned'[58] himself (as he called it) and went to live with his mistress, the famous *salonniere,* Julie de Lespinasse. As well as his work in mechanics, d'Alembert was also a literary figure and joint editor of the famous *Encyclopedie* with Denis Diderot. His introduction to the *Encyclopedie* was an important document in the Enlightenment and a manifesto for the *philosophes*. According to the *Dictionary of Scientific Biography*, d'Alembert was slight and had a rather high-pitched voice, but this only enhanced his reputation as a mimic, a wit, and a brilliant raconteur. He died in 1783, aged 66 years.

We return now to d'Alembert's ideas. The seeming triteness of $F - m\mathbf{a} = 0$ is explained by the fact that the especially simple scenario it describes is perverse in its simplicity. How can the virtual tremors of a whole *system* be manifest when that system is composed of just one particle and one applied force? (We would, likewise, not expect to be enlightened by applying, say, a factory management method to a factory with just one employee and producing just one item; or expect to appreciate the utility of the technique known as linear regression by taking the average of just two data points.) If we consider a scenario where there are many masses and a multiplicity of forces (some applied externally, some internal, and some 'reversed-inertial'), then d'Alembert's Principle makes sense, and is an essential analytical tool.

But why is the Principle so profound, reaching across most of physics? According to d'Alembert, a 'mass undergoing an acceleration' constitutes a force—but it doesn't matter how that acceleration has arisen: it could be consequential upon the application of a force ($-a = -F/m$); or it could 'merely' be due to the acceleration of the reference frame; or some combination of these. Stated another way, there is no way of telling—absolutely—if the effects one is experiencing are due to forces, or to the (motion of the) reference frame.[59] But, by the very fact of looking at a whole system, we are freed up from obscuring influences, such as the choice of this or that frame of reference. What are these whole-system parameters? They are the energies.

Lagrangian Mechanics

A synthesis of the Principles of Least Action, Virtual Work, and d'Alembert was brought about by Lagrange, and his *Analytique mechanique*[60] was the pinnacle of eighteenth-century mechanics (Hamilton—see Chapter 13—referred to it as a 'scientific poem'[61]). It was published in Paris in 1788, 101 years after Newton's *Principia*, one year after Mozart's opera *Don Giovanni*, and one year before the storming of the Bastille. Certainly, there are parallels between the elegance, depth, and beauty of Lagrange's work and Mozart's work, and certainly Lagrange's ideas were revolutionary. However, Lagrange's style in the *Analytical Mechanics* was most extraordinary: whereas d'Alembert shunned force and just about tolerated mass, Lagrange appeared to shun physics altogether. He proudly announced in the preface that the work contained no figures and no geometrical or mechanical arguments; there were, instead, 'solely algebraic operations'.[62] The work is as close to pure mathematics and as divorced from physical considerations as a work in mechanics can possibly be.

Joseph-Louis Lagrange (1736–1813) was an Italian from Turin. He eventually moved away from Italy, never to return, and worked first in Berlin and then in Paris (upon his arrival he lodged at the Louvre, at the invitation of Louis XVI). He worked for the Paris Royal Academy of Sciences until this was dissolved in 1793 by the new revolutionary government; he continued at the Commission of Weights and Measures, and subsequently for the Bureau des Longitudes at the new Institut National (the first meeting was in the Salle des Cariatides, in the Louvre in 1796); finally, he lectured at the Ecole Normale (it only lasted three months and 11 days) and at the Ecole Polytechnique.

Curiously, despite a lifetime's correspondence and collaboration with Euler, the other leading mathematician of the age, Lagrange and Euler never met. Euler, as we have seen (Part II) had a hectic family life, permanently surrounded by children and grandchildren: Lagrange, like Euler, was happily married, widowed, and then remarried, but expressly desired not to have any children, as they would interfere with his work. Euler and Lagrange did have something in common, though (as well as mathematics)—they both had good powers of self-preservation when it came to the tumultuous political, social, and economic upheavals of the eighteenth century. On being invited to the Berlin Academy of Sciences, Euler was presented to the queen mother of Prussia. She took great interest in the conversations of illustrious men, but could never draw more than monosyllables from Euler. When he was asked to explain his reluctance to talk, Euler replied 'Madam, it is because I have just come from a country [Russia] where every person who speaks is hanged'.[63] Lagrange had also formulated a prudent rule of conduct: 'one of the first principles of every wise man is to conform strictly to the laws of the country in which he is living, even when they are unreasonable'.[64] Laplace, a leading mathematician of a slightly later age, also shrewdly managed to keep in favour with whoever was in power on either side of the French Revolution. By contrast, some major experimentalists and engineer–scientists of the time—Lavoisier, Lazare Carnot, and Priestley—had a less detached attitude to the real world, and sometimes suffered the extreme consequences (see Chapter 9). Maybe this points to a difference between the personalities of theoreticians and experimentalists.

Lagrange founded a new energy mechanics—a way of solving mechanical problems that did not rely on Newton's '$F = ma$'. Lagrange wanted to dissociate his mechanics from Maupertuis' Least Action, or even Euler's version (which Lagrange called Euler's 'beautiful theorem'[65]), and go back to a more fundamental starting point—Johann Bernoulli's Principle of Virtual Work. This Principle, as we have seen (in the previous section), could be extended from statics to dynamics via d'Alembert's Principle. Now it's strange to say, but d'Alembert's Principle, while it relates to dynamics, still only applies at *one* instant. This can be remedied by applying d'Alembert's Principle over and over again at successive increments of time. Unfortunately, it would take too long (and too much mathematics) to explain, in detail, how this is accomplished,[66] but the outcome is momentous, and of great relevance to energy, and so we give it in synopsis.

Lagrange proceeds as follows. He defines the functions T and V (he was the first to give them these denominations), representing the kinetic and potential energies, respectively. As well as the usual energy conservation requirement, that $T + V$ is constant,[67] the *difference* between kinetic and potential energy, $T - V$, now gains in importance. Now, this new function, $T - V$, later called 'the Lagrangian', has to be a minimum when averaged over time; and the equations ensuring this minimum condition are known as 'the Lagrange Equations of Motion'[68] or 'the Lagrange Equations', for short (see Fig. 7.3). Despite Lagrange's aim to by-pass Maupertuis, it turns out that his method *is*, generically, a 'Principle of Least Action'.

Lagrange changed the world-view. He not only brought the energy functions, T and V, into physics for the first time, but he modelled reality in a totally new and more general way: in addition to Johann

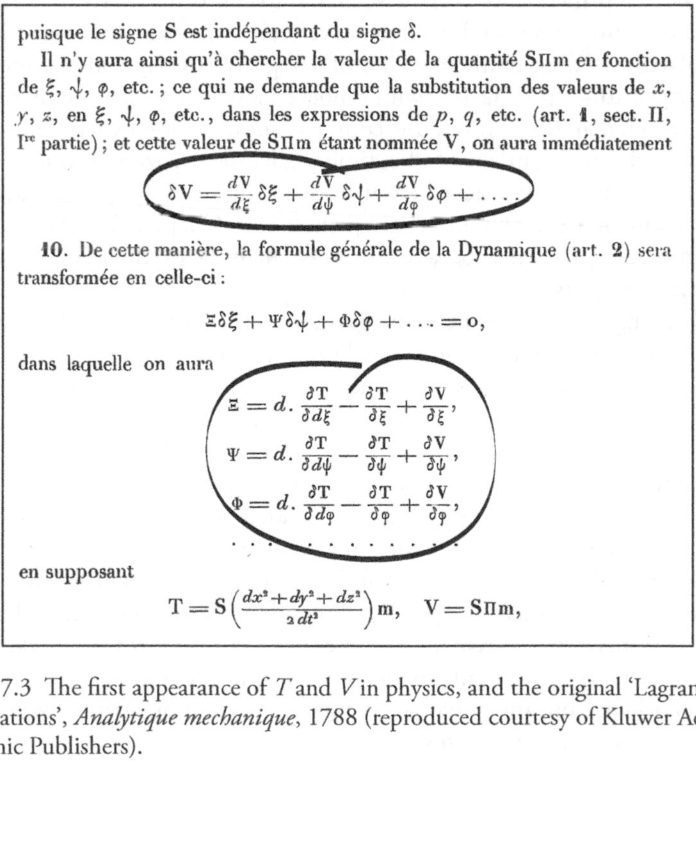

---

puisque le signe S est indépendant du signe δ.

Il n'y aura ainsi qu'à chercher la valeur de la quantité SΠm en fonction de ξ, ψ, φ, etc. ; ce qui ne demande que la substitution des valeurs de x, y, z, en ξ, ψ, φ, etc., dans les expressions de p, q, etc. (art. 1, sect. II, Iʳᵉ partie) ; et cette valeur de SΠm étant nommée V, on aura immédiatement

$$\delta V = \frac{dV}{d\xi}\delta\xi + \frac{dV}{d\psi}\delta\psi + \frac{dV}{d\varphi}\delta\varphi + \ldots$$

10. De cette manière, la formule générale de la Dynamique (art. 2) sera transformée en celle-ci :

$$\Xi\delta\xi + \Psi\delta\psi + \Phi\delta\varphi + \ldots = 0,$$

dans laquelle on aura

$$\Xi = d.\frac{\partial T}{\partial d\xi} - \frac{\partial T}{\partial\xi} + \frac{\partial V}{\partial\xi},$$

$$\Psi = d.\frac{\partial T}{\partial d\psi} - \frac{\partial T}{\partial\psi} + \frac{\partial V}{\partial\psi},$$

$$\Phi = d.\frac{\partial T}{\partial d\varphi} - \frac{\partial T}{\partial\varphi} + \frac{\partial V}{\partial\varphi},$$

$$\cdots\cdots\cdots$$

en supposant

$$T = S\left(\frac{dx^2 + dy^2 + dz^2}{2\,dt^2}\right)m, \quad V = S\Pi m,$$

Fig 7.3 The first appearance of T and V in physics, and the original 'Lagrange equations', *Analytique mechanique*, 1788 (reproduced courtesy of Kluwer Academic Publishers).

Bernoulli's particles, Lagrange allowed 'lever-arm', 'spinning top', 'oscillating spring', 'loaded beam', and so on; and instead of 'displacement of a particle', there could be 'angle turned', 'strain in a material', 'change in a coefficient', 'change in voltage',[69] and so on (these are called 'generalized coordinates', denoted, q); and, finally, instead of 'force', there was 'generalized force'. Each new system was modelled afresh, taking account of the 'degrees of freedom'—the set of independent 'motions' that characterize the system (a beam *bends*, a lever-arm *rotates* about the fulcrum, and so on). This modelling of the system is where the physical intuition, the physical nous, enters the problem. The advantage of Lagrange's generalized world-view was that prior knowledge of the physical set-up (e.g. the *symmetry* of a spinning top) could be exploited, and constraints could be built in (e.g. one could have a '*taut* cord', or a '*rigid* rod').

But what of energy? As far as energy is concerned, the important development was that force was generalized to such an extent that, in effect, it melted away altogether, and was replaced by the potential energy function, V. Also, the motions (of the generalized particles) were taken care of by the kinetic energy function, T. What Lagrange showed is that the system really was more than the sum of its elemental, Newtonian parts:[70] and it was the combined energy function, $T - V$, that captured these whole-system attributes.

Let's get a feel for how energy enters 'least action' by minimizing the time average of $T - V$ in three simple examples.[71] (One important point is that $T - V$ must be minimized between *definite* end-conditions: this may seem like a mere technicality but, if time goes on indefinitely, then it is evident that no minimum can be defined.) Consider, first, the two extreme cases where (a) there is only potential energy and (b) there is only kinetic energy. In case (a), T is zero, and therefore we have no motion—we are in static equilibrium. We still have our starting requirement that V is minimized—but this is exactly what we expect in static scenarios: the potential energy is at a minimum (cf. Daniel Bernoulli's analysis of flexed metal bands; Part III). Case (b) is even more interesting: here, we have no potential energy, and therefore the particle is 'free'. We already know what the resulting motion will be from Newton's First Law: the particle's speed, v, will be a constant, and so will its direction. However, we can get this result in a novel, non-Newtonian, way by minimizing T through time, between the definite end-conditions. As the positions at the start and end of the motion are given, we can see at once that the direction must be constant (any changes in direction would increase the distance to be travelled, and so use up more T). It is a bit more subtle, but we can also show that any deviation away from the mean speed (while travelling in a straight line) would also use up more T. The particle could wait a long time at the starting position, using up no T at all, but then it would have to make an almighty dash just before the time was up, in order to get to the final position on time; or it could accelerate like mad at the beginning, but then it would have to dally, as it mustn't arrive at the end position early. We can think of an infinity of other possible variations, but we are always constrained by the fact that the start and end times are predetermined. It is ultimately this fixity of the end-conditions which determines that the minimal T occurs when v is constant (and equal to the mean speed[72]). Finally, we look at case (c), where both T and V are present. For example, consider a ball thrown up against gravity. The ball rises up quickly, at first, using up lots of T, but we know that V (gravity, in this example) increases with height, so if we can get to a high enough region, then the large V will compensate the large T (remember that we are minimizing T *minus* V). Overall, we have a compromise between getting to a region with a large V, but not using up much T along the way.

An extra observation is that not only does action get minimized, but also the interchanges between T and V are always smooth and gradual. This can be seen in the motion of a swing (the swing's speed slowly decreases to zero as it achieves its maximum height); a spring (its speed increases smoothly as its potential energy decreases smoothly, and vice versa); and so on for countless other scenarios. This smooth interchange is a feature of Lagrangian mechanics:[73] T and V must be mathematical functions satisfying certain criteria of being continuous, finite, and so on, and their interchange is therefore always smooth.[74]

One more, extraordinary, feature of the method of least action is that the Lagrange Equations (the equations that follow on from the minimization of the time-integral over $T - V$) always have exactly the same form.[75] This is true even though the choice of generalized coordinates varies from problem to problem (it is not even unique for a given problem), and the respective functions for T and V must be worked out afresh each time. To repeat, the functions T and V are different for each new problem, and yet the Lagrange Equations, taken as a whole, remain exactly the same. The invariance of a set of equations is astounding and puzzling, but it results from the fact that we are using a minimization principle. This sort of principle has the remarkable attribute that it winnows away all the unwanted model-dependent features, and all those dependent on the reference frame. Some analogies can be found in physical geography: the Khyber Pass is the shortest route through the Hindu Kush from Peshawar to Kabul, and this is true whether a Mercator's or a Peter's projection is used. Likewise, Mount Everest is the highest mountain, and it doesn't matter whether the scale is in metres or feet. The 'shortest route' or 'highest peak' are superlatives ('extremal' or 'optimized' features) and these are preserved whatever coordinate system is used, and even for different maps.[76] Yet an analogy always has something wrong with it: in physical geography, the extremals occur in a real, physical landscape, whereas in Lagrangian mechanics, they occur in a landscape that is imagined, virtual.[77]

There is still one more puzzle to answer: Why is it exactly '$T - V$' that must be minimized? Why is it not '$T + V$' or some other combination?[78] Now we must appreciate that the T-function determines the *motions*, while the V-function relates to the applied *forces*. But d'Alembert introduced an ambiguity between the *motions* of a reference frame and inertial *forces* (see the section on d'Alembert's Principle), and this translates into an ambiguity between T and V, a blurring of the distinction between kinematics (masses in motion) and dynamics (forces generating and altering motion):

(1) Take the case of a bead constrained to move on a curved wire (ignoring friction). The bead has only one degree of freedom (motion along the wire) and so only one coordinate is needed—the distance moved along the wire from some fixed point. However, as the bead twists and turns on its way, forces are needed to keep it on the wire, and the tighter the curving of the wire, the greater is the force required to keep the bead on. These forces are not explicit (they don't show up in the function V); rather, they are implicit in the shape of the wire. Although I have called them 'forces', they show up as an extra term in T, in addition to the usual $\frac{1}{2}mv^2$. In summary, it is the *curvature* in the wire that leads to *an extra term in T*.

(2) On the other hand, there are scenarios in which there is now a link between V and 'curvature'. Consider the case of a meteor passing near a planet: it could have its path deflected, or go into orbit, or even be drawn down towards the planet; the stronger the gravitational potential, V, the greater is the *curving* of the meteor's path.

The T and V components are getting mixed up, and are sort of bootstrapping each other into existence. We said earlier[79] that d'Alembert's Principle anticipated Einstein's Principle of Equivalence; now we are finding that Lagrange's mechanics anticipates Einstein's Theory of General Relativity, the essence of which is captured in the slogan 'matter tells space how to curve; curved space tells matter how to move'.[80] But as there is this feedback between T and V, we have the danger of each reinforcing the other endlessly, and this could lead to a runaway growth in the total action. This would be highly unphysical—it could lead to perpetual action (as bad—in fact, the same—as perpetual motion; Chapter 3). T and V *must* therefore act in opposition; they must each act such as to limit the effect of the other, on average, over time. Thus it is $(T - V)$, the *difference* between T and V, that must be minimized through time.

Overview

In this chapter, we have seen the evolution of mechanics over 101 years, from Newton's Laws of Motion in 1687 to Lagrange's *Analytical Mechanics* in 1788, via Euler's *Mechanica* in 1736. Leibniz introduced *vis viva*, the kinetic energy in all but name, and potential energy slipped in over time. Daniel Bernoulli championed 'energy' before it was fashionable to do so. He developed the kinetic theory of gases and, remarkably, managed to derive an expression for the energy (the 'live force') in the gas.

The 'principle of the conservation of live force' (really, the conservation of kinetic, T, and potential, V, energy together) emerged as an important tool in the analysis of mechanical problems. For example: the height attained by a pole vaulter could be linked to the athlete's speed during the approach; the range reached by the debris from an explosion could be predicted from the 'strength'[81] of that explosion; and (modern-day) teasers could be answered, like the one about the weight of a lorry of homing pigeons before and after the pigeons have left their perches.[82] It is hard to see how these examples could be solved using forces. However, a force analysis was still useful in certain cases, especially where directions had to be determined (e.g. the directions of three taut cords supporting a mass in static equilibrium).

By the end of the eighteenth century, it was understood that the energy method is made up from *two* principles: the aforementioned conservation principle and an optimization principle (the minimization of the time average of $T - V$). While the conservatory principle leads to a knowledge of what the 'blocks' of energy are (the actual formulae for T and V), the minimization principle leads to something new: an understanding of *why* the mechanical energy is transformed between its various 'blocks', and of the striking qualitative feature—that this transformation is usually *smooth*.[83] For example, a child on a swing, an oscillating spring, a falling apple, and a spinning and bobbing gyroscope all demonstrate a smooth and continuous conversion between kinetic and potential energies (cf. the frontispiece to this book, Poussin's *Dance to the Music of Time*). Together, these conservatory and optimization principles make up the energy or systems view. Let us compare and contrast this systems view with the Newtonian force view.

In the Newtonian case, we consider an individual particle and watch it as it goes here and there, exposed to absolute, externally applied forces. In the energy approach, we look at a *whole system*, and the scalar functions T and V (the kinetic and potential energies, respectively), with appropriate boundary conditions, are the prescriptions that completely determine the progress of the system through time. Which approach is better; which is more fundamental?

Of course, where they are applicable, then both approaches work. For example, consider a cyclist going round the cambered track at the velodrome. In the Newtonian method, we carry out the vector sum of forces, and we must take into account the pull of gravity, the reaction of the tyres against the surface of the track, and the 'centripetal force' as the cyclist takes the curve. In the energy approach, we find that, in order for T and V to best counteract each other on average through time, the cyclist must climb the track to an intermediary height, somewhere between the highest and lowest extremes of potential and kinetic energy.

It is chiefly with regard to the systems aspects that the two approaches differ. In the Newtonian outlook, the problem has been broken down into the participating mass-points and the forces acting on them. There are no precepts such as 'a bicycle'. These Newtonian concepts—masses and forces—feel more elemental and intuitively comprehensible than kinetic and potential energy; and when a handful of mass-points and forces are considered together, isn't this all there is to a system? Not necessarily. In the energy mechanics, the masses don't simply inhabit space and wait for a force to come along; rather, the sharp distinction between masses and forces on the one hand, and space and time on the other, has softened. New precepts blending massy and space–time aspects, the energy functions T and V, are now the ingredients that make up the system. But there is a price to pay for the increased versatility of the systems approach; it is that these new concepts, T and V, the 'blocks' of energy, must be redefined for each new system, and have lost the intuitive immediacy of the forces and mass-points.

All of this feels unnecessarily complicated in the case of, say, two masses attracting each other by a central force—and so it is. Likewise, as children, we have need only of counting-numbers, and real numbers ($\sqrt{2}$, -10.333, 6.02×10^{23}, etc.) would be an obscuring complication. Newton's attempt at reducing all physical phenomena to just three laws of motion was possibly the most important single leap forward that has ever happened in physics, and if we had only to deal with a small number of slow-moving neutral particles then, yes, Newton's vectorial mechanics would answer the case. But we shall find (in the subsequent chapters) that our physical world—with high speeds, moving charges, continuous media, statistical assemblies, microscopic particles, and very large gravitating masses—is better served by the systems or energy view. Let's survey the advantages and disadvantages of this energy view.

First, as we have already said, the energy approach may be the only method that easily provides an answer (cf. pole vaulting, explosions, and lorry loads of pigeons).

Second, the problem may be modelled[84] so as to incorporate prior knowledge of the system, such as pre-existing symmetries, or built-in constraints. This represents an enormous simplification, opening up the number of problems that can be solved.

Third, in the systems approach,[85] effects are always transmitted by small, *local* variations, and yet the boundaries can always be made sufficiently extensive.[86] Certain philosophical problems ('action-at-a-distance') are thus avoided. This brings me on to teleology. This really is a red herring in both the energy and force views: the system is no wiser in 'choosing the right path' than is Newton's individual particle in 'knowing how to solve' $F = ma$ in order to 'know where to move to next'.

Another philosophical problem concerns compliance with the Principle of Relativity—discovered by Galileo and reasserted by Newton (see Chapter 3), and later, in a more general form, by Einstein (see Chapter 17). In order for a meaningful mechanics to emerge, something absolute about motion has to be extracted. Newtonian mechanics deals with this by looking at the *change* in the motion (i.e. the acceleration) of a body as it is subjected to a force. As a change involves a comparison of motions, then this change can be determined absolutely, even while the motions themselves are relative. But we shall find that the Newtonian stratagem doesn't always work: sometimes accelerations cannot be determined absolutely. However, there is another route to the absolute.

An outstanding feature of Lagrangian mechanics is that the 'least action' for the given scenario is an invariant quantity—it doesn't change with the reference frame.[87]

As regards disadvantages, the systems approach lacks intuitive simplicity: forces have become 'generalized', and space and time may be melded together, or may be 'massy', or chopped into infinitesimal chunks. One of the hardest adjustments to make concerns the evident directionality of motion—how can mere scalars, T and V, hope to capture this directional quality? But upon deeper reflection, we see that direction *is* accounted for—in the directions implicit in T and V, and in the kinematical constraints. For example, billiard balls *approach each other* in a head-on collision, an apple falls *towards* a lower gravitational potential, and the bead *twists* and *turns* as it moves along the *curved* wire. In other words, there are directions, but all are defined relative to something within the system; none are absolute. (There is also the direction in time, either forwards or backwards, but that is another story—for the time being.)

Another disadvantage is that T and V must be in 'functional form', and so dissipative effects (friction, air resistance, etc.) can't easily be accounted for. In Newtonian mechanics, these dissipative effects *can* sometimes be treated, by modelling them as one 'dissipative force'.[88]

Some problems are endemic to both approaches. For example, both the force and energy views share the problem of 'circularity'. Force and mass are introduced together (in Newton's Second Law), and so the definition of one is dependent on the definition of the other. Likewise, in the systems view, we have the problem of defining which effects are inside the system and which are outside the system. Resolution in both views comes from the innumerable experimental observations where the consistency of the initial definitions is supported.

To summarize, the Newtonian approach considers the viewpoint of an *individual* (particle or body) as it goes along its path in life, subject to all the 'slings and arrows' (forces) of 'fortune'. This approach is more intuitive but brings in metaphysical problems of absolute force, space, time, and mass. In the energy view, the *systems* view, T and V are not absolute but their *difference* is an absolute invariant, when it is minimized through time, subject to certain end-conditions. Thus both energy *and* time are the determinants of 'what happens'.

The blocks of energy, T and V, introduced in this chapter are both types of *mechanical* energy. Classical mechanics has to do with 'matter in motion' (levers, projectiles, pulleys, springs, etc.), and ignores everything that is warm–fuzzy, wet–sticky, and so on. For example, we can predict the range of the debris from an explosion, but not explain the chemical cause of that explosion. There is a sense in which mechanics only bites off what it can chew—it deals with systems that can be modelled in a certain way (functional forms, smooth changes in the coordinates, etc.). The puzzle is that these concepts, T and V, end up having a more cosmic role outside the narrow confines of classical mechanics. Staying within the realm of mechanics, a case could be made—a rather poor case—that we could get by with Newton's Laws, and without even the concepts of T and V. It is only when we start to look outside mechanics and discover other 'blocks of energy' that we come to realize that energy is an indispensable concept. So, for now, we leave the arcane world of algebraic manipulation, the scratching of quill on paper, and look for some other 'blocks' in the fizz, bang, and crackle of chemical reactions, gases, electricity, and light, and in the belching and clanking of engines.

8

A Tale of Two Countries: the Rise of the Steam Engine and the Caloric Theory of Heat

In France, this is a tale of one city, Paris, but in Britain it is a tale of the whole country, and especially of the Scottish university cities of Glasgow and Edinburgh, and the new provincial centres in the Midlands, Manchester, and Cornwall.

Our main characters are James Watt and Henry Cavendish in Britain, and Lazare Carnot and Antoine Lavoisier in France. As far as Lavoisier was concerned, it was the worst of times (he lost his head to the guillotine), but for French science it was the best of times. Science was not merely a by-product of the French Revolution; it was the *chief cultural expression of it*, ahead of, say, achievements in the arts, music or literature.[1] The list of French scientists and mathematicians from this period (from the Revolution to the Restoration) goes on and on: Lagrange, Laplace, Monge, Condorcet, Jussieu, Lamarck, Cuvier, Saint-Hilaire, Bichat, Lavoisier, Berthollet, Biot, Poisson, Gay-Lussac, Fourier, Coulomb, Coriolis, Navier, Poncelet, Arago, Ampère, Fresnel, Legendre, Galois, Lazare Carnot, Sadi Carnot, Magendie, Cauchy, and others.

The scientists in these two countries could hardly have been more different. The French were salaried professionals and worked in the new French Institute and later in the École Polytechnique. The British were individuals—from rich gentleman-scientists to poorly paid laboratory technicians. They worked from home in their own private laboratories or on site at mills, mines, and in the new 'manufactories'; and they formed their own societies to discuss 'matters scientific and philosophical';[2] for example, the Lunar Society in the Midlands (so-called because the members met when there was a full moon and they could travel safely in their carriages). Paris was nevertheless the scientific hub, and everything French was the touchstone of style. For example, Matthew Boulton, Watt's backer and business partner, welcomed the 'lunatics' to his 'Hotel d'amitié sur Handsworth Heath'.[3]

While the French were in the throes of revolution (political and social), something unplanned and unprecedented was also happening in Britain—industrialization. The pattern of life that had persisted for hundreds of years was changed in just a few decades and forever. It was all to do with engines and the generation of power.

Part I: Engines

In Lancashire, for example, virtually every stretch of every river and tributary had a water-wheel on it by the end of the century. These were to drive the new spinning and weaving machines of the burgeoning textile industry. But in summer millponds can dry up (yes, even in the North of England) and soon steam engines were being used—not to drive the textile machines themselves, but to pump up the water to feed the water-wheels. In the Cornish copper and tin mines, steam engines were again used as pumps, but here water was the enemy, and the engines were to stop the mines from flooding. In 1777 there were just two Watt engines in Cornwall; by 1800 there were 55.[4]

Perpetual motion machines had been shown to be impossible, but these proliferating new machines *could* keep on running provided that they had a source of power. Wind, water, horse, manual, and heat (from burning wood or coal) were all exploited. (In one case, even a Newfoundland dog was used to turn a wheel.) While all these 'fewels' were understood as a source of 'power', the idea of some abstracted essence common to all was still not apparent. Boulton was a new breed—a venture capitalist and a very advanced thinker. He alone had the understanding that power was soon to be of huge *economic* significance. 'I am selling what the whole world wants: POWER'[5] he wrote to the Empress Catherine of Russia.

Watt

The young James Watt (1736–1819) loved tinkering with all the instruments in his father's shipwright's workshop on the banks of the Clyde near Glasgow. He knew what he wanted to be when he grew up—a maker of instruments, particularly mathematical instruments. After an arduous apprenticeship in London, Watt returned to Glasgow for health reasons, but he couldn't obtain employment as the guilds in Scotland were too strong. Only one door was open to the young apprentice, and that was instrument-maker at the University of Glasgow. So it was, around 1760, that Watt came into contact with Joseph Black, professor of natural philosophy in Glasgow at this time (see Chapter 6).

Black soon came to appreciate Watt's natural talent for any technical problem, and their relationship quickly developed into one of professional co-operation and friendship rather than professor and subordinate. As the student, Robison, recalled: 'Dr. Black used to come in [to Watt's rooms] and, standing with his back to us, amuse himself with Birds Quadrant, whistling softly to himself . . .'[6]

Watt's duties included preparing apparatus for lecture demonstrations. One commission was to bring to working order a benchtop model of a Newcomen engine (the workings of the Newcomen engine are described in Chapter 5). Watt became engrossed in this particular project: he not only got the engine to work but started to wonder at all the factors limiting the functioning of the little engine.[7] Why didn't it work as well as the full-sized engine?

Watt observed that after only a few cycles the model engine slowed down, became very hot, and soon ran out of steam and stopped altogether. He thought deeply about these symptoms and read widely. From his reading of Boerhaave, especially, he realized that the model engine, being small, had a large ratio of surface area to volume, and so the surface of the cylinder would have a disproportionate effect on cooling the steam as compared to the full-size engine (Boerhaave had written about how such geometric factors could influence the rate of cooling).

These were impressive conjectures but Watt proceeded to check them out. He determined that a cubic inch of boiling water turned into about a cubic foot of steam, and he also measured how much water was boiled away in a typical run of the engine. Thus he found that the engine used up many times more steam than was required merely to fill the volume of the cylinder. He presumed that this excess was needed just for the purpose of warming up the cold cylinder at the end of each cycle.

Perhaps making the cylinder out of a different material would lead to less waste of steam. Watt therefore determined (what would later be called) the heat capacity of various materials (iron, copper, tin, and wood), and experimented with wooden cylinders as well as the brass one. The wooden cylinders did remain at a more even temperature than the original brass cylinder, but they were mechanically unsatisfactory (the wood was prone to warping and cracking). It can be noted that, at this stage, Watt had never heard of Black's concept of specific heat capacity.

Another way to economize (on steam and hence coal) would be to avoid cooling the cylinder any more than needed just to cause condensation. Watt therefore investigated the optimum amount, and temperature, of condensing water required. In so doing, he effectively determined the latent heat of condensation of steam—but, at this stage, Watt hadn't heard of Black's concept of latent heat.

However, Watt found that when he operated the engine with this minimum of cooling water, the power of the stroke was severely curtailed. Again, after critical thinking and reading (on Cullen's experiments on the boiling of tepid water in *vacuo*; see Chapter 6), Watt realized what the problem was: warm water left in the cylinder after condensation would boil and generate a back-pressure, vitiating the vacuum, and reducing the strength of the power-stroke. Once again, Watt diligently followed up this line of enquiry. He used Papin's 'digester' (see Chapter 4) to find the law of the increase of steam pressure with temperature, plotted the results as a graph—a very rare procedure in those days—and extrapolated back to find the pressure corresponding to the temperature of the leftover condensing water.

At this point, Watt faced a dilemma. For economy, the cylinder needed to be kept as hot as possible (as hot as the steam), while for maximum power, it needed to be as cool as possible (cooled down once every cycle): but how could the cylinder be kept hot and cold at the same time?

Then, in 1765, in a moment of deep reflection (apparently, he was walking across Glasgow Green on a Sunday afternoon),[8] Watt suddenly thought of the solution: he had to divert the steam from the working cylinder into another cylinder; and then this *separate* cylinder, where the steam would be condensed, could be kept permanently cold. Legend has it that Watt's 'eureka moment' was consequent upon his first hearing of Black's theory of latent heat, but Cardwell demonstrates that this is nonsense.[9] It seems that Black and Watt simply pursued their own researches, at the same place (the University of Glasgow) and at the same time, but quite independently of each other (at least, during these early years).

The advantages of separate condensation would not only be an enormous saving in fuel (coal for the boiler), but a more powerful engine (the vacuum was better), and one where the interval between power-strokes was much reduced, as there was now no need to heat up and cool down one and the same cylinder every cycle. Thus, in 1764, was born Watt's revolutionary 'condensing steam engine' (see Fig. 8.1). (We shall not detail the history of its manufacture.)

Watt didn't leave it at that, but continued to improve his engine. He diverted some of the steam coming from the boiler, and used this steam—rather than cold atmospheric air—to drive down the piston at the end of each cycle. Even more noteworthy, he continuously throttled down the amount of steam allowed to leave the main cylinder and be fed into the separate condenser—so that, at the end of the stroke, every last wisp of steam pressure had been exploited. This became known as Watt's 'method of expansive operation' and, in the hands of Sadi Carnot, it was to become a crucial thought experiment in the founding of thermodynamics (see Chapter 12).

In order to put 'expansive operation' into practice, Watt needed to continuously monitor the steam pressure—but everything inside the engine was hidden from view. So Watt commissioned his assistant, John Southern, to come up with a solution. Southern's solution was ingenious: a pen was held in a clamp that moved with the piston, as the latter rose or fell vertically, due to the expanding or contracting volume of steam; at the same time, the clamp was moved horizontally by a small spring-loaded piston that measured the steam pressure. The pen was poised over a fixed sheet of squared paper, and simultaneously traced out these vertical and horizontal motions. What resulted was a curve that showed the variation of volume with steam pressure—this curve became known as the 'indicator diagram'.

What now seems like a simple chart-recording was, in fact, a revolutionary new practice of inestimable significance. More 'analogue to graphical' than 'analogue to digital', it was a mathematization of the fundamental processes of the engine (see below). Boulton and Watt kept this innovation top secret, but all the competitors were desperate to know how the indicator diagram was generated. The cotton-spinning magnate George Lee complained 'a physin can give no good account of ye state of his Patient's health without feelg his Pulse, & I have no Indicator',[10] and also 'I am like a man parch'd with thirst in the Expectation of relief, or a Woman dying to hear (or tell) a Secret—to know Southern's Mode of determining Power'.[11]

In later years, Watt further improved his full-sized commercial steam engine by making it 'double-acting' (both up and down movements of the piston constituted a power-stroke) and, in the mid-1780s, able to provide rotative power directly (although this design was not fully satisfactory). Both advances required Watt to invent exceedingly clever new

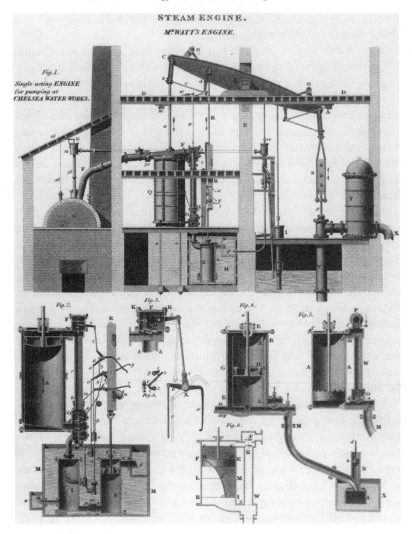

Fig 8.1 Watt's single-acting steam engine, Chelsea, end of the eighteenth century, from Rees' *Cyclopaedia* (Science Museum, London/SSPL).

mechanisms: the 'parallel motion' and the 'sun and planet gearing'. In his retirement, he told his son that he was 'more proud of the parallel motion than any other mechanical invention I have ever made'.[12]

It is clear that Watt was a genius and no mere mechanical maestro of the sort that can make anything work just by walking into a room (I don't believe in such sorts anyway)—he always tried to *understand* the underlying scientific principles. Nevertheless, it is fair to say that Watt's motivation was always to improve the efficiency and economy of any design or process rather than to study science for its own sake. When working on textile bleaching using chlorine, for example, Watt was horrified that the inventor of the process, the French chemist Berthollet (1748–1822), was 'making his discoveries publick'.[13] Berthollet replied that 'Quand on aime les sciences, on a peu besoin de fortune'.[14] In his memoirs, Watt admitted that he had not had a day free of worries about money until his retirement, aged 80.

(Watt did make one significant contribution to pure science—the idea that water was not elemental. Black, on the other hand, was of the opinion that water was essentially an 'earth', as it could be reduced to this by evaporation of the liquid parts.)

The revolution brought about by Watt's engine required more than the invention itself, but also new people (venture capitalists), new social and financial structures, and so on. Watt needed to define quantities such as 'efficiency' and 'duty' in order to prove that his engines were better than the competition, and also to know how much to charge the customer. He looked back to Parent's 1704 treatise on water-wheels (see Chapter 7, Part II)—and in doing this he was almost certainly the first to apply the theory of water-engines to a heat-engine. (Watt taught himself French, German, and Italian specifically for the purpose of being able to read foreign texts.)

Parent had defined the efficiency of a water-engine as the percentage of water that could be returned to the starting height. For his steam engine, Watt defined the efficiency (also called the 'duty') as the ratio between the 'work done' to the fuel (bushels of coal) used up. But how should the 'work done' be determined? Watt stuck with common engineering practice, and used the specific example of weights lifted through a given height (in fact, pound-weights lifted through a foot), a measure that had been initiated by Descartes over 100 years earlier (see Chapter 3). Finally, Watt introduced a new measure—for the *rate* of doing work—the horsepower, which he set at 33,000 foot-pounds per minute. (To commemorate Watt, we now have the SI unit of power called the watt, which is defined as one joule of energy per second.) The owners of a Watt steam engine enjoyed free installation and servicing, but had to pay a 'royalty' equal to a third of the savings on their previous fuel bills; for example, the cost of hay for horses.

A competitor of Watt's was the Cornish engineer Jonathan Horn-blower (1753–1815), inventor of the compound steam engine (similar to Watt's engine, but with additional—'compounded'—cylinders). In 1792, Hornblower applied to Parliament for an Act to extend his patent. He asked Davies Gilbert (1811–54), a Cornish MP and also a mathematician (and later president of the Royal Society), to help him. Gilbert (known as 'Giddy') showed that the area under the graph of pressure versus volume (in other words, the integral, $\int P dV$) was equal to the work done by the engine. Alarmed, Watt wrote to Southern: 'they have brought Mr Giddy, the high sheriff of Cornwall, an Oxford boy, to prove by *fluxions*, the superiority of their engine; perhaps we shall be obliged to call upon you to come up by Thursday to face his fluxions by common sense'.[15]

To recap, 'Giddy' (Gilbert) had shown that $\int P dV$ (the area under the 'indicator curve') was proportional to the work done by the engine (P is the pressure of the steam, and dV is a tiny incremental change in the volume; an 'integral' may be thought of as a sum of separate increments, $P_1 dV + P_2 dV + P_3 dV + \ldots$). This was an equivalent but completely different measure of 'work done' from the standard contemporary one of 'weight times height'. The formulation, $\int P dV$, would reappear in the classical thermodynamics of the nineteenth century, but this wasn't the first time it—or something very similar—had been derived: it had been discovered once before, by Daniel Bernoulli, almost 60 years earlier, but had passed by unnoticed (see Chapter 7, Part III).

Remarks

Amontons' and Newcomen's engines were known as 'fire-engines', but by the time of Watt, the term 'fire' was being replaced by the more prosaic term 'steam'. It would still be another 40 years before the steam engine would be recognized as an engine that ran on *heat* rather than on steam.

On a human note, did the inventors and improvers of the steam engine, all of them men, draw any parallels between the reciprocating motion of the piston (and the consequent 'creation' of work) and the sex act? Apparently, yes, such comparisons were made on rare occasions,[16] but, for the most part, the correspondence[17] between Watt and Boulton was concerned with engine design, patent infringements, spies, future prospects for Watt's son, travel arrangements, polite enquiries regarding health, and so on.

This heat-engine was a tricky engine to get into production. One could say that the steam engine needed the Industrial Revolution as much as the Industrial Revolution needed the steam engine—in order to provide the coal and the high-precision and high-strength cylinders and pistons, for example. Also, it needed the visionary outlook of Boulton, who was prepared to risk a fortune with no profits for over a decade, as well as the inventive genius of Watt.

However, it still seems strange that this steam or heat-engine, which was so influential in the development of physics, let alone in the indus-trialization, economics, and politics of the world, should have arisen in just one place—Britain—and at just one time in history. By contrast, machines using water-, wind-, or muscle-power have cropped up inde-pendently, all over the globe, time and time again. Some have suggested that the steam engine originated in Britain rather than in France because the French had peasants to do the work (before the Revolution) and the aristocrats were only interested in clever devices as 'toys'.[18] Others have argued that England ran out of streams and forests but had a plentiful supply of coal, and so coal-fired steam power was particularly attractive.[19] These arguments are persuasive, but they don't tell the whole story. I be-lieve that there is another reason—a *physics* reason—why the evolution of the steam engine was unique and serendipitous. It is that the steam engine is really a heat-engine, and getting work from heat is an inherently difficult thing—one might almost say, against nature. In a heat-engine, we shall find that the separate random motions of trillions upon tril-lions of molecules must be harnessed and turned into the coordinated motion of a bulk object. Nature has to be cajoled into doing this, and can only do it to a certain extent. This is the substance of the Second Law of Thermodynamics, which emerged from the work of Sadi Carnot (see Chapter 12) and was formally discovered by Clausius and Thomson (see Chapter 16). (However, presumably the laws of thermodynamics would still have emerged even if the steam engine had never been discovered . . .)

What is the 'block of energy' connected with the steam engine? Some formulae arrive in physics as major pronouncements by their discov-erers (for example, Newton—via Euler—defines force as 'ma', Leibniz declares that *vis viva* is 'mv^2', and so on). In the case of an engine driven by the expansion of steam, or other 'elastic fluid', the expression for the work done is '$\int PdV$'. This expression was implicit in Watt and Southern's indicator diagram (it is the area under the curve of pressure against volume); it had been 'derived' by Daniel Bernoulli over 50 years earlier; and re-derived by Giddy—an unknown person today—at the end of the eighteenth century. By this disjointed route, the 'block' for the 'work done by a gas expanding against a piston' arrived. As energy is a difficult concept, so its avatars, heat and work, are difficult.

It is ironic that these 'unnatural' heat-engines are now the most common sort of engine, the world over. It is even more ironic that the most prevalent type of heat-engine is the reciprocating cylinder engine (in cars), when what is required is direct *rotary* motion. The inefficiency is scandalous and we can only wonder what Watt would have made of it: for example, in my station wagon, running at 100 kilometres per hour, the pistons come to a dead halt and then reverse direction around 2,000 times per minute.[20]

Lazare Carnot

The engineer–scientist Lazare Carnot (1753–1823) is a rare personage in that he is equally famous as a scientist and as a statesman. He was the 'Organizer of Victory' in the wars of the French Revolution and, later, served under Napoleon, and had to go into exile when the Bourbons came back into power.

In his *Essai sur les machines en general*[21] of 1783, Lazare Carnot was the first to attempt to abstract the idea of a machine, or engine, to its idealized essentials. He took the conservation of *vis viva* ('live force') as his guiding physical principle. The chief exponent of this principle had been Daniel Bernoulli, who especially considered examples drawn from hydrodynamics (see Chapter 7, Part III); and Carnot always used the *water*-powered engine as the test case.

From the conservation of *vis viva*, Carnot developed his Principle of Continuity for the most efficient mode of operation of a machine: the water had to enter the machine smoothly, continue with no turbulence, and leave the machine with as little speed as possible (else this speed could have been exploited to yield yet more power). For example, in the case of an overshot water-wheel, the water should enter the wheel with as little shock to the buckets as possible, continue with no spray or splashing, and merely trickle out at the end.

Lazare[22] sought to analyse machines in terms of the interactions of the 'corpuscles' (infinitesimal elements) of which they were composed. His analysis had shades of Johann Bernoulli's Principle of Virtual Work about it (see Chapter 7): he considered virtual motions (he called them 'geometric motions') that were any mathematically allowed motions, infinitesimal and reversible. (As these motions were infinitesimal, then the Principle of Continuity was satisfied automatically.) The total work done by the machine was the sum of the *vis viva* of its 'corpuscular' parts. Significantly, the idea of *reversibility* was brought into physics for the first time; Lazare defined it as any 'geometric motions'[23] where the 'contrary motion is always possible'.[24] The idea of reversibility would be crucial in the work of Lazare's son, Sadi Carnot (see Chapter 12).

We have said that the engineers and the mathematicians moved in different worlds—well, Lazare Carnot was the crossover point. For the first time, the *engineer's* concept of work (usually 'force times distance' but sometimes the more specific 'weight times height') began to be incorporated into physics. The process was completed by Coriolis (1792–1843) who, in 1829, formally defined work as the integral of force over distance. Thus the recognition of 'work', from Amontons and Parent through to Coriolis, via Watt, and Lazare Carnot, had taken over 125 years. Was this the first time that an engineering concept had been integrated into the body of physics?

The engineers' 'work' is a 'block' of energy. But even today, it sits strangely in physics, relating as it does to such man-made artefacts as machines, and having an unusual provenance with no alternative Latin name such as *vis* or *vis viva*. (I think all of us can agree, however, that 'work' is a four-letter word.)

Part II: Theories of Heat

'The French . . . formulate things, and the English do them',[25] writes Gillispie in *The Edge of Objectivity*. We have already seen an example of this with regard to the science of engines—Lazare Carnot formulated the general theory of machines and Watt built the steam engine. Turning now to chemistry, we find the same trend: the English 'pneumatic chemists', Priestley (1733–1804) and Cavendish (1731–1810), discovered new gases (oxygen, hydrogen, and others), while the French scientist Lavoisier (1743–94) incorporated their discoveries into a new system of chemistry, and set up the chemical naming conventions we still use today. We are interested in Lavoisier because of his historic work on heat, *Memoire sur la chaleur* (*Memoir on Heat*, 1783),[26] which he carried out jointly with the French mathematical physicist Laplace (1749–1827).

Heat was a topic on which virtually every French philosopher wrote a treatise, from Voltaire and Mme La Marquise du Châtelet, to Jean Hyacinthe de Magellan (grandson of the Portugese explorer), to Jean-Paul Marat, the French Revolutionary who was assassinated in his bath. The picturesque term 'fire' was used more often than the scientific term 'heat'. The Montgolfier brothers' first hot-air balloon flight occurred in the same year as Lavoisier and Laplace's Memoir. Yet Joseph Montgolfier still felt able to say that their balloon ascended because it was filled with fire, the lightest of the four [Aristotelian] elements.[27]

Lavoisier

Antoine-Laurent Lavoisier (1743–94) was a Parisian of comfortable middle-class origins. His mother died when he was five years old, and he was brought up by an adoring maiden aunt. Following in his father's footsteps, he initially studied law, but a family friend and geologist tempted him into geology and mineralogy, which led him into his passion—chemistry.

Lavoisier seems always to have had a great capacity for work. When only 19, he cut himself off from social life, lived on a diet consisting largely of milk, and devoted himself for several months to scientific investigations.[28] His wife tells how, in later years, Lavoisier worked at his beloved chemistry from 6 to 8 a.m. and then again from 7 to 10 p.m., while the rest of the day was taken up with working as a private tax collector (a 'tax farmer'), and with his many civic commitments (e.g. supervising the production of gunpowder), and other scientific studies (he was on committees reporting on ballooning, the water supply of Paris, mesmerism—with Benjamin Franklin, invalid chairs, and the respiration of insects). On one day a week, his 'jour de bonheur',[29] Lavoisier devoted himself entirely to his own researches.

Lavoisier's wife, herself the daughter of a tax farmer, assisted Lavoisier with his work—translating the works of Priestley and Cavendish into French, recording the results of experiments and making detailed illustrations.[30] Their apartment and very well-equipped laboratory was at the Arsenal in Paris (see Fig. 8.2).

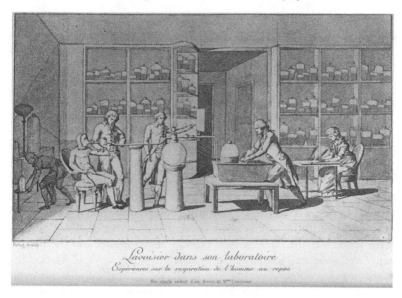

Lavoisier dans son laboratoire
Expériences sur la respiration de l'homme au repos

Fig. 8.2 Lavoisier and Mme Lavoisier at work on experiments on respiration (courtesy of the Edgar Fahs Smith Collection, Kislak Center for Special Collections, Rare Books and Manuscripts, University of Pennsylvania).

Pierre-Simon Laplace (1749–1827) was one of the major French mathematicians of the Revolutionary period—he claimed he was *the* major one (modesty was not his strong point). (However, he did say this in 1780–81, before Lagrange arrived in France, in 1787.) Laplace's chief contributions were in astronomy and in probability theory. In astronomy, his work was written up in his monumental *Traité de mécanique céleste*, published in five volumes between 1799 and 1825. The American mathematician Bowditch translated this work into English; and whenever Laplace had written 'It can easily be shown that . . .',[31] Bowditch knew he had hours of work ahead of him. Laplace's most famous saying arose when Napoleon asked him why God did not appear in the *Traité*: Laplace replied 'I have no need of that hypothesis'—however, the story is apocryphal.[32] In probability theory, Laplace became famous for his deterministic views: he speculated that if we could exactly describe the state of the physical world at any one time, then the laws of physics would be able to completely determine the past, and completely predict the future. As regards his work in physics, we have met Laplace before, in connection with potential function theory (see Chapter 7, Part II), and we shall meet him again when he starts the Societé d'Arcueil with his neighbour, Berthollet, and when he corrects Newton's calculation of the speed of sound (see Chapter 10).

Let us now return to the earlier project, the *Memoir on Heat*. The work was a collaboration, but it was a continuation of Lavoisier's overall programme of research on heat, and so we shall continue in Lavoisier's voice alone. Lavoisier was interested in heat because he wanted to explain combustion and respiration (were they linked?), and vapours and 'airs' (were *they* linked?). Remarkably, even while Watt was carrying out his extensive experiments on water and steam, it had still not been established that all substances can exist in three states—solid, liquid, and vapour. It was Lavoisier who made this explicit, and who saw that it was the 'quantity of heat' within a substance that determined which of these three states was the current one. He saw that vapours and the permanent 'airs' were not fundamentally different—they were all gases. (The *Memoir* was the first time that the word 'gas' gained currency. It came from the Greek *kaos* via Van Helmont in the seventeenth century—see also Chapter 5).

In the *Memoir*, Lavoisier starts by describing the two main theories of heat[33]:

Scientists are divided about the nature of heat. A number of them think it is a fluid diffused through nature . . . Other scientists think that heat is only the result of the imperceptible motions of the constituent particles of matter.

Lavoisier attempted to be impartial, and proposed only principles that (he thought) were universal and didn't depend on which heat theory had been adopted:

whichever one [hypothesis] we follow, in a simple mixture of bodies the quantity of free heat remains always the same.

But, as with the elastic collisions in Chapter 7, there was some circularity in the definitions (what is a *simple* mixture, and what is *free* heat?). Lavoisier continued with his reversibility principle:

If, in any combination or change of state, there is a decrease in free heat, [then] this heat will reappear completely whenever the substances return to their original state; and conversely, if in the combination or in the change of state there is an increase in free heat, this new heat will disappear on the return of the substances to their original state.

This appears incontestable, but it did lead, eventually, to a false assumption—that in the cyclic operations of a heat-engine, all the processes could be reversed. This will be discussed more fully in Chapter 12. ('Free heat' means the heat that can be detected by a thermometer; in other words, the heat that isn't latent or combined.)

How should heat changes be measured? The usual 'method of mixtures' wasn't always appropriate (for example, as when trying to determine the amount of heat emitted by a guinea pig over 10½ hours). Laplace devised the ice calorimeter, whereby the subject of the experiment was placed within an ice 'bomb', and the amount of ice that was melted within a given time was proportional to the amount of heat emitted. The design relied on an understanding of latent heat, but it appears that Lavoisier and Laplace knew nothing of Black's work. (They learned of specific and latent heats from Magellan's translation of Crawford's work—see Chapter 6.) The ice calorimeter was very successful—the start of a new field, now known as calorimetry: in fact, it *had* already been invented by Black, but he graciously acknowledged that Laplace had discovered it independently. (Nevertheless, I can't see how the ice bomb could have been of use in those 'endothermic' cases where heat is absorbed rather than emitted.)

Lavoisier and Laplace carried out three main categories of investigation: the measurement of specific heats, the heats of chemical reactions, and the heat evolved on combustion and in respiration.

In the case of the aforementioned guinea pig, the poor creature had to go on a starvation diet, remain in a small basket held at freezing temperature, and have only three inputs of fresh air in 10 hours and 36 minutes. In later experiments (to determine the gases emitted while exhaling), a further[34] insult to injury occurred when the guinea pig was forced through a trough of mercury on its way to the bell jar. At least, the conclusions were important: the exhaled gas was 'fixed air' rather than the phlogistonists' 'vitiated air', and respiration was thus shown to be a form of combustion.

This, and all the other ice-calorimeter experiments, helped Lavoisier to arrive at his famous conclusion—that all burning, whether by combustion, 'calcination' (rusting), or respiration, is *oxidation*. In all cases, a component of the air, oxygen, is used up. Hence, Lavoisier had debunked the phlogiston theory (see Chapter 4). However, he wrongly thought of oxygen as the 'acidifying principle' ('oxygen' means 'acid' in Lavoisier's Greek lexicon of chemical terms).

As we have said, Lavoisier claimed to be impartial on the question of the nature of heat, but he was, privately, totally committed to the idea of heat as a special fluid. In fact, he needed the heat-fluid as the expansive agent in boiling and evaporation, especially where these processes occurred in a vacuum. (Boiling, evaporation, and even sublimation had been found to occur in the evacuated receiver of the air pump. Where air *was* present, then some liquid could bind with it and become . . . airborne.)

Lavoisier's views had thus advanced to the point at which he saw heat as instrumental in changes of state, and in the elastic properties of gases. He saw a gas as a combination of a gaseous 'base' and the subtle matter of heat. (It could be argued, therefore, that Lavoisier was still a phlogistonist of sorts, as he had, in a sense, merely transferred the 'phlogiston' from the burning substance to the flammable gas.) Some of Lavoisier's ideas were apparently inspired by an obscure article, 'Expansibilité', written anonymously by a government minister, Turgot (1727–81), and appearing in Diderot's *Encyclopaedie*[35] (the *Encyclopaedie* was a banned book). What *is* clear is that Lavoisier's ideas on heat and the gaseous state were completely interdependent.

Only five years later, Lavoisier was unequivocal on the nature of heat. In his famous treatise, *The Elements of Chemistry*,[36] published in 1789, he listed 'caloric' (the subtle heat-fluid) and 'light' as two of his 'primitive elements'. However, taking heat as a subtle fluid *did* prejudice Lavoisier into making an unwarranted but largely unconscious assumption—that the total heat is always conserved. An *actual* fluid (albeit subtle) may be transmitted here and there, sometimes free, sometimes combined, but it is never created or destroyed. But, in the operations of a heat-engine, we shall find that the total heat is *not* conserved; it is instead the even more subtle stuff, *energy*, that is conserved.

When Lavoisier was at the height of his powers, his life was cut short as he fell prey to the Terror and to the guillotine (on 8 May 1794, in the Place de la Révolution, now called the Place de la Concorde). Lavoisier was a committed revolutionary but he came under suspicion due to his work as a 'tax farmer'. Lagrange said 'Il ne leur a fallu qu'un moment pour faire tomber cette tête, et cent années peut-être ne suffiront pas pour en reproduire une semblabe' ('It took but a moment to make this head fall, but even a hundred years may not be enough to produce a like one').[37]

We shall discuss in a while the reasons why the caloric or subtle-fluid theory of heat was so appealing, and became so entrenched, but let us turn first to the tale of a singular and singularly gifted individual, whose whole life was nothing but science.

Cavendish

Henry Cavendish (1731–1810) was born into one of the richest and best-connected families in England: his father was the son of the second Duke of Devonshire, his mother was Lady Anne de Grey, daughter of the Duke of Kent. Henry Cavendish inherited a fortune of over a million pounds, but he was completely uninterested in money. Once, when his banker arrived uninvited at the house, wanting only to offer financial advice, but disturbing Cavendish's privacy and train of thought, Cavendish said 'If it [the money] is any trouble to you, I will take it out of your hands. Do not come here to plague me'.[38]

A career in politics was open to Henry, but he was as uninterested in politics as he was in money. Fortunately, his family gave him free rein to pursue his passion for science. He converted his large residence on Clapham Common (and also the family apartments on Great Marlborough Street) into laboratories, and carried out the experiments on which his fame rests: the discovery of inflammable air (hydrogen), the identification of the dew formed when Priestley ignited a mixture of inflammable air and common air (the dew was water), 'weighing' the Earth, and many others. He was meticulous in his experimenting, and achieved unparalleled precision for those times.

Less well known is that Cavendish was also a brilliant theoretician. A hundred years after his death, Maxwell was sorting through Cavendish's electrical researches and found that Cavendish had already discovered Ohm's Law, Coulomb's Law, the concepts of electric potential, electric capacitance, and many other results. Also, in 'pneumatics', Cavendish had anticipated Dalton's Law of partial pressures, and Charles's Law relating the volume and temperature of a gas.

It may have been Cavendish's loathing of publicity, and his eccentric personality, that led to his keeping so much material unpublished. The recent trend in retrospective diagnosis is to be deplored, but I have to admit that, in Cavendish's case, he could be the prototype for Asperger Syndrome. There are many anecdotes relating to his extreme shyness; for example, he communicated with his house-servants by notes, and is said to have had an extra staircase put in to avoid encountering the maid. Even with scientific colleagues, Cavendish was painfully shy. A fellow of the Royal Society recounts how Cavendish was introduced to a visitor as a celebrated philosopher, and then forced to listen to a flattering speech: 'Mr. Cavendish answered not a word, but stood with his eyes cast down, quite abashed and confounded. At last, spying an opening in the crowd, he darted through it with all the speed of which he was master, nor did he stop till he reached his carriage, which drove him directly home'.[39]

In 1969, some papers from the Cavendish estate at Chatsworth House came to light for the first time.[40] Here was uncovered Henry Cavendish's 43-page work, 'Heat', probably written in 1787.[41] Cavendish's views were against the contemporary orthodoxy, the subtle-fluid theory of heat, and this may provide another reason why he didn't publish them: 'I think Sir Isaac Newton's opinion, that heat consists in the internal motion of the particles of bodies, much the most probable'.[42]

But what was this internal motion? Even while he was a staunch Newtonian, Cavendish realized that it was not the Newtonian but rather the Leibnizian quantification of 'motion', in other words, $\frac{1}{2}mv^2$, that answered to the needs of heat. He took a body as being made up of an 'inconceivable number'[43] of interacting microscopic particles (actually, too small to be seen in the microscope) and tacitly assumed that the usual laws of mechanics would apply on this microscopic scale. He proceeded to an analysis reminiscent of Daniel Bernoulli's—not the latter's kinetic theory, but his interactions of celestial bodies.[44] According to Cavendish, the microscopic particles had 'active *vis viva*' (which would affect a thermometer) and 'inactive *vis viva*' (due to their configuration—a measure of the 'latent heat' of the body). Both Bernoulli and Cavendish took the conservation of total *vis viva* as the overriding principle.

Thus Cavendish had outlined the first kinetic theory not related to a gas. He was able to use his theory to explain all the phenomenology of heat. First, for the communication of heat between identical bodies at different temperatures, he saw that the motion of particles in the hotter body would be slowed down, while those in the cooler body would be speeded up, until the 'particles of both come to vibrate with the same velocity'.[45] (Nevertheless, for different bodies, such as lead and copper, Cavendish was hazy about what was being equalized, and couldn't bring himself to admit to an equality of active *vis viva*—kinetic energy—for particles of very different masses.)

Second, Cavendish saw that the different particle configurations, and the different strengths of particle interactions, within different materials, meant that the proportions of active to latent heat would be specific to each type of material—this explained Black's concept of specific heat. Likewise, a change of configuration during a change of state or chemical reaction explained the latent heats of fusion and vaporization, and the 'heats of reactions'.

Finally, Cavendish also explained the heating effect of electricity to his partial satisfaction, but admitted that 'it is an effect which I should not have expected'.[46]

On the other hand, in one class of heat phenomenology, that due to radiant heat, Cavendish readily admitted a substance theory of heat. Referring to the experimental investigations of Scheele, and of de Saussure (see Chapter 6), Cavendish was convinced by the parallels between heat and light, and therefore took radiant heat to be made up of rays of 'heat-particles', by analogy with Newton's rays of 'light corpuscles'. Cavendish also knew of Michell's experiments to determine the momentum of light (Reverend John Michell, 1724–93), and used Michell's results to calculate the 'work done by light'. He found that the light falling on a thin vane of copper sheet, of area 1½ square feet, would cause the vane to rotate, and do work at the rate of approximately 2 horsepower.

Returning to Cavendish's kinetic theory, one can detect echoes of Prevost's 'Theory of Exchange' (see Chapter 6), although, in fact, Cavendish's work was some three or four years earlier. Cavendish wrote:

strictly speaking, the . . . [active and inactive *vis viva*] . . . must be continually varying, & can never remain exactly the same even for an instant. Yet as the number of vibrating particles, even in the smallest body, must be inconceivably great, & as the *vis viva* of one must be increasing while another is diminishing, we may safely conclude that neither of them can sensibly [macroscopically] alter.[47]

Cavendish had therefore realized that the process of temperature equalization was really a *dynamic* one. He was tantalizingly close to, but had not quite reached, the idea of *average* speed, or *average* kinetic energy.

Daniel Bernoulli's work was not picked up by the scientific community, and Cavendish's work was hidden from it. The motion or kinetic theory of heat would have to be discovered afresh. In the meantime, we return to the question of why the fluid theory held such a grip.

The Caloric Theory of Heat

The eighteenth century was the heyday of the subtle fluid; it was invoked to explain gravity, electricity, magnetism, phlogiston, and heat. (As remarked in Chapter 5, in the case of electricity we could argue that the subtle fluid has never gone away.) Cavendish was an adherent of the phlogiston theory, but instead of thinking of phlogiston as a subtle fluid, he identified it directly with his 'inflammable air' (hydrogen). (His version of phlogiston was, therefore, less subtle than before, as hydrogen does have weight and other measurable properties.) In the case of heat, the idea of a subtle fluid was exceptionally rich and fruitful in explaining almost all the phenomena:

- Black's 'equilibrium of heat'—the temperature of bodies becomes equalized in the same way that water (a real ponderous fluid) finds a common level.

- Irvine's theory of heat capacity (see Chapter 6) drew on the analogy of a tank holding water to describe a body containing heat-fluid.

- There were analogies made between water-powered engines (water-wheels and column-of-water engines) and heat-engines: in the first case, the water fell between two heights; in the second case, the heat-fluid flowed between bodies at a hotter and a lower temperature (see Chapter 12).

- From the supposed property that the heat-particles within the heat-fluid are self-repelling (and this had the authority of Newton behind it; see Chapter 5) arose the idea that extra heat would lead to expansion. (Curiously, Cavendish, using his motion theory, didn't find this in the least bit obvious. He could see that the heating of a body would lead to changes in the average separation of the constituent particles, but why shouldn't this lead to contraction as often as expansion?)

- The class of phenomena known later as 'adiabatic' expansions and compressions (see Chapter 10) were suggestively explained using the subtle-fluid theory. (For example, as a gas was compressed the heat-fluid was squeezed out, like squeezing water out of a sponge: the heat so liberated became 'sensible'—detectable by a thermometer.)

- In changes of state, extra heat-fluid was required in order to cause expansion (from solid to liquid, and from liquid to gas) but also to overcome the cohesive forces binding particles together within the solid or the liquid. We have already seen how Lavoisier *required* the heat-fluid in order to explain the expansibility of gases, and changes of state.

- For heat being transmitted across a void or from the Sun, the fluid theory had a more ready explanation than the motion theory. The subtle fluid could be transmitted across the void, perhaps as 'rays of caloric', whereas the void, by definition, contained no constituent particles, whether these were in motion or not. (We have seen that Cavendish resorted to rays of heat-particles in this instance.)

There was really only one category of phenomena that the fluid theory struggled to explain. We'll reveal this in the work of the colourful character, Count Rumford, in the next chapter.

Overview

In this second half of the eighteenth century, we have witnessed the formulation of the abstract theory of machines (by Lazare Carnot), the building of a very real machine (Watt's steam engine), and the emergence of the caloric theory of heat (Laplace and Lavoisier). As usual, the kinetic theory (Cavendish) was developed in parallel but remained hidden from view.

The caloric, or subtle-fluid, theory was mostly taken up by chemists rather than physicists. This led to a subtle error, almost a subconscious one—that the conservation of heat was unquestionably true, and was paramount. In chemistry, substances can be variously ground down, boiled away, mixed together, chemically combined, and so on, but always there is the assumption that the starting substances can be recovered if the processes are gone through in reverse. Heat was hypothesized as a material substance, so it was natural to assume that it too would be conserved in all processes. (As we have seen, the conservation of heat was a central tenet of Lavoisier and Laplace's *Memoir*.) We shall find, however, that in the operations of a heat-engine, the total heat is *not* a conserved quantity (see Chapter 12).

We have observed that the steam engine was discovered in just one part of the world, at just one time, and for the fundamental *physical* reason that it is hard to tame heat in this way.

Likewise, the motion or kinetic theory of heat was not readily taken up because the ideas were *difficult*; that is, statistical in nature. 'Heat' is statistical, partly because there are a very large number of microscopic particles involved (there are around 10^{24} atoms in an apple, a mug of tea, or a bucket-full of air) but also because we are dealing with statistical *processes* (Prévost's 'mobile equilibrium', and the as-yet undiscovered 'mean-free-path', 'random-walk', and so on). By contrast, 'weight' is not statistical: the weight of a litre of hydrogen *is* just the sum of the weights of the individual gas molecules (this was one way in which molecular weights were first determined).

Cavendish, and Daniel Bernoulli, made the assumption that the laws of mechanics, first discovered on the macroscopic (everyday) scale, would be applicable on the microscopic scale. They appreciated that *average* microscopic velocities were involved, although neither stated this explicitly. The problem was—they simply didn't know how to proceed: what were the speeds when not at the average? How should this be treated mathematically?

We all know how counterintuitive results can come out of statistics—for example, there is the surprisingly high chance of two people at a party sharing a birthday. Some very counterintuitive and unexpected results began to emerge for heat, as we shall begin to discover with the remarkable researches of Rumford.

9

Rumford, Davy, and Young

Rumford

It was Rumford who rumbled the caloric theory of heat. His famous experiment on the boring of cannon went straight to the Achilles's heel of the caloric theory—*the generation of heat from friction.* The caloric theory was, remarkably, able to account for virtually all the varied phenomena of heat (see the end of Chapter 8), and was adopted by the 'Laplacian school', and most of the rest of the scientific community. Only in the case of frictional heating were its solutions forced and *ad hoc.* But such a successful and well-entrenched theory was not going to be overthrown by one experiment, and that carried out by an outsider to the scientific community. Rumford was really an outsider to all communities, as a biographical sketch will show.

Count Rumford (1753–1814) was born Benjamin Thompson, in Woburn, Massachusetts, a small rural town near Boston, into a family of farmers. His father died when Benjamin was only 18 months old, but his grandfather and uncle left him some land and a small allowance—so he was hardly destitute, despite the script of Cuvier's eulogy: '. . . leaving his grandson almost penniless. Nothing could be more likely than such a destitute condition to induce a premature display of talent'.[1]

Benjamin did show talent—an early aptitude for science, self-education, and self-promotion. He read Boerhaave's *The Elements of Chemistry* (see Chapter 5) when only 17, and this gave him a life-long interest in the study of heat. At only 19, he married a 33-year-old widow who was wealthy and well-connected; and at the young age of 20 he became a major in the army on the loyalist (pro-British) side. This resulted in his precipitate departure, first from his home in Concord (formerly, Rumford), New Hampshire, and subsequently from America: 'I thought it absolutely necessary to abscond for a while, & seek a friendly Asylum in some distant part'.² Thus Rumford was one of the world's first asylum seekers and global citizens. From 1776 onwards (i.e. after the American War of Independence) he lived in England, Germany (Bavaria), and France, had only one return trip to America (to recruit soldiers for his regiment, the King's American Dragoons), and never saw his first wife again.

While the youth of today are enjoined to consider flexibility as the single most important job qualification, they had better not try to follow Rumford's example. Starting out as a soldier (a major in America, a colonel in England, and a major-general in Munich), Rumford became a spy (for just about every pairing between America, England, Bavaria, and France) and a profiteer. He also was, at various times: a statesman and diplomat; a nobleman (knighted in England, made Count of the Holy Roman Empire in Bavaria—he chose to be the Count of Rumford—and a knight of the order of St Stanislaus in Poland); a philanthropist and social reformer (in Germany he founded the 'English Garden' in Munich, put beggars to work, reformed the army, and initiated workhouses, soup kitchens, a 'house for ladies', and a home for illegitimates of noble birth); an educationalist (founding the Royal Institution in London)³; and a designer and inventor (of invisible ink, lamps, candles, a carriage-wheel, a frigate, a life-belt for a horse, the Rumford fireplace, and many other appliances especially connected with heat—see below).

Rumford was also a ladies' man: he married and separated twice (his second wife was Lavoisier's widow); had a daughter from his American wife and in addition at least two illegitimate children; and had many mistresses (his favourites were Lady Palmerston in England and the sisters Countess Nogarola and Countess Baumgarten in Bavaria).

On top of all this, Rumford was a scientist, chiefly in the areas of heat and light. His interest in science was twofold—practical and philosophical. On the practical side, Rumford was always motivated to improve the 'economy of human life',⁴ and he invented the coffee percolator, cooking range, roaster, double-boiler, Rumford stove, and, famously, a smokeless fireplace, still the best design after 200 years.⁵ He also researched lighting, the 'science of nutrition', fabrics (he came up with tog ratings), and clothing (he looked somewhat eccentric in the Paris winter, wearing a white hat and coat—to better reflect the 'frigorific' rays). On the philosophical side, Rumford designed scientific instruments (the photometer, difference thermometer, and standard candle), and contemplated the nature of heat and light.

From early on, Rumford had misgivings about the substance theory of heat. While carrying out some experiments on gunnery and explosives (in England, from 1778 onwards), he made the astute and surprising observation that the gun barrel was *hotter* when fired *without* a bullet inside. Moreover, he drew the right but not obvious conclusion: when there was a bullet, then its bulk motion carried away the excess heat. (With the benefit of hindsight, we can see that Rumford's intuition was pregnant with meaning.)

It wasn't until some ten years later, in 1789, that Rumford again returned to his beloved heat studies, and carried out the experiments that have since become legendary in the history of physics. While supervising works at the arsenal in Munich, Rumford noticed that, in the manufacture of cannon, the brass barrel—and especially the metallic chips and swarf—all became very hot when a new cylinder was being bored.* His interest piqued, Rumford set about designing and conducting his own experiments—'for the express purpose of generating Heat *by friction*'.[6] He therefore arranged for a deliberately blunt borer to be held fixed against a solid cannon waste-head, while the latter was turned on its axis by two horses (Fig. 9.1). The steel borer was 'forcibly shoved (by means of a strong screw) against the bottom of the bore of the cylinder'.

Rumford found that the temperature of the brass cannon, after 960 revolutions, went up by an impressive 70 °F. Now the caloric theory, in Irvine's version of it (see Chapter 6), accounted for this increase in temperature by a decrease in the heat capacity of the brass chips (as compared with the heat capacity of the solid brass). However, the metal chips weighed only 837 grains Troy (approximately 1.9 oz), and Rumford asked 'Is it possible that the very considerable quantity of Heat . . . which actually raised the temperature of above 113 lb of gun-metal at least 70 °F . . . could have been furnished by so inconsiderable a quantity of metallic dust? And this merely in consequence of *a change* of its capacity for Heat?' Furthermore, Rumford noted that some earlier experiments had shown that the heat capacity of gun-metal was '*not* sensibly changed' by being reduced to the form of metallic chips.

However, Rumford was careful not to jump to conclusions, and conscientiously considered other possible explanations. For example, he wondered if the heat could have originated from the air within the

---

* It was easier to make cylinders true, and of uniform diameter, by boring rather than casting: also, 'rifling' wasn't common practice until the invention of smokeless gunpowder, well into the nineteenth century.

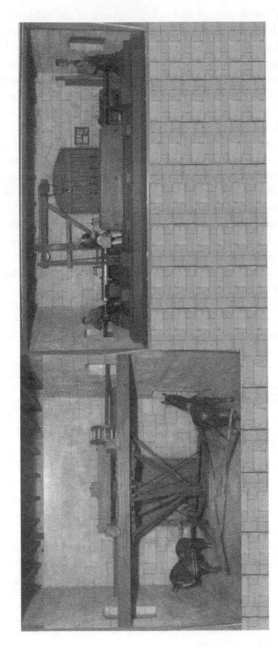

Fig. 9.1 A model of Rumford's cannon-boring experiment, 1789 (courtesy of the Rumford Historical Association, Woburn, MA).

newly drilled cavity: 'I . . . endeavoured to find out whether the air did, or did not, contribute anything in the generation of it [the heat]'. He checked this by submerging and encasing the whole cannon in a box filled with water (the water displaced the air), before proceeding with a repetition of the experiment ('shoving' the blunt borer up against the rotating cannon barrel, and so on).

Rumford did repeat the experiment—but with a difference: he invited a crowd of onlookers, and took the opportunity to continue the tradition of experiment-as-public-demonstration (Galileo had dropped weights from the Leaning Tower of Pisa, and Pascal had constructed lengthy barometers of wine and of water at the port of Rouen; see Chapter 4). Imagine the suspense as Rumford and the assembled crowd watched and waited, for two and a half hours, after which time the water 'ACTUALLY BOILED!' Rumford, clearly relishing the spectacle, writes: 'It would be difficult to describe the surprise and astonishment expressed in the countenances of the bystanders, on seeing so large a quantity of cold water heated, and actually made to boil, without any fire'.

But what was the explanation? No changes in heat capacity had occurred, the air had not participated, and neither had the water (Rumford was careful to note that the water had remained chemically unaltered). Another possibility was that heat was somehow being *supplied* to the cannon and borer (say, from the water, or via the shaft) at the same time as it was being emitted to the surroundings. But Rumford held that the simultaneous emission and receipt of a material substance was absurd. (For this reason, he rejected Prévost's theory of exchange; see Chapter 6.) However, the crux of the argument for Rumford was the fact that the heat generated by friction was 'inexhaustible', and so clearly it could not be a '*material substance*'. In fact, nothing could supply these prodigious quantities of heat 'except it be MOTION'.

One other bit of phenomenology that spoke against the materiality of heat was the fact that heat was weightless, or very nearly so. The prior results were inconclusive, and plagued by the extra weight of water condensing on to cold materials, by convection currents, and by relative changes in buoyancy as bodies cooled. Rumford cleverly got around these difficulties—by comparing the weights of two or more bodies all at the *same* temperature. For example, he compared the weights of water, 'spirit of wine' (alcohol), and mercury, as these were simultaneously cooled down past 32 °F. This was a telling experiment, as the water, upon freezing, lost a large quantity of 'latent heat' as compared to the alcohol and the mercury. Another telling experiment was one in which equal weights of water and mercury were cooled from 61 °F down to 34 °F, their weights being compared in a balance all the while. The water and the mercury remained 'in *equilibrio*',[7] despite the fact that around 30 times more loss of heat was implicated for the water. Rumford concluded that heat was indeed weightless, to within the limits of the sensitivity of his balance, 'an excellent balance, belonging to his Most Serene Highness the Elector Palatine Duke of Bavaria'.[8]

In spite of all this, Rumford's cannon-boring experiments cannot be said to have ousted the caloric theory, except in retrospect. Rumford didn't help his cause as he ran a largely negative campaign (i.e. against the caloric theory rather than for the motion theory) and also had some quirky views that were easy to rebut.

However, Rumford did further the cause of the motion theory in his choice of lecturers for the Royal Institution in London,[9] which he founded in 1800: Humphry Davy and Thomas Young. These two young men, mostly unknown to science, supported the motion theory of heat, and advanced ideas relating to 'energy'. We shall briefly outline their careers.

Rumford's Protégés: Davy and Young

The Cornishman Humphry Davy (1778–1829) is famous for his miners' lamp; the discovery of potassium, sodium, chlorine, and other elements; his inhalation of laughing gas; his associations with the Romantic poets Coleridge and Wordsworth; and, finally, for *his* choice of protégé at the Royal Institution, the experimenter of genius, Michael Faraday (see Chapter 15).

When only a teenager, Davy had devised an experiment in which he rubbed two blocks of ice against each other continuously, until, eventually, they melted. He satisfied himself that 'It has thus been experimentally demonstrated that caloric or matter of heat does not exist'.[10] He satisfied *himself* but not twentieth-century historians of science, who all conclude that Davy couldn't possibly have justified this inference from this experiment[11] (the pressure would have lowered the melting point, and the ambient air temperature would also have caused melting). Rumford, certainly, was impressed, and even more so by his young protégé's lecturing style. Davy was a natural performer, who pitched the content of his lectures at the right level for his fashionable audience and also played to the ladies (Fig. 9.2).

Fig. 9.2 'New Discoveries in Pneumatics!' —Rumford, Davy, and Young at the Royal Institution, (Science Museum London/SSPL). (Young applies the gas, Davy holds the bellows, Rumford stands by the open door.)

Rumford's other new lecturer was the polymathic genius Thomas Young (1773–1829). Young was a child prodigy of the best kind—'the kind that matures into an adult prodigy'.[12] At age two he was a fluent reader, by six he had read through the Bible twice, and by 13 he was teaching himself Hebrew, Chaldean, Syriac, Samaritan, Arabic, Persian, Turkish, and Ethiopic. He was later a scholar of Greek and Latin, an Egyptologist, and first decipherer of the Rosetta Stone.

In physiology, Young discovered how the eye focuses (by changes in the curvature of the lens rather than the cornea—and not by moving the lens backwards and forwards as in a telescope); and that the eye detects only three primary colours. In engineering, Young is known for 'Young's modulus', and in physics for 'Young's slits' (which, according to Feynman, brings out the one true mystery of quantum mechanics[13]). Let us now return to 1800, and to the development of the concepts of light, heat, and energy.

Young was the first to explain the curious fact that two sources of light can be combined to yield . . . darkness. He saw that when two or more light sources are superimposed, they 'interfere', and the resulting patterns of light and dark (for example, as occur in 'Young's slits') show that light is a *wave*. At this stage, Young thought only in terms of a pressure wave, similar to a sound wave, but when the experiments of the French physicist Malus, in 1809, showed that light could be 'polarized' (after reflection or transmission through Iceland spar), Young suggested that this could be explained if light was a *transverse* wave, like the standing wave in a violin string. This was a 'mathematical postulate',[14] but was it physical? Young had his doubts. At this time (early nineteenth century), the idea of waves in empty space was unthinkable, so some sort of medium or ether was implicated. Now, the faster the waves, the more rigid the ether would have to be; but the speed of light was known to be very fast, perhaps infinite, and so the 'luminiferous ether' would have to have the contradictory properties of being 'not only highly elastic, but absolutely solid!!!'.[15] (These are Young's exclamation marks.)

All the above refers to light, but Young was one of the first to suggest that heat and light were really the same thing. As light was a wave (an 'undulation'), then so was heat. This was music to Rumford's ears (almost literally, as Rumford likened the focusing of heat radiation to the tuning of a sound wave to a given note). Young went further and posited that the different colours of light were part of a spectrum in which the frequency of vibration determined the colour. He predicted that heat radiation would occur at lower frequencies than the red end of the 'visible spectrum', and also that invisible high-frequency radiation would be found beyond the violet end of the spectrum. These predictions (in his paper of 1801) occurred almost simultaneously with the experimental findings of William Herschel (in 1800) and Ritter (in 1802), who discovered infrared and ultraviolet radiation, respectively. (Herschel (1738–1822) was astonished to see the thermometer rise as it went beyond the red end of the visible spectrum; Ritter (1776–1810) witnessed the blackening of silver chloride just after the violet end of the spectrum.)

Connections

All this happened around 1800. The other major achievement in physics at this time was Volta's pile. The Italian physicist Alessandro Volta (1745–1827) stacked alternating discs of zinc and silver, separated by pieces of felt soaked in brine, into a 'pile'. This resulted in the first continuous source of current electricity—a battery. Davy rushed into this new field, and was one of the first to show that a chemical reaction (as opposed to the mere contact of dissimilar metals) was essential in producing the current. He also showed that the reverse was true; in other words, that electricity caused chemical reactions to occur and, ultimately, that all chemistry was electrical in origin.

As with heat and chemistry (see Chapter 8, 'Lavoisier') and heat and light (above), there were now links between electricity and chemistry. Soon there was found to be a connection between electricity and heat, as in the Seebeck and Peltier effects. (Seebeck (1770–1831) found, in 1821, that when the two ends of a metal bar were held at different temperatures, a current flowed from the hot end to the cool end; Peltier (1785–1845) found the reverse effect, in 1834—in other words, a heat difference was generated when dissimilar metals were connected in an electrical circuit.)

Another pairing was electricity and light, and it had long been known that they were linked: in 1752, Benjamin Franklin (1706–90) had shown that lightning carried electricity (and lived to tell the tale); and highly charged bodies caused sparking. Also, all three of electricity, light, and heat could be generated by friction. All these connections led Young to speculate that the 'luminiferous' and electrical ethers were really one and the same.[16] However, one of Young's predictions was soon shown to be wrong—that electricity and magnetism would never be linked;[17] as, in 1820, the Danish physicist Oersted (1777–1851) happened to notice that a magnetic compass needle flickered when a nearby electrical circuit was switched on. (Oersted, and many others—for example, Ampère and Faraday (see Chapter 15)—went on to explore such effects.)

Did the connections between heat, light, chemistry, electricity, and magnetism imply that there was some aspect that was common to all? Young was in no doubt. He lasted only two years at the Royal Institution—his lectures were too abstruse—but in 1807, his masterful two-volume *A Course of Lectures in Natural Philosophy and the Mechanical Arts*[18] was published. Here, he proposed the Greek word 'ενεργεια' (energy) in place of 'living force'. He was the first to use this word in physics since Johann Bernoulli had used it in 1717 (see Chapter 7, Part IV). Although Young introduced the term solely to describe what we would now call *mechanical* energy, Young's coinage was soon applied to diverse 'energies', as in:

the relation of electrical energy to chemical affinity is . . . evident [Davy][19]

different parts of a solar ray, dispersed by a prism, possess very unequal energies . . . [Biot][20]

Soils . . . which act with the greatest chemical energy in preserving Manures [Davy][21]

iodine had less 'energy' than chlorine but more than sulphur' [Gay-Lussac][22]

While the meaning of the term didn't exactly coincide with our modern meaning, the increasing use of the word 'energy' indicated that a new player was soon to enter the physical stage.

Overview

Physicists have long been impressed by Rumford's perceptive insights after his observations of everyday practice at the arsenal in Munich, but now the history makes us even more impressed: he didn't simply observe, but planned and carried out a series of thoughtful experiments, fulfilling Newton's guideline—that Nature must be 'teased' into revealing her secrets. However, Rumford's cannon-boring experiment cannot be said to have ousted the caloric theory (except in retrospect). This theory was well entrenched, and it was no good simply showing up a problem with it—an alternative theory had to be put forward. Nevertheless, it is surprising that there weren't more tell-tale cracks within the caloric theory, as it was, in fact, false. Who, for example, could not be swayed towards the motion theory of heat, after seeing the quick wrist-action needed to light the new friction matches? These were invented in 1827, by John Walker, around 30 years after Rumford's experiments. (Walker called them 'Congreves' and they were later marketed as 'Lucifers'.)

Young (following on from the earlier work of Scheele; see Chapter 6) had shown that light and radiant heat were the same sort of stuff, and that, in both cases, the radiation was wave-like rather than being made from rays of 'corpuscles'. The accidental discovery of the thermoelectric effect by Seebeck enabled more precise experiments to be carried out (for example, by Bérard, Melloni, Forbes, and Delaroche[23]) and these experiments amply corroborated Young's ideas. (It should be noted that while Young made suggestive propositions, the true wave theory of light was developed soon after by the brilliant French theoretical physicist Fresnel (1788–1827) in 1816; and in 1835, Ampère (1775–1836), the 'French Faraday', developed a wave theory of heat.)

But now there was a new dichotomy—instead of a split between the caloric and motion theories of heat, there was a split between two kinds of heat, radiant and ordinary. A possible bridge between these heats lay in the ether—the ether supposedly carried the waves but was itself, perhaps, made up of tiny particles. Then, in Einstein's landmark Special Theory of Relativity, in 1905, the (heat and light) waves were found *not* to need an ether. Also, after the several works of Planck, Einstein, de Broglie, Compton, and others in the twentieth century (see Chapter 17), it was understood that radiation is *both* a wave and a particle (and also, there is no ether).

There was increasing evidence, as we have seen, of the links between heat, light, chemistry, electricity, and magnetism—all different 'energies'. However, there was still one kind of energy that was not linked to these others. In 1800, the engineer Richard Trevithick (1771–1833) built the first high-pressure steam engine, heralding the age of locomotion and the mass use of heat-engines. Trevithick and his son, Francis, were unable to rouse Rumford's enthusiasm for these engines. Francis comments (somewhat patronizingly?), that 'Count Rumford was not quite up to the idea of the new steam engine . . . still he gave his opinion of a proper fire-place for the boiler of the steam carriage'.[24] Despite greatly enhancing the status of the motion theory of heat, Rumford never made a conclusive (in other words, quantitative) connection between the *microscopic* motions in heat and the *bulk* motion generated by a heat-engine (although his gunnery experiments had been suggestive). This is the final link that still had to be made, the link between heat and *work*.

10

Naked Heat: the Gas Laws and the Specific Heats of Gases

Because it's a gas.

The Rolling Stones

Prologue

The gas laws and specific heats of gases occupy a somewhat dusty corner in physics, but the dust is well and truly blown away when we put the physics into its historical context. We can then appreciate the crucial role of gases in the development of *two* major advances in nineteenth-century science—the theory of energy and the atomic theory.

Briefly, how the physics of gases helped the atomic theory is as follows.

John Dalton was the founder of the atomic theory—the theory that atoms are the ultimate constituents of matter, and that there are *different* atoms for the different chemical elements (oxygen, hydrogen, gold, sulphur, carbon, etc.). Now Dalton, in common with most other atomists at this time (around 1800), initially assumed that atomic *size* was the characteristic that determined chemical identity. The atomic size, according to Dalton, was given by the size of the envelope of caloric surrounding each hard atomic kernel. To understand Dalton's motivations, we must appreciate that he had a totally *static* view of a gas: the individual gas-atoms were at fixed positions in a 3D lattice, and were as close to each other as possible, consistent with the repulsive force acting between neighbouring envelopes of caloric (see Fig. 10.1).[1] So, rather than the 'passion fruit pip jelly' model of Chapter 5, we now have a new analogy for a gas—polystyrene shapes filling a packing case. The shapes stack differently, depending on whether they are large or small, and depending on whether they are, say, spherical or squiggly.

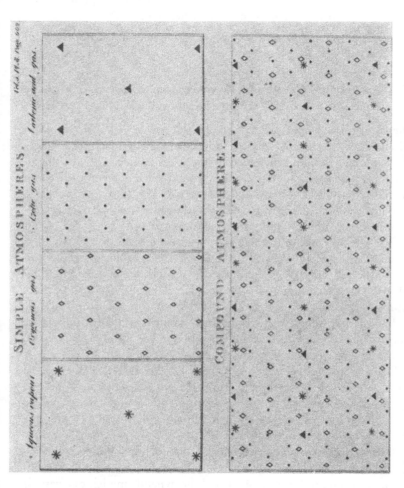

Fig. 10.1 Dalton's 'Simple and compound atmospheres', 1810 (courtesy of the Manchester Literature and Philosophical Society—the 'Lit and Phil').

However, there were some everyday observations that puzzled Dalton. The first was the observation that the different components of the atmosphere did not separate out into different layers, with the lightest layer at the top.[2] Other careful observations showed Dalton that:

(1) All gases increase in volume to the same extent for a given increase in temperature (when the pressure is held constant).

(2) For a mixture of gases, the pressure of any one gas is completely independent of the presence of the other gases (Dalton's famous 'law of partial pressures').

(3) The solubility of a given gas, in a given liquid, depends only on the gas pressure (the higher the pressure, the more soluble the gas).

These *physical* as opposed to chemical results made Dalton re-think his initial ideas. He realized that gas pressure and solubility were purely mechanical effects, and therefore atomic size, atomic shape, and chemical affinity were all irrelevant. In fact, from the first observation, Dalton surmised that all gases have atoms of exactly the *same* size at a given temperature—in other words, there was the same amount of 'caloric atmosphere' surrounding each atom,[3] whatever the gas.

What, then, *was* the atomic property characterizing an element, and differentiating one element from another? Once again, the physics of gases provided the clue. Dalton found that for different gases, the solubility *did* vary according to the type of gas: the heavier the gas, the more soluble it was. Drawing together all these observations from physics, and combining them with results from chemistry (chiefly, the relative weights of chemicals combining in compounds[4]), Dalton founded an atomic theory based, for the first time, on *weight*.

Dalton's atomic theory, his ideas on solubility, and his law of partial pressures have all stood the test of time. However, what is astounding and baffling to us moderns is how Dalton's ideas could have survived his totally static, and therefore utterly wrong, conception of a gas—see Fig. 10.1 again, and especially Fig. 10.2.

That his theories came through is because Dalton stumbled across—or, rather, his profound physical intuition led him to—the one salient feature common to both the static and the dynamic models of a gas: *the temperature alone is decisive*. It determines the caloric per atom (static model) or, as we now understand, the average* kinetic energy per gas

* The notation ‹ *x* › means 'the average *x*'.

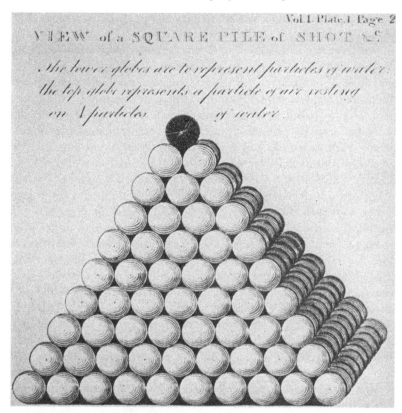

Fig. 10.2 'View of a square pile of shot', 1802 (courtesy of the Manchester 'Lit and Phil'). (A particle of air rests on 4 particles of water.)

particle, $\langle\tfrac{1}{2}mv^2\rangle$ (in the modern dynamic model). The gas temperature is oblivious to details such as the particle's mass or the particle's average speed; only the average particle *energy* is important.

One remarkable consequence of this result is Avogadro's Hypothesis, formulated by Avogadro (1776–1856) in 1811. It states that at a given temperature and pressure, a certain volume of gas contains exactly a certain number of molecules.[†] It doesn't matter what the type of gas

---

[†] Avogadro realized that the *molecule* was the basic unit, never mind that it could be made up from one, two, three, or more atoms.

is, or whether the molecules are shaped like polo mints, Maltesers, or Smarties—there is always the same number.‡

Now if we can just find out *what* that certain number of molecules is, then we'll be able to weigh the gas and find out what the weight of just *one* molecule is. Some ten years after Avogadro's death, that number was discovered, and was named after him.[5] But, wait a moment. If the caloric per molecule (in today's terms, the average kinetic energy of a molecule) is *the* key to gassy behaviour, and is blind to any gas-specific atomic details, such as atomic weight, how could we ever find out anything specific to this or that type of gas?[6] In other words, what effects could possibly cleave apart the 'm' and the 'v^2' in the locked casket of $\langle\frac{1}{2}mv^2\rangle$?

It turns out that there are certain effects that do offer a key to this casket. One such effect we have cited already—it is the dependence of solubility on the type of gas. Another effect is the rate of diffusion of gases. Different smells traverse a room at different rates, and this rate does depend on *specific* properties of the smelly molecules. (The transport of smells is, moreover, hard to explain with a static model.) One final effect to mention is the speed of sound in different gases. This speed is related to the average molecular speed, $\langle v \rangle$, in the given gas. In a light gas, such as helium, the molecular weight, m, is small and so the $\langle v \rangle$ is correspondingly higher.[7] The speed of sound should therefore be higher in helium than in air, and so it is. (This is the reason why people who have inhaled helium talk in squeaky voices.)

That the physics of gases could lead to two such major scientific advances (the atomic theory and the theory of energy) is due ultimately to the fact that gases allow heat-energy to show up in its most naked state (as we have seen in earlier chapters).

Part I: The Gas Laws

Starting with the result, we have

$$PV = nkT \qquad (10.1)$$

where P and V are the pressure and volume of the gas, respectively, n is the number of particles, T is the temperature, and k is a constant (Boltzmann's constant; see Chapter 17). This law was first discovered

‡ The explanation is given in Chapter 18, 'Which is more fundamental, temperature or pressure?'

empirically (see below) and then derived from the kinetic theory for an idealized gas. A gas is ideal when its particles are point-sized, don't interact, and bounce off the container walls elastically. In reality, the gas particles do have a finite size, and do have a small attraction for each other, especially when close together; and so there is a better convergence between theory and experiment when the gas density is low. In the limit, as the density goes to zero (in other words, when there's no gas at all!), the match is perfect (see Chapter 18, 'Impossible Things').

This law (Equation 10.1) has four variables, so holding any two constant we have a relationship between the other two. This is summarized in Table 10.1.

We have already described the work of Boyle, Mariotte, and Amontons in the seventeenth century (see Chapter 4) and shall now continue with the work of Dalton, and Gay-Lussac at the start of the nineteenth century.

John Dalton (1766–1844) was a Quaker and village schoolmaster in the Lake District. In 1793, he moved to the smallish town of Manchester, to teach at the Manchester Dissenting Academy. He remained in Manchester for 51 years as a teacher, scientist, and meteorologist—he recorded the local rainfall, temperature, barometric pressure, and other weather details every day of his adult life. He never married, and his only activities outside science were weekly bowls, an annual trip to the Lake District, and regular church attendance. Even so, he was affable and had a few close friends (for example, the Reverend Johns), and many scientific contacts in the newly formed Manchester Literary and Philosophical Society (the 'Lit and Phil').

Manchester was transforming itself as one of the centres of the Industrial Revolution, and Dalton was the first in a line of scientists from

Table 10.1 The ideal gas laws

	Relationships	Discoverers
n and T fixed	$PV \propto$ constant	Boyle, Mariotte
n and P fixed	$V \propto T$	Amontons, then Dalton, Gay-Lussac
n and V fixed	$P \propto T$	Amontons, then Gay-Lussac
P and T fixed	$V \propto n$	Avogadro's Hypothesis, also implied in Gay-Lussac's work

Manchester, the 'Manchester school', later to include Joule (see Chapter 14), Rutherford, and others. Despite this, science in England at this time was still mostly practiced by rich and/or eccentric amateurs. This was in sharp contrast to France where, since the French Revolution, a new breed, the professional scientist, was emerging. This contrast is brought out in the following quote from Roscoe in 1826:

M. Pelletier, a well-known savant, came to Manchester with the express purpose of visiting the illustrious author of the Atomic Theory. Doubtless, he expected to find the philosopher well-known and . . . lecturing to a large and appreciative audience of advanced students. What was the surprise of the Frenchman to find, on his arrival in Cottonopolis, that the whereabouts of Dalton could only be found after diligent search; and that, when at last he discovered the Manchester philosopher, he found him in a small room of a house in a back street, engaged looking over the shoulders of a small boy who was working his 'ciphering' on a slate. 'Est-ce que j'ai l'honneur de m'addresser à M. Dalton?' for he could hardly believe his eyes that this was the chemist of European fame, teaching a boy his first four rules. 'Yes' said the matter-of-fact Quaker, 'Wilt thou sit down whilst I put this lad right about his arithmetic?'[8]

(As well as the atomic theory, Dalton is famous as the discoverer of (his own) colour blindness—'Daltonism'.)

Dalton's most relevant contribution to the gas laws was his fourth 'Experimental Essay',[9] read to the 'Lit and Phil' in 1801, and in which he had set out to determine the change in volume of a gas upon heating. Dalton followed good experimental practice, as best he could (for example, he first dried the gas and the glassware using sulphuric acid), and his findings confirmed the investigation of Amontons, 100 years earlier, that all gases expand to the same extent, for the same increase in temperature. In detail, he found that the gas increased its volume from 1,000 units to 1,376 units, between 32 °F and 212 °F. This regular increase in volume was, for Dalton, a testimony to the regular increase in the volume of caloric surrounding each separate gas particle, as the gas was heated.

At almost exactly the same time as Dalton was carrying out this research, Gay-Lussac, in Paris, was conducting the same investigations, and coming to the same conclusions. But Gay-Lussac's experiments were altogether more thorough and professional. This was unsurprising, given that Gay-Lussac was working in a different milieu, an environment that was indeed more professional, and where Gay-Lussac was a scientist with a job description and a salary. In fact, Joseph-Louis Gay-Lussac (1778–1850) was, according to his biographer,[10] both a scientist and a bourgeois. He was a protégé of the chemist Berthollet, and became a member of the French Institute, First Division, at the young age of 28. Berthollet had recently founded a science society at his house in Arcueil on the outskirts of Paris, and had some rooms converted into laboratories. When the Marquis de Laplace bought the house next door, Berthollet arranged for a door to be put in connecting their two properties, and the new 'Societé d'Arceuil' now had the authority of the great French mathematician at its head. Some other members were Biot, Poisson, and the German explorer Humboldt. The young Humboldt and Gay-Lussac became friends and carried out some experiments jointly; for example, measuring the combining volumes of oxygen and hydrogen (see below), and going on a balloon ascent to measure the proportion of oxygen at different altitudes.

It was chiefly at Laplace's instigation that Gay-Lussac carried out his early investigations into the physical properties of gases (Laplace, as astronomer, needed to know the factors that affected the refraction of starlight in the atmosphere). Besides, Gay-Lussac had recently been appointed to the French Institute as a physicist, so he had to find something physical to investigate rather than pursue his beloved chemistry.

Gay-Lussac's research into the expansivity of gases was, as we have said, altogether more professional than Dalton's. He was able to determine that the expansion coefficient was 1/267 per degree Centigrade, and that this applied to oxygen, nitrogen, carbon dioxide, air, ammonia, sulphuric ether, and moist air. When Gay-Lussac conscientiously acknowledged some desultory earlier experiments of Charles, the law of the uniform expansion of gases was mistakenly attributed to Charles ever after (apart from in France).[11] In any event, Gay-Lussac—as chemist rather than as physicist—achieved lasting fame with a completely different sort of volume relationship. In experiments carried out with Humboldt, he found that hydrogen and oxygen gases combined to form water in a specific ratio by volume: 1.00 volume of oxygen combined with 1.99 volumes of hydrogen. He could have left it at that, but simplicity argued the case for a volume ratio of 1 : 2. This supposition was given extra confirmation by Gay-Lussac's later experiments, from 1807 onwards. He found that whenever gases were the reactants or products in a chemical reaction, the gas volumes were always in a *simple* ratio. For example, 1 volume of ammonia was made from 1 volume of nitrogen and 3 volumes of hydrogen.

This 'law of combining volumes of gases', along with Avogadro's Hypothesis, gave wonderful support to Dalton's atomic theory—but

Dalton rejected this helping hand, as he couldn't accept the concomitant idea that more than one atom of the same kind could be found in an elementary gas molecule. (For example, for Dalton, ammonia's composition was NH and not NH_3.) So, rather than being grateful for this endorsement of his theory, Dalton complained of the 'French disease of volumes'.[12] Avogadro's Hypothesis was only accepted 49 years after it was formulated and, unfortunately, four years after Avogadro had died.

Gay-Lussac's researches also helped to bring out the relationship $P \propto T$, which, along with Boyle's Law and Charles' Law (so-called), all confirmed the ideal gas law of Equation 10.1.

We might now ask which are the most important thermodynamic parameters in this equation, $PV = nkT$. Not V or n, as surely no truly fundamental attribute could depend on something so arbitrary as the sample size, whether by volume or by weight.[13] That leaves P or T. Black (see Chapter 6) had found that bodies all strive to reach an equilibrium temperature, but this also applies to pressures. T demonstrates direction (hot bodies cool down) but the same is also true for P (a gas will try to expand rather than contract).

It turns out that T, after all, is the more fundamental thermodynamic parameter. First, every body has a temperature, while only gases have pressures. Second, and more crucially, two gases at the same pressure are not yet in equilibrium—not until their temperatures have been equalized as well (see Chapter 18). Pressure is a meaningless concept when it comes to individual molecules, whereas we shall find that temperature straddles the macroscopic and microscopic worlds—for example, T is a macroscopic thermodynamic parameter occurring in expressions such as Equation 10.1, but it also relates to the energy of an individual, albeit average, molecule. For all these reasons, temperature wins out over pressure as regards how important and fundamental it is.

Part II: The Specific Heats of Gases

Laplace had an idea for improving Newton's famously wrong estimation of the speed of sound in air (Newton's result was 10% too low) and, for this, Laplace needed to know more about the specific heats of gases—and so he encouraged Gay-Lussac to measure them.

From Chapter 6, the specific heat is the amount of heat needed to cause a one-degree rise in temperature for a given weight of substance. This refers to heat *directly* applied; for example, by a flame, hot coals, a hot-water jacket, and so on. However, in a gas, a temperature rise can also happen in a completely different way—by sudden compression of the gas (Fig. 10.3). The rise in temperature can be so dramatic that tinder may be ignited (Mollet[14] demonstrated such 'fire pistons' in Lyons in 1804).

So, we have the direct heating of a gas (say, by flame or 'heat-jacket') and the indirect heating by a sudden change in volume. In the caloric theory, adopted by Dalton, Gay-Lussac, and the 'Laplacian school' (Laplace, Biot, Poisson, and others), these two processes were thought to be equivalent, and this led to no end of difficulties, as we shall see. As in Part I, we shall outline the modern interpretations first.

The cases of indirect heating or cooling are called *adiabatic*, meaning that no heat is transferred between the system and its surroundings—in other words, the system is enclosed by insulating walls. (The walls of a 'dewar' or thermos flask are a good approximation to such 'adiabatic walls'.)

In the case of direct heating, obviously a transfer of heat across a system boundary does occur. In one special case,[15] however, this does not lead to a change in temperature. This is the case where, at exactly the

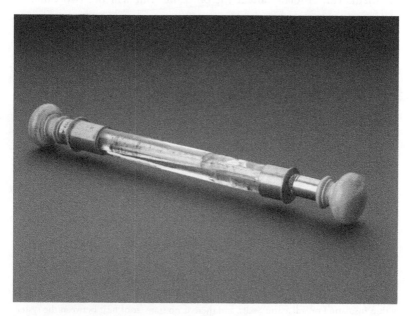

Fig. 10.3 A French demonstration model of a fire piston, early nineteenth century (Science Museum, London/SSPL).

same time as the gas is being directly heated, the gas is also expanding and pushing a piston back against the 'surroundings'.* We can imagine ideal conditions in which the temperature rise from direct heating *exactly compensates* the cooling of the gas as it pushes against the piston. The volume, *V*, of the gas[16] increases, and its pressure, *P*, falls, but *the temperature remains constant*. This is called the *isothermal*[†] case. For a given weight of gas, at a given temperature, all the possible *P* and *V* pairs join up and define a curve, known as an 'isothermal', as in Fig. 10.4(a).

We have met this isothermal case before. If we put *T* = constant into the ideal gas equation (Equation 10.1), we find that *PV* = constant (the curves are hyperbolas), and this is also the same thing as Boyle's, or Mariotte's, Law. Boyle had discovered this law experimentally—and with no direct heating or moving pistons. He had slowly varied the pressure of the gas ('air') by gently raising one arm of a J-shaped glass tube filled with mercury (see Chapter 4), and then noted the accompanying adjustment in the volume of the trapped gas. (Boyle was careful to try to keep the temperature constant all the while.)

We can begin to appreciate how the details of all these various experiments can crucially affect the outcome. But will the two different

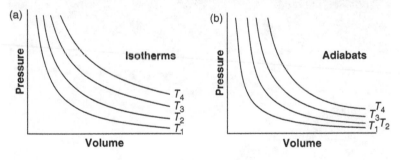

Fig. 10.4 (a) Isotherms and (b) adiabats for an ideal gas.

---

* 'Surroundings' means the air, foam, sponge, vacuum, and so forth on the outside surface of the piston.
  † We repeat these definitions for future reference. 'Isothermal' means 'at constant temperature', and usually means that the system makes 'thermal contact' with an external heat source through 'diathermal walls'. 'Adiabatic' means that the system is contained within insulating or 'adiabatic' walls, and there is no transfer of heat between the system and the exterior. In both cases, there may still be 'work done' on or by the system, by or on the surroundings.

scenarios—'compensated expansion' and Boyle's experiment—really map out one and the *same* curve on the *PV* graph? The 'compensated expansion' is an ideal set-up, never to be exactly realized in practice (see Chapter 18, 'Impossible Things')—in fact, I'm not sure if there has ever been any attempt to check it experimentally. Nevertheless, for theoretical reasons, we believe that the same curve would indeed be traced out in these two scenarios. The reasons are as follows.

In modern terms, the pressure of a gas arises out of the enormous number of collisions of gas molecules against the walls of the container (around a billion per second per square centimetre for air under standard conditions). Now consider a sample of gas at one temperature, and imagine progressively 're-housing' it in larger and larger containers. (Imagine that the container walls simply melt away, and the gas 'finds itself' in a new, larger-void container.) There is no piston, and no work is done. The amount (the weight) of gas is constant, and therefore, as the volume of the container is increased, the density of gas molecules (and hence the gas pressure) decreases proportionately; therefore, from *purely geometric considerations*, we must have PV = constant. So, one of the curves in Fig. 10.4(a) is traced out.

Repeating the 're-housing' process, but now at a new, higher, temperature, the gas pressure will be correspondingly higher ($P \propto T$, as in Table 10.1), but, once again, we'll find that PV = constant, and so an isotherm higher up in Fig. 10.4(a) will be followed.

If the gas is not simply re-housed but actually expands *against a piston*, its volume will increase, its pressure will decrease, *and* its temperature will fall. However, *if* external heat is supplied, at just the right steady rate so that the temperature of the gas stays constant, then its pressure will be maintained at just such a value that an isotherm (at this new, higher, temperature) will again be traced out.

Now for adiabatic changes (the gas pushes back against a piston, but the temperature fall is not compensated for by contact with a heat reservoir), then the pressure decreases not only for geometric reasons, but also because the gas is getting cooler. We therefore have the *steeper* curves of P against V (see Fig. 10.4(b)). These are called *adiabatic* curves.

Let us now return to our nineteenth-century narrative.

Gay-Lussac commented that the results of his experiments were not conclusive, as the vessels and thermometers absorbed large amounts of heat relative to that absorbed by the gas itself. Also, the response time of the thermometers was very slow. These considerations were to dog all determinations of the specific heats of gases. Nevertheless, Gay-Lussac tentatively suggested the following trends:

(1) The specific heat for *different* gases is inversely proportional to the molecular weight[17] of the gas (for example, the lightest gas, hydrogen, has the greatest specific heat).

(2) For a given volume of a *given* gas (in fact, he only showed it for air), the specific heat is less as the pressure is less.

(3) For a given weight of a *given* gas, the specific heat increases with increasing volume.

The first two suggestions were correct and the last was wrong—but none of them make any sense until we specify the experiments more exactly. In (1), the amount of gas was determined by the sample always having a given weight—but for a lighter gas (say, hydrogen) this implied a greater number of gas particles. In (2), lower pressures were achieved by releasing some gas; therefore the sample amount became progressively smaller. In (3), at last, the weight of gas was specified—but the third suggestion was wrong. Gay-Lussac kept changing his mind about it, but it was almost forced upon him by the prevailing caloric theory. According to this theory, the subtle heat-fluid, caloric, wrapped itself around each gas molecule like a miniature atmosphere (cf. Dalton's assumptions at the start of this chapter). For a given weight of a gas, then, the larger the volume of the container, the greater the physical space was around each molecule, and the more caloric atmosphere could be accommodated.

The Specific Heat Capacity of the Vacuum

From this wrong supposition came the following extrapolation *ad absurdum*: as the specific heat increases with increasing volume, then the specific heat capacity of the vacuum is the largest of all. The eighteenth-century German polymath Lambert (1728–77) (see also the end of Chapter 6) believed this, and he suggested that suddenly reducing the volume of a void should have a heating effect.[18] De Saussure, Swiss Alpinist and natural philosopher (see Chapter 6), agreed. There were also historical antecedents to this view in the work of Descartes (see Chapter 3) and Boerhaave (see Chapter 5).

Two French scientists, Clément and Desormes, argued the case in a highly original way. They had entered into a competition of the French Institute, First Class, on the specific heats of gases, in 1812. A large spherical vessel was partially evacuated and then left to regain thermal equilibrium. The stopcock was then opened for a fraction of a second, and outside air allowed to rush in. Clément and Desormes claimed that their partially evacuated flask could, in effect, be considered as a *mixture* between 'air at atmospheric pressure' and 'a void'. The sudden injection of yet more air caused compression and adiabatic heating of the pre-existing components. The relative specific heats of the void and the air could then be found as these two mixed and came to equilibrium (in other words, using the usual 'method of mixtures'; see Chapters 5 and 6).

You may wonder why they didn't test Lambert's idea more simply— by compressing a vacuum. But a vacuum was hard to obtain and to hold, especially in an adjustable container. Then Gay-Lussac came up with a method that was most ingenious. He prepared a Torricellian vacuum (the cavity at the top of the column of mercury in the glass tube; see Fig. 4.1); and then he simply tilted the barometer this way and that, so that its size was altered. He found no changes in temperature as the size of the void was varied. (In the modern view, the vacuum, having no mass, has no assignable specific heat capacity.)

Returning now to the specific heats of gases rather than of voids, another pair of French scientists, Delaroche and Bérard, carried out a series of investigations that were the last word in accuracy and professionalism at the time. They had entered the same competition as Clément and Desormes (these two pairs were, in fact, the only entrants) and had come first and won the prize of 3,000 francs. Delaroche and Bérard had gas flowing at a constant rate through a spiral copper tube in a water bath. In this way, a large volume of gas was used, and the uptake of heat was maximized (so that measurement errors were reduced). After calibration, they could work out how much heat had been imparted to how much 'fixed' volume of gas, and measure the corresponding temperature increase. Once they had determined the specific heat of a number of different gases, they set out to test the dependency of specific heat on volume. They tested just one gas ('air') at just two 'volumes' (in fact, pressures) and found that, taking the specific heat to be one unit at atmospheric pressure, it was 1.2396 units at a pressure of 1 metre of mercury.

From just these two data points, Delaroche and Bérard thought they had confirmed Gay-Lussac's proposition (3), and with some hubris they wrote:

Everyone knows that when air is compressed heat is disengaged [the temperature rises]. This phenomenon has long been explained by the change supposed to take place in its specific heat; but the explanation was founded upon mere supposition, without any direct proof. The experiments which we have carried out seem to us to remove all doubts upon the subject.[19]

But they were wrong; and this prize error was to reinforce the caloric theory and retard acceptance of the true dynamic theory of heat. Still, it is hard to agree that it was really 'one of the most significant ones [errors] ever made in the history of science'.[20] The error was perpetuated in the work of Sadi Carnot, but one has the feeling that Carnot was not totally taken in by it (see Chapter 12).

Apart from the comparison between Boyle's experiment and the case of compensated isothermal expansion, another troublesome duo was the case of a gas *expanding* due to direct heating, and a gas heated by adiabatic *compression*. Was the heat really the same, and was the expansion in the first case the inverse of the compression in the second case? Biot thought yes. Gay-Lussac had experimentally found the relationship $V \propto T$ (Part I), and had measured the expansion coefficient of all gases to be 1/267 for a one-degree rise in temperature. Poisson, using theory, had shown that an adiabatic compression of 1/116 caused a temperature rise of one degree. (By this stage, 1800 and after, the French scientists were using the Centigrade scale, now convergent with the Celsius scale.)

Another conundrum was the case of the free expansion of a gas.

The Free Expansion of a Gas

Is a gas expanding against a piston against 'the surroundings', the same as a gas expanding into a vacuum? (Imagine that the piston has simply dissolved away, and outside there happens to be a vacuum: in other words, it's the same purely geometrical, isothermal, expansion we mentioned earlier.) Today, this is called the 'free expansion' of a gas. The investigation of this effect was a particularly confusing episode spanning over 50 years. Here are four experiments all giving different outcomes.

(1) First was the experiment of 1807, in which Gay-Lussac's aim was to measure the relative specific heats of gases. He used two 12-litre glass flasks, each with its own thermometer, and connected by a pipe fitted with a stopcock. The apparatus was dried thoroughly with anhydrous calcium chloride, and then pumped out to as good a vacuum as was possible. The test gas was introduced into one of

the flasks, and everything was left for 12 hours to ensure stable conditions. The stopcock was then opened and the test gas allowed to rush from the first flask into the empty flask. Air, hydrogen, carbon dioxide, and oxygen were tested in turn, all at roughly atmospheric pressure, and also air at two other, lower, pressures. The stopcock was specially designed with a variable orifice so that different types of gas would flow through the pipe at roughly equal rates (otherwise the lightest gas, hydrogen, would flow the fastest).

What Gay-Lussac found was that the temperature of the first flask always fell, and the temperature of the second flask always rose, by an almost equal amount. (For example, his actual experimental results show that for air at atmospheric pressure, the temperature in the first flask fell by 0.61 °C, and the temperature in the second flask rose by 0.58 °C.) The biggest temperature rises and falls were noticed with hydrogen, then air, then oxygen, and finally, carbon dioxide. Gay-Lussac tentatively concluded that the magnitudes of these temperature changes, and hence the specific heats themselves, were in inverse proportion to the gas weights (so hydrogen, the lightest gas, would have the largest specific heat).

(2) We have already described the competition entry of Clément and Desormes, carried out in 1811/12 (published in 1819). Air from outside was allowed to expand into a partially evacuated vessel. The temperature in the vessel rose slightly.

(3) In 1844, Joule repeated Gay-Lussac's experiment but with copper vessels instead of glass flasks (see Chapter 14). There is no evidence that he knew of Gay-Lussac's work. He found *no* temperature changes upon expansion of the gas from one vessel into the next.

(4) In 1852–53, Joule and Kelvin (see Chapter 16) together carried out a similar experiment, but this time with the gas expanding through a cotton-wool porous plug or 'throttle'. The temperature fell. Later (1861), Joule found that the higher the starting temperature, the less marked was the cooling. Later again, in the twentieth century, it was found that above a certain critical temperature, the 'inversion temperature', the temperature *rises* as the gas expands.

Leaving you, for the moment, to devise your own answers to these divergent outcomes (all will be explained at the end of the chapter), let us return to the vexed question of specific heats.

C_p and C_V

There was yet one more troublesome duo in specific heat studies—one that has great relevance to energy. We have talked of the compensated expansion of a gas as it tracks an isothermal curve, but there is another kind of compensation possible—that required to keep the *pressure* rather than the temperature of the gas constant as the gas expands. (We could call this isobaric expansion.) Thus, two kinds of specific heat can be defined: C_V, the specific heat at constant volume (the heat needed to raise the temperature of a given weight of gas while its volume is held constant); and C_p, the specific heat at constant pressure (the heat needed to raise the temperature of a given weight of gas while its pressure is held constant).

Crawford, in 1788 (see Chapter 6), was attempting to explain animal heat, and was the first scientist to make any measurements of the specific heat of a gas. He measured the temperature rise after direct heating, but contained the gas first in an extensible bladder, and then, in later experiments, in a brass vessel of fixed dimensions. He therefore inadvertently obtained rough measures of C_p and C_V—but, of course, he had no idea that there were two types of specific heat.

The importance of C_p is that it is explained in utterly different ways in the caloric and modern theories. In the caloric theory, the excess of C_p over C_V is due to the 'latent heat of expansion' of the gas: as the gas expands, there is more space for the caloric, and so the capacity of the gas (to 'hide heat') goes up, and the amount of 'sensible' heat (heat detectable by a thermometer) goes down. In the modern theory, the heat capacity of the gas is *not* dependent on its volume. Rather, the expanding gas does mechanical *work* in pushing the piston, and this work is at the expense of the heat-energy in the gas. This recognition—of the actual conversion of heat into work—heralded the discovery of 'energy', and will be covered in Chapter 14.

The remarkable thing is that while the qualitative explanations are radically different, the quantitative agreement between the two theories is total. In both theories, $C_p = C_V + $ 'heat of expansion'. Whether this 'heat of expansion' goes to increase a hidden reserve within the gas, or whether it is lost as it is converted into work, makes no difference whatsoever to the thermometer readings. Moreover, this 'heat of expansion' serves exactly the same role in both the caloric and modern theories: $C_p - C_V$ is a constant in both theories; C_p/C_V is another constant, called γ; the adiabatic curves in Fig 10.4(b) are given by PV^γ = constant in both theories; and finally, the correction to Newton's calculation of the speed of sound is $\sqrt{\gamma}$ in both theories.

This last was famously worked out by Laplace. In 1816, he published a brief note in which he had closed the gap between Newton's value and the experimental value, but he didn't give his reasoning. In 1823, in the last volume of his *magnum opus*, the *Mécanique Céleste*, he explained all. (It would be ungenerous to say that in his first paper he looked around for a correction that would fit, and then in the intervening years worked out how to justify it.) This work was the reason why Laplace had asked Gay-Lussac to investigate the specific heats of gases in the first place. The value of γ was measured by Gay-Lussac and Welter in 1822, using a modification of Clément and Desormes' apparatus (experiment (2) above—and ever after referred to as Clément and Desormes' Method). Their value of 1.37 brought Laplace's calculations into better agreement with experiment—so the latter's work could be considered as a *resounding* success.

Dulong and Petit's Law

Before we leave the subject of specific heats, we must look at the influential work of Dulong and Petit with respect to solids. In 1819, Dulong and Petit measured the specific heat capacities of a number of solids, mostly metals ('*bismuth, plomb, or, platine, etain, argent, zinc, tellure, cuivre, nickel, fer, cobalt, soufre*'[21]). They also knew the relative atomic weights of these solids (from the work of chemists such as Berzelius). They then stumbled upon a curious correlation—the product of the specific heat and the relative atomic weight was roughly constant—and realized this meant that the specific heat *per atom* was a constant. They elevated this result to the status of a law:[22]

The atoms of all simple bodies have exactly the same capacity for heat.

(This law, while not exact, gave wonderful support to the atomic theory—but Dalton, as usual, was not to be helped.) Dulong wondered whether there might be a similar law in the case of gases. (The year after the original law was promulgated, Petit died of tuberculosis, aged only 29.) Dulong therefore carried out a comprehensive review of the specific heats of gases, in 1829, and ended up extending Dulong and Petit's Law to the case of gases. He also managed to show that, as the specific heat per gas molecule was the same, whatever the gas (oxygen, nitrogen, etc.), the specific heat for a given *volume* of gas was the same, whatever the gas (assuming the same conditions—of pressure and temperature). This is exactly what Dalton and Gay-Lussac had suspected all along (and it also ties in with Avogadro's Hypothesis).

The modern explanation is as follows: the specific heat is determined solely by the average kinetic energy per atom, and is therefore unrelated to the type (*mass*) of the atom. This explains Dulong and Petit's Law for solids. The same is true when the 'atom' is in fact a gas molecule, and this then explains Dalton's and Gay-Lussac's findings. However, this explanation is only an approximation to the truth. When the gas molecule is polyatomic (for example, carbon dioxide, water vapour, or methane) then there are more avenues through which heat can be taken up apart from purely kinetic. The specific heat is therefore larger in such cases (and this is why the so-called 'greenhouse gases' contribute to global warming).

So it seems that the specific heat of gases, like the temperature, straddles the macroscopic and microscopic worlds. On the one hand, the specific heat is a macroscopic thermodynamic parameter, telling us about a bulk property of the gas—how much heat it absorbs per degree for a given mass of gas. On the other hand, it tells us about certain details of the actual molecules, such as whether they are made from one, two, or more atoms. (Later, in the twentieth century, anomalies in specific heats were a window into the quantum world; see Chapter 17.)

Before we finish this section on the specific heats of gases, we mustn't forget to tie up the loose ends of experiments (1)–(4) above. In modern terms, we explain the findings as follows.

Free Expansions, Again

First, in Gay-Lussac's experiments, the variations in temperature that he observed were due to the fact that he didn't wait long enough after opening the stopcock for equilibrium conditions to be achieved. The effect was more marked for the lighter gases, as these have a higher average speed, and therefore travelled to the second flask more quickly (despite the specially designed valve).

Joule, on the other hand, measured the average temperature across both vessels—as there was just one water bath surrounding these; furthermore, he waited until all temperature fluctuations had ceased.

In Clément and Desormes' experiment, as with Gay-Lussac's, the conditions hadn't settled down to a steady state. Also, as there was some air already in the vessel, the expansion wasn't even approximately 'free'.

The temperature change following the Joule–Kelvin expansion is tricky to explain (and without talking about enthalpy), but is entirely due to the fact that no gas is truly ideal (there *are* interactions between the molecules). First, the gas 'does work on itself' pushing itself through the 'throttle': the molecules repel each other as the gas is squashed up, and the temperature may rise. Second, as the gas emerges into the larger volume, its molecules are further apart, and there is therefore some cooling (the molecules attract each other at these larger separations, and so resist the expansion).

Theoretically, in a true free expansion of a gas (no intermolecular forces, and no work done pushing back pistons or pushing back other gases), the temperature remains *constant* whatever the volume or pressure of the gas. This counter-intuitive result cannot be explained in the caloric theory. In modern classical thermodynamics, we say that the 'internal energy' of an ideal gas depends only on the temperature (or, equivalently, it doesn't matter *where* you are on a given isotherm). Once again, the temperature is appearing as *the* defining thermodynamic parameter. There is another puzzle—why is there a direction involved? It seems that it *does* matter where you are on a given isotherm, because the gas is always trying to reach a lower pressure and a larger volume. This is mysterious and can only be answered by reference to a new abstract quantity—entropy. We shall give the explanation near the end of Chapter 18, in the section 'Difficult Things'.[23]

Real Experiments—Some Anecdotes

The experiments we have been describing all point to the crucial influence of the experimental conditions. To give you a flavour of the ingenuity of these early nineteenth-century experiments, consider this rather appealing example—Gay-Lussac's method of determining vapour pressure. Gay-Lussac blew a hollow glass teardrop (such as may be seen today in the so-called Galileo thermometer), weighed it, filled the bulb completely with the relevant volatile liquid, sealed the neck with a flame, and then re-weighed it. This full teardrop was then introduced into a Torricelian barometer so that it rose up through the mercury and bobbed into the vacuum. Heat was then applied from outside the barometer (using

a 'burning glass'?) until the teardrop exploded and the vapours were released. Taking account of the volume of floating glass shards—including
the volume of displaced mercury, and of the thermal expansion of the
mercury and the vessel—the change in the mercury level was a measure
of the pressure of the alcoholic vapours.

Many of these early experiments were fraught with danger. In 1804,
Gay-Lussac ascended higher in a balloon than ever before, 7,016 metres,
a record that was not matched for half a century afterwards. He complained of nothing more than a headache. In 1808, he was temporarily
blinded by an explosion with the newly discovered potassium. He recovered but his eyes were permanently affected, and he subsequently always
wore protective goggles. In 1811, Dulong lost a finger and the sight of
one eye during his discovery of nitrogen trichloride. He resumed his
investigations only four months later. Needless to say, all chemists ran
the risk of breathing in noxious fumes (for example, Scheele checked the
smell and taste of his new discovery, hydrogen cyanide).

Overview

One thing that we have learned is that the minutiae of the experiments
were crucial, not just with regard to precision but with regard to what
was actually being measured. How did Boyle maintain a constant temperature (when checking PV = constant) and Gay-Lussac a constant
pressure (when measuring the expansion coefficient of a gas)? Were the
specific heats of gases measured by reference to a constant weight or a
constant volume of gas, and were such 'volumetric' specific heats then
measured at constant volume or at constant pressure? Were the conditions diathermal or adiabatic, and was the gas simply 're-housed' or did
it expand against a piston against the external air? How good were the
vacuums? Was the expansion fast or slow, did the piston slide freely,
what was the heat capacity of the vessels, and what was the time lag of
the thermometers? How is a water bath maintained at constant temperature, and is the specific heat of the water a constant at all the relevant
temperatures?

The technology required was also increasingly sophisticated: accurate
glassware, brass vessels to sustain high pressures, good-quality pumps
and seals, the collection and storage of gases, accurate thermometers,
and so on.

All the while, the gas molecules themselves were still only conjecture.

As regards theoretical advances, this chapter has been long and difficult, but it will be sufficient to hold on to just two findings. First, the temperature T is *the* defining thermodynamic parameter for a gas. Second, when a gas does work in expanding or contracting against a piston, heat is added to or subtracted from the latent heat-reserve in the gas (caloric theory), or is *converted* to or from the work (modern theory of energy).

That there could be such a *conversion* between the mechanical energy of the piston and the heat energy of the gas was not dreamt of in anyone's philosophy at the time. All, that is, except for one philosopher— Pierre Louis Dulong (1785–1838) (the same Dulong as in 'Dulong and Petit's Law'; see above). In a letter to the great Swedish chemist Berzelius, in January 1820, he wrote:

I can show that, in making the volume of a gas or vapour vary suddenly, one produces changes in temperature incomparably greater than those which result from quantities of heat developed or absorbed [merely by supposed changes in heat capacity], if [it] wasn't [for the] heat *generated* by the movement. Rumford had already used more or less this line of reasoning . . .[24]

If Dulong's significant *aperçu* had been noticed, it might have speeded up the course of physics by 25 years—but no response to this letter has ever been found. We can at least console ourselves with the knowledge that Dulong was, by all accounts, a very nice man. He was a doctor who treated his poor patients in Paris for free, and even paid for their prescriptions out of his own pocket. This was despite the fact that Dulong himself always struggled financially, had four children to raise, and suffered from chronically ill health as well as his work-related injuries. He died in 1838, aged 53.

All the progress in empirical knowledge described in this chapter— the gas laws, the specific heats of gases, and the determination of γ— provided the crucial evidence to take Dulong's vision further. Two men in particular, Sadi Carnot and Robert Julius Mayer, were to make great use of the data on the specific heats of gases. Their work (see Chapters 12 and 14, respectively) would lead to the formulation of thermodynamics (Carnot) and the concept of energy (Mayer).

We started this chapter with a look at how the physics of gases helped to give birth to two scientific revolutions: the atomic theory and the theory of energy. The atomic theory was at a more advanced stage in the opening decades of the nineteenth century, but the energy theory had leap-frogged ahead by mid-century. The main reason for the changing fortunes of the atomic theory was philosophical doubts about the reality of atoms. The positivist philosophy of Comte (1798–1857) was then at its height. (Comte was proclaiming the impossibility of ever knowing the composition of stars even as experimentalists were examining the first stellar spectra.) It is puzzling that this positivist outlook should not also have hindered the idea of energy. After all, it's not as if anybody actually sees a blob of '$\frac{1}{2}mv^2$' coming towards them. As in Feynman's allegory, the 'blocks' of energy would only emerge through quantitative formulation, experimental corroboration, and the honing of the idea of energy itself.

11

Two Contrasting Characters: Fourier and Herapath

We have already made comment on the different approach to science in England and France during the post-revolutionary period, and have seen the contrasts between contemporaries such as Watt and Lavoisier (see Chapter 8) and Dalton and Gay-Lussac (see Chapter 10). Now (1812–1822), once again, appear two characters who epitomize their respective national characteristics: the English eccentric amateur, John Herapath, and the French professional mathematical physicist, Joseph Fourier.

Fourier

Joseph Fourier (1768–1830) came from humble origins, but through sheer talent managed to rise in the fields of mathematical physics, government office, and literary accomplishment. The son of a poor tailor in Auxerre, he was orphaned at eight, and would have been lost to science but for a lady who was touched by his especial gentleness of manner.[1] She recommended him to the Bishop of Auxerre, who arranged for Fourier to be brought up by Benedictine monks. From early on, Fourier developed an interest in mathematics—he collected candle-ends by day so that he could secretly pursue his studies by night. Later, despite his gentle disposition, and bolstered by testimonials, the young Fourier applied for permission to join the artillery, but was rejected: 'Fourier, not being of noble birth, cannot enter the artillery, not even if he is a second Newton'.[2] However, after playing an active part in the French Revolution (in Auxerre), Fourier was at last received into society, and rewarded by an appointment, first at the École Normale, and then at the École Polytechnique. (During the Terror, Fourier's gentle manners worked against him, and he was criticized for being too lenient, and even arrested for his part in helping an innocent man escape the guillotine.[3]) Fourier's abilities were soon spotted by Napoleon, and he was asked to accompany Napoleon on his Egyptian campaign (in 1798): Fourier performed his duties well, and after only three years he was virtually governor of half of Egypt. (Later, Fourier's introduction to the 'Description of Egypt' was so admired that he was admitted to the prestigious French Academy.) Upon Fourier's return from Egypt in 1802, Napoleon appointed him Prefect of Isère, and he remained there, in Grenoble, for the next 13 years.

It was while he was Prefect of Isère that Fourier first became interested in the subject of heat (in the 1820s). Like Boerhaave (see Chapter 5) and Lavoisier (see Chapter 8), Fourier attributed heat with truly cosmic significance:

Heat, like gravity, penetrates every substance of the universe, its rays occupy all parts of space.[4]

it influences the processes of the arts, and occurs in all the phenomena of the universe.[5]

Newton's Law of Gravity and Laws of Motion explained both the motions of the heavenly bodies and Earth-bound motions. Likewise (thought Fourier), the varied phenomena of heat should be bound by simple mathematical laws, and should apply equally in the celestial and terrestrial domains. Fourier especially wanted to explain the temperature of the planets, the heat radiated by the Sun, and the cold of outer space:

How shall we be able to determine that constant value of the temperature of space, and deduce from it the temperature which belongs to each planet?

What time must have elapsed before the climates [on Earth] could acquire the different temperatures which they now maintain?

From what characteristic can we ascertain that the Earth has not entirely lost its original heat?[6]

He was also the first to appreciate the influence of the atmosphere on the surface temperature of the Earth (what we now call the greenhouse effect).

Despite his cosmic ambitions, Fourier concentrated on just one aspect of heat—its *transmission*, whether by conduction (through the body of a planet) or radiation (across the empty space between the Sun and the planets). The English natural philosopher Halley (see Chapter 4) had already demonstrated the 'sine law' for the intensity of radiation emitted from a surface, but Fourier showed that this was a *necessary* law. In a remarkable piece of reasoning, he argued that if the sine law didn't apply, then perpetual motion would follow:

Bodies would change temperature in changing position. Liquids would acquire different densities in different parts, not remaining in equilibrium [even] in a place of uniform temperature; they would be in perpetual motion.[7]

This was probably the first time that the impossibility of perpetual motion had ever been used in a thermodynamical setting.[8]

Fourier's magnum opus was *The Analytical Theory of Heat* of 1822, and he was not backward in coming forward, claiming to have 'demonstrated all the principles of the theory of heat, and solved all the fundamental problems'.[9] The work was based on his earlier (1812) prize-winning entry for a competition on the modes of transmission of heat. Now Biot, a member of the Societé d'Arcueil, and an adherent of the 'Laplacian school' (see Chapter 8), had carried out experiments (in 1804) on the transmission of heat along a bar. The iron bar was held hot at one end, cooled at the other end, and left until a steady state had been reached. This meant that, at any position along the bar, the rate of heat loss by (1) radiation through the surface and (2) conduction along the bar was exactly matched by the quantity of heat supplied continuously at the hot end. Fourier's masterstroke was to submit Biot's experimental set-up to mathematical modelling. As regards (1), Fourier took it that the rate of heat loss by radiation from the surface was proportional to the excess temperature (i.e. Newton's Law of Cooling; see Chapter 5). But as regards (2), he modelled the process in a completely novel way[10]: he imagined a cross-section through the bar, and defined the 'flux' of heat across it (the amount of heat crossing unit area per unit time). One of Fourier's physical assumptions was that Newton's Law of Cooling, of a sort, applied for heat transmission *within* the bar as well as for heat losses from the surface. (In other words, the heat flux at any distance, x, along the bar, was proportional to the temperature gradient, $\partial T/\partial x$, at that distance.) Finally, from the twin requirements of continuity of heat flux, and conservation of total heat, Fourier arrived at his famous heat-conduction equation.[11]

One striking feature of Fourier's heat-conduction equation is that it is the first equation in physics that is different for time running forwards and time running backwards (i.e. if '$-t$' is substituted for 't'). However, there is no evidence that Fourier, or any of his contemporaries, noticed this.

Apart from its correctly describing heat conduction, the equation was enormously important for physics, as it led Fourier on to one of the greatest discoveries in mathematics—the 'Fourier series'. This is a method for decomposing a wave of any shape into a sum of elemental (sinusoidal) components.

The heat-conduction equation is also intriguing in a number of other ways, but the discussion is rather technical, and has been relegated to the endnotes.[12] From our point of view, the most interesting thing about Fourier's achievement is that the equation is totally modern in approach, and, indeed is still used today to describe the conduction of heat through a solid.

Despite the comment at the end of the previous chapter, the influence of positivism wasn't all negative. Fourier was a positivist, and so he didn't commit himself to the caloric theory (he encountered some enmity from Biot and Poisson for this, but Laplace himself remained aloof[13]); and he always insisted that every mathematical statement or variable had to have a direct physical meaning, and be capable of measurement. He therefore examined the clumps of physical constants occurring in the exponents[14] of his Fourier series, and realized that they had to correspond to something real—the actual physical 'dimensions' (such as 'time', 'space', 'mass', etc.). He proceeded to carry out the first 'dimensional analysis' in physics, and to develop a theory of units. This was a major advance, possibly the first since Galileo,[15] in the mathematical representation of physical quantities. (Comte went on to adopt Fourier as the leading promoter of positivism in the physical sciences.)

Napoleon's downfall also marked the nadir of Fourier's fortunes. The coming of the Restoration, and the end of the Hundred Days, left Fourier deprived of the Prefecture, in disgrace, and almost destitute. Fortunately, a friend and former pupil, the Prefect of Paris, managed to arrange eminently suitable employment for Fourier, as the director of the Bureau of Statistics in Paris. Fourier retained this position until he died in 1830. By a strange irony, Fourier's death was due in part to overheating—he had a habit of wrapping himself up 'like an Egyptian Mummy'[16] and living in airless rooms at an excessively high temperature. (Possibly, he suffered from an underactive thyroid gland.)

Herapath

John Herapath (1790–1868) was the archetypal English eccentric. He was largely self-educated, and was acquainted with the major works of mathematical physics, from Newton's *Principia* onwards, and with the recent experimental investigations of Dalton and Gay-Lussac (Herapath had taught himself French as well as physics). His style was verbose and opaque, full of made-up words such as 'Numeratom', 'Voluminatom', and 'Megethmerin', set out in old-fashioned scholia, propositions, and lemmas, and also he used proportions instead of calculus.

Yet Herapath, working outside the mainstream scientific community, was one of the founders of the kinetic theory of gases. In fact, there seems to be a correlation between these two groups. First, there was Daniel Bernoulli in Switzerland in 1738 (see Chapter 7, Part III), 50 years later came Cavendish (see Chapter 8), and then came Herapath in 1816, and Waterston in 1845. All worked in isolation (in Cavendish's case, he also worked in secret) and all were unaware of the work of their predecessors. We shall come to the sad neglect of Waterston's work later (see Chapter 14).

Like Fourier, Herapath came to heat studies via gravitation. He sought to provide a mechanistic explanation of gravity, and proposed that a subtle 'aether', made up from 'gravific' particles, became rarefied by the high temperatures near to celestial bodies. The reduced density of the aether then allowed gravity to hold sway. All this brought him to the connection between temperature and particle velocity, and on to his kinetic theory of gases.

Herapath rejected repulsion as the cause of elasticity in a gas, and hit upon the true kinetic theory, wherein particles move freely of each other, and are in 'projectile motion', as opposed to some sort of vibratory motion. However, he soon came across an obstacle that baffled him and 'was a shock I had hardly philosophy enough to withstand'.[17] This was the inconvenient truth that his particles had to be both perfectly hard and yet undergo perfectly elastic collisions. He eventually resolved this dilemma (to *his* satisfaction, at any rate) by making mv rather than mv^2 the fundamental measure of motion. This led him on to Boyle's Law, and to the correct association between gas pressure and particle speed ($P \propto v^2$), but also to the incorrect $T \propto v$ and $PV \propto T^2$.

Despite the 'unbeatenness of the track'[18] that he followed, Herapath was disappointed that his ideas were not better received—even by Davy, who had supported Rumford in his stand against the caloric theory (see Chapter 9). Now Davy accepted the idea of heat as motion of the constituent particles, but he thought in terms of a localized vibration or rotation rather than the more mechanistic picture of a gas of randomly moving non-interacting particles. This may have been due to his affiliation to the 'Romantic' movement in England and France ('*Naturphilosophie*' in

Germany) through the influence of his friends—chiefly, the poet Samuel Taylor Coleridge. This new Romantic movement, as applied to physics, promoted the idea of mutual influences, ebbs and flows, and quasi-organic evolutions, rather than austere mechanism. As well as this, Davy may simply have found Herapath's style hard to take, and his archaic mathematical proofs impossible to follow. There were also some glaring mistakes in Herapath's work. In any event, Davy informed Herapath that the paper would not be published in the *Philosophical Transactions* (Davy had recently become the President of the Royal Society, and the *Philosophical Transactions* was its journal). He was mindful to advise Herapath to withdraw his paper, otherwise it would become the property of the Royal Society, and could not be returned to the author.

So, Herapath sent his paper instead to the *Annals of Philosophy*, where it was published in 1821. It was almost completely ignored, although a certain anonymous character, 'X', made some pertinent criticisms in a short reply.[19] Five years later, Herapath attacked Davy, and the Royal Society, through the letters pages of *The Times* newspaper. Davy never responded, but when Davy resigned from the presidency soon afterwards, Herapath took it as a victory for himself.

Herapath then retired from the fray. Soon, the new era of railways provided him with an opening, and in 1835 he became editor of his own *Railway Magazine* (he had 11 children to support). This gave him a forum to continue publishing his scientific ideas, and in 1836 he presented a calculation of the speed of sound, which included a calculation of the *average* speed of a molecule in a gas. This was the first time that this had ever been done (priority is usually wrongly given to Joule).[20]

Herapath's work went unnoticed until, in 1848, Joule was trawling the literature to find some support for his radical new departure in physics (see Chapter 14). Later still, Maxwell (see Chapter 17) acknowledged Herapath's pioneering work. At least Herapath—unlike Avogadro (see Chapter 10) and Waterston (see Chapter 14)—lived long enough to see the vindication of his ideas with the publication, in 1860, of Maxwell's famous kinetic theory of gases.

Overview

Fourier's heat conduction equation would be of prime importance to Thomson in his meshing together of Carnot's work and the new thermodynamics (see Chapter 16). Herapath's work, as we have seen, was eventually recognized by Joule and by Maxwell. Another contemporary of Fourier and Herapath, the 'French Faraday', André-Marie Ampère (1775–1836), had yet another approach to heat: he considered its propagation as a wave, in his paper of 1832. (This wave theory applied to ordinary heat; it is distinct from the theory of radiant heat.)

However, none of these three approaches to heat—its transmission via conduction, or as a wave, or the kinetic theory—considered the connection between heat and *work*. This was all the more surprising because this was the heyday of the steam engine: all over Europe, and the East Coast of North America, but especially in Britain, steam engines were pumping water out of mines, powering industry, and transporting people and goods in trains and on boats. These were all examples of the performance of work from heat. It was an engineer rather than a physicist who finally investigated this connection, and, in so doing, brought in an overarching new principle, and initiated the discipline of thermodynamics. We shall cover this in the next chapter.

12

Sadi Carnot

Often, the greatest leap forward occurs with a totally new approach (as when the principle of the economy of nature joined the principle of the conservation of nature's resources) or when utterly different fields of enquiry are found to have something in common (Newton embracing the orbiting of the Moon, and the falling of an apple, in a single set of laws). Contemporaneous with the French 'Laplacian school' (chiefly, Laplace, Biot, and Poisson) and other French scientists, such as Ampère, Fresnel, Dulong, and Fourier, came a revolutionary new departure. It was, though, a very quiet revolution, as we shall see. All the workers listed above were physicists, whereas the new departure came from a young engineer, Sadi Carnot (1796–1832)—contrary to common perception, the influence from engineering to physics is rare; the spin-off usually goes the other way.[1]

Laplace and his followers tried to understand inter-particle forces (between 'heat-particles' and matter-particles); and Fourier asked what laws of heat transmission were necessary to ensure equilibrium in the motion of celestial bodies. But Carnot was the first to ask about the connection between heat and motion on an everyday, engineering scale—not motion on a microscopic or astronomical scale, but the sort of motion that occurs in the workings of a heat-engine. He asked: what, if any, are the limits to the efficiency of a heat-engine?

We can see in this question a concern with engineering but also with social issues such as economics. On both counts, Sadi's chief influence was his father, Lazare Carnot, a mathematician and engineer, and a leading statesman during the time of the French Revolution (see Chapter 8).

Lazare named his son Sadi after a medieval Persian poet and moralist, Saadi Musharif ed Din. There are not many anecdotes from Sadi's short life, but all show him to be a person with high moral standards. For example, when the Buonaparte and Carnot families were on an outing near a lake, and Napoleon was throwing stones to splash his wife and other ladies in a boat, Sadi, aged four, ran up and shook his fist at Napoleon, shouting: 'You beastly First Consul, stop teasing those ladies!'[2] (Fortunately for physics, Napoleon just laughed.) Later, as a young adult, Sadi drew up a list of rules of good conduct for himself: 'Say little about what you know, and nothing at all about what you don't know . . . When a discussion degenerates into a dispute, keep silent . . . Do not do anything which the whole world cannot know about'.

Coming from a famous, political family, Sadi was motivated to improve the economic and political standing of France. He understood the huge importance of the heat-engine:

The study of these engines is of the greatest interest, their importance is enormous, their use is continually increasing, and they seem destined to produce a great revolution in the civilized world.

He also understood the particular importance of the steam engine to England:

To take away today from England her steam-engines would be to take away at the same time her coal and iron. It would be to dry up all her sources of wealth, to ruin all on which her prosperity depends, in short, to annihilate that colossal power. The destruction of her navy, which she considers her strongest defence, would perhaps be less fatal.

Now steam engines had been improved steadily, from the time of Savery onwards, and Sadi readily acknowledged the enormous debt of the world to the eighteenth-century English engineers—Newcomen, Smeaton, 'the famous Watt', Woolf, and Trevithick. Sadi asked himself the following question: was the work that could be obtained from a heat-engine, the 'motive power of heat', unbounded, or was there an assignable limit to the improvements that could be made, 'a limit which the nature of things will not allow to be passed [exceeded] by any means whatever'?

The laws of operation of machines not using heat (the simple machines such as the lever, pulley, and inclined plane; and more complicated machines powered by water, wind, or muscle) were well known, and ultimately attributable to the laws of mechanics (noted Carnot). Sadi's own father, Lazare, had initiated the fundamental theory of such machines. Sadi's genius lay in his isolation of the one law relevant to all these examples, and his realization that this law would apply to *all* machines, whether mechanical or *heat*-driven. He set out a simple argument, using no mathematics, and couched in the style of the ancient

Greeks—the syllogism. His very argument was ancient—he appealed to the age-old law of the impossibility of perpetual motion.

His argument was as follows: hypothesize an ideal engine that is the best possible—it yields the maximum amount of work. As it is ideal, it is also reversible (it can be run equally well in reverse). Now imagine that there could exist another engine—say, the 'superior' engine—which produces *more* work than the ideal engine. If the ideal engine was run in reverse, the 'superior' engine could totally counteract it—it could replace all the work consumed and return the world to the identical starting conditions—and there would still be some work-capacity left over. This leftover work could then be used to generate a perpetual motion. But perpetual motion is impossible—and therefore the initial premise (that a 'superior' engine exists) is ruled out.

In succinct syllogistic form, the argument runs as follows: perpetual motion is not possible; a 'superior' engine would permit perpetual motion; therefore a 'superior' engine is impossible. In other words, *there can be no engine better than an ideal engine.*

So far, this result is not too surprising—after all, we hypothesized the ideal engine as the best possible one in the first place. However, more can be teased out of the argument. There can be no engine better than an ideal engine; but can there be a worse one? Obviously, in the case of ordinary engines (real, actual engines), these are all worse than the ideal engine, but could one ideal engine be worse than another? For example, suppose that ideal engine IE_{worse} is worse than ideal engine IE. Then IE would be better than IE_{worse}. But our original argument prohibits this— *no* engine, ideal or not, can be better than the ideal engine.

So, combining the two results, no ideal engine can be better or worse than another ideal engine. We finally come to the remarkable conclusion: *all ideal engines perform equally well*, and this will be true whatever the details of their construction, whether they use this fuel or that, and so on.

Let us analyse this argument.

First, we must be sure that we are only comparing ideal engines of the same 'size' even though they might work in different ways. For example, if we were to employ IE ten times over, then it would be equivalent to one ideal engine ten times larger, and would obviously do ten times as much work. We shall define the 'size' of an ideal engine a bit later on.

Second, we note that Sadi was careful to define 'perpetual motion' and show that it could be of two kinds. It was 'not only a motion susceptible of indefinitely continuing itself after a first impulse [has been] received', but also 'the action of an apparatus, of any construction what-ever, capable of creating motive power in unlimited quantity'. (This was a subtlety that had already been appreciated by Leibniz; see Chapter 3.) By ruling out *both* kinds of perpetual motion, Carnot was bringing in what would become the Second Law of Thermodynamics as well as what would become the conservation of energy and the First Law of Thermodynamics.

Sadi was thorough, and mindful to justify his premise of the impos-sibility of perpetual motion. As we saw in Chapter 2, this is one of those truths that has always been challenged even while it has always been accepted. Sadi says:

The objection may perhaps be raised here, that perpetual motion, demonstrated to be impossible by mechanical action alone, may possibly not be so if the power either of heat or of electricity be exerted; but is it possible to conceive the phe-nomena of heat and electricity as due to anything else than some kind of motion of the body, and as such should they not [also] be subjected to the general laws of mechanics?

This didn't mean that Sadi accepted the motion theory of heat—in fact, he adopted the mainstream caloric theory—but that he assumed that, at the most fundamental level, all processes were mechanical. Sadi was also convinced by the fact that, experimentally, *all* attempts at perpetual motion had failed: 'Do we not know, besides, *a posteriori*, that all the attempts made to produce perpetual motion, by any means whatever, have been fruitless?'

Finally, we note that Sadi's syllogism *required* an ideal engine to have the property of being reversible.

While Sadi's argument is admirable, our admiration is tempered by the fact that we want to know: what *is* an ideal heat-engine?

Sadi was influenced, of course, by his father's seminal work on the efficiency of mechanical engines (see Chapter 8, Part I) and he wanted to find similar strictures for heat-engines. Lazare had specified condi-tions for optimizing the performance of any machine, in his Principle of Continuity. By this Principle, power should be transmitted smoothly, continuously, and without percussion or turbulence.

Take, for example, the case of a water-engine driven by a fall of water—the scenario so often considered by Lazare. We can imagine two water-wheels, A and B, and, once again, use the argument of the impos-sibility of perpetual motion to state that B can't generate more work than A; otherwise, B could be run in reverse, employed as a 'water-raiser', and be used to drive A, perpetually. But this doesn't clinch the proof—perhaps B has better-shaped scoops and does, in fact, work more efficiently than A?

For a complete proof, we must first idealize the water-wheel, A, to be the best possible—in other words, it must be an *ideal* water-wheel: the falling water must meet the scoops with zero relative velocity, so that there are no losses due to 'shock'; there must be no friction at the bearings; the water must trickle out at the bottom with zero speed; and there must be no loss of water due to spray or splashes; and so on. Only then can we use it in our argument, and assert that, yes, this idealization *is* the best possible, no water-wheel could work better, and so a perpetually acting machine is ruled out. Behind the details of this idealization of the water-wheel lies the deeper understanding that it is the 'falling of weights' (the weight of water through a given height) that is the source, sole determinant, and measure of the work done by any gravity water-engine.

Now, getting back to Sadi, he asked: What is the deeper understanding behind the motive power of *heat*? Or, in other words, what is the idealization in the case of a *heat*-engine? It is here that Sadi Carnot's genius truly came to the fore. He understood, for the first time, the quintessential nature of a heat-engine: *it is a machine that does work as a result of heat falling between two temperatures.*

It is hard to be impressed enough by the simplicity of this, 200 years later. Listing the multifarious varieties of contemporary heat-engine, it was far from obvious that they all shared this underlying essential mechanism: there were steam-driven paddle-wheels, contemporary versions of Amontons' fire-wheel (see Chapter 4), Newcomen's atmospheric engine (see Chapter 5), Watt's condensing engine (see Chapter 8), Trevithick's high-pressure steam engine, compound engines, engines using the expansive properties of air, or alcohol vapour, combustion engines, and buoyancy engines. We could also include heat-engines in the more abstracted sense, such as the cannon and the hot-air balloon, or take examples from geology (e.g. volcanoes) or meteorology (e.g. the water-cycle). Carnot saw that in *all* these 'engines', the overriding feature was that a certain quantity of heat flowed between two temperatures. But which two temperatures?

Another crucial, fundamental, and simple understanding came in with Carnot: heat always flows from a high temperature to a lower temperature. This knowledge was as old as the hills, but who had thought to raise awareness of it and bring it into physics? Even Black (see Chapter 6), who understood that separate bodies will lose or gain heat until they achieve an equilibrium of temperature—even he did not bring out the fact that heat always flows *from* hot *to* cold. With Carnot, *direction* (of a process) came into physics for the very first time.

Let's get back to the idealized heat-engine and see how Carnot makes it approachable in reality. After all, a water-wheel may be nothing but a machine driven by a flow of water between two heights, but a real water-wheel must have the best design of scoops, the most friction-free bearings, and so on. Carnot's considerations led him to three conditions.

With regard to the performance of work, Carnot saw that when heat flows it can do work by causing a change in volume. For example, a gas could be heated, made to expand, and thereby push a piston that could be linked to a weight-raising machine. He did see the possibilities of other sorts of action—such as chemical change, which then resulted in work—but he chose to concentrate on the most commonplace effect, a change of volume. He had been on a fact-finding tour of actual heat-engines in use across Europe. These were mostly driven by steam, and so he was aware of the special connection between gases (e.g. vapours) and heat, and of all the contemporary research into the physics of gases (see Chapter 10).

Now, Carnot stressed that any heat-flow that occurs but does *not* result in a volume change is wasted—the potential for doing work was there but was not exploited. Therefore, in an ideal heat-engine, there should be no direct (i.e. thermal) contact between bodies at different temperatures.

The final feature of Carnot's ideal heat-engine was that it should be reversible—his syllogistic argument required this feature.

Altogether, the requirements for his ideal heat-engine were as follows:

(1) A heat-flow from a high temperature to a lower temperature.

(2) No *direct* heat-flow between these temperatures (i.e. no thermal contact between bodies at different temperatures).

(3) Only a reversible heat-flow between these temperatures.

These multiple requirements appear incompatible, as we know that heat only flows where there is a temperature gradient, and then only in one direction, from hot to cold. Carnot's remedy is novel, simple, and ingenious—and, furthermore, it will become the basis of the whole approach of modern thermodynamics. In a nutshell, it is that, in an ideal heat-engine, every temperature gradient is infinitesimal, and every heat-flow shall occur only by infinitesimally small steps. This guarantees that each such step will be individually reversible. Direct contact between bodies at different temperatures is allowed—but only where the temperature difference is infinitesimal. (In fact, such temperature differences *must* be allowed; otherwise no heat will flow.) Heat will still only flow from hot to cold, but the temperature gradient, being infinitesimal, can always be reversed at will.

We have already acknowledged the great influence of Lazare Carnot on his son. It was Lazare who inspired Sadi to consider the utility of machines for the glory of France, and for all mankind: 'If real mathematicians were to take up economics and apply experimental methods, a new science would be created—a science which would only need to be animated by the love of humanity in order to transform government'. Also, Lazare's work on the theory of generalized machines, and his Principle of Continuity, were undoubtedly starting points for Sadi. However, perhaps the greatest legacy of Lazare to Sadi was the concept of *reversibility*. Lazare had introduced the idea of infinitesimal 'geometric motions' (see Chapter 8), and further stipulated that these were reversible ('the contrary motions were always possible'[3]). Sadi incorporated this idea of infinitesimal reversibility, but added a brilliant extension to it: for a succession of *individually* reversible, infinitesimal processes, one could arrive at a *macroscopic* process that was still reversible—the idealized heat-engine.

All this is rather abstract. We turn now to Sadi's own description of how such an ideal heat-engine would run. This comes straight from Sadi's historic book, *Reflections on the Motive Power of Fire*,[4] a work embarked upon after a visit to his father[5] in exile in Magdeburg in 1821. Sadi never saw his father again, and the book was published in 1824, one year after Lazare's death. Sadi first sets the scene (our Fig. 12.1 is straight from his work, and the instructions are given verbatim):

let us imagine an elastic fluid, atmospheric air for example, shut up in a cylindrical vessel, abcd (Fig. 1) [our Fig. 12.1], provided with a moveable diaphragm or piston, cd. Let there be also two bodies, A and B,* kept each at a constant temperature, that of A being higher than that of B.

Sadi takes the starting position of the piston to be ef and the cylinder to be initially isolated from the heat reservoirs A or B. He describes the following cyclic series of operations (his numbering has been left intact):

---

* A and B are reservoirs of heat at temperatures T_A and T_B, respectively.

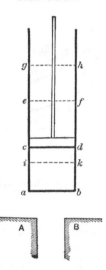

Fig. 12.1 Carnot's diagram of his cycle, in *Reflections on the Motive Power of Fire*, 1824 (by permission of Dover Publications, Inc.).

(3) . . . The piston . . . passes from the position ef to the position gh. The air is rarefied without receiving caloric, and its temperature falls. Let us imagine that it falls thus till it becomes equal to that of the body B; at this instant the piston stops, remaining at the position gh.

(4) The air is placed in contact with the body B; it is compressed by the return of the piston as it is moved from the position gh to the position cd. This air remains, however, at a constant temperature because of its contact with the body B, to which it yields its caloric.

(5) The body B is removed, and the compression of the air is continued, which being then isolated, its temperature rises. The compression is continued till the air acquires the temperature of the body A. The piston passes during this time from the position cd to the position ik.

(6) The air is again placed in contact with the body A. The piston returns from the position ik to the position ef; the temperature remains unchanged.

(7) The step described under number (3) is renewed, then successively the steps (4), (5), (6), (3), (4), (5), (6) . . . and so on.

This is the famous Carnot cycle. It is most easily understood by means of a pressure against volume graph, or '*PV* diagram' (see Chapters 8 and 10), as in Fig. 12.2. (Carnot himself never used such an 'indicator

Energy, the Subtle Concept

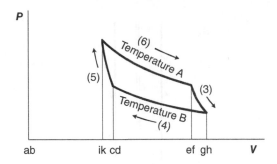

Fig. 12.2 Carnot's cycle shown on a graph of pressure versus volume.

diagram'. It was used to describe Carnot cycles for the first time by Clapeyron in 1834; see Chapter 16.)

The path from ef through gh, cd, ik, and back to ef again—roughly, a parallelogram—traces the various 'states' (pairs of *P* and *V* values) of the air enclosed in the cylinder as it expands or is compressed. As we know from Chapter 10, the air temperature rises during the compressions and falls during the expansions.[6] The whole cycle constitutes the ideal heat-engine. (We can see the influence of Watt's steam engine: section (3) is the same as Watt's 'expansive operation', except that in the latter case the amount of steam is continuously throttled down.)

There are many qualifications that must be added to this sparse outline. We are to understand that the 'bodies', A and B, are infinite reservoirs of heat. They stay at the temperatures T_A or T_B however much heat is added to or taken away from them. We are also to assume that the piston moves without friction or leaks; the heat transmission between cylinder and reservoir is perfect; and the thermal isolation of the cylinder, when not in contact with a reservoir, is total. Moreover, all the operations must be reversible, and must therefore be carried out smoothly and gradually (in effect, Lazare's Principle of Continuity must be adhered to for heat-engines as well as for mechanical engines). Thus, for compressions, the external pressure applied to the piston must be increased slowly and continuously so that it is always only marginally above the pressure inside the cylinder. Likewise, for expansions, the pressure outside the cylinder must be continually adjusted so that it is at all times only marginally less than the internal pressure. Sadi doesn't explicitly mention these provisos. He also tacitly assumes that the temperature and pressure can be monitored continuously without compromising their values in any way.

With these idealizations, we see that during the long sections ik to ef, and gh to cd (i.e. when the cylinder makes contact with a heat reservoir), the air in the cylinder remains at the constant temperature T_A or T_B, respectively. These sections are therefore *isothermal*, and follow the Ideal Gas Law, $PV = RT$ (see Chapter 10). (Sadi would have described this as Mariotte's Law.)

In the short sections ef to gh, and cd to ik (see Fig. 12.2), the cylinder is isolated from the reservoirs and does not lose or gain heat to/from the surroundings: in other words, the change is not 'compensated', and the temperature therefore does change by a finite amount. But this finite change occurs by infinitesimal steps, smoothly, continuously, and slowly, as the expansion or compression proceeds. The internal air pressure also changes (smoothly, continuously, and slowly) but for *two* reasons: (1), the usual volume change and, in addition (2), the temperature change just mentioned. The path followed is thus the *steeper* 'adiabatic' curve (see Chapter 10). (Sadi knew that the curve was steeper, but not that it followed PV^γ = constant, where $\gamma = C_p/C_V$, although this relationship had already been worked out by Poisson a year or two earlier; see Chapter 10 again.)

We now ask how much work this ideal engine does. We could gear up the piston to raise and lower weights but, in fact, we already know the work output for the volume change of an ideal gas. It is $\int P dV$, the area under the 'PV curve' (see the discussion on Giddy in Chapter 8, Part I). In going all the way round the Carnot cycle, the total work is given by the *area enclosed* by the cycle (the area of the 'parallelogram' in Fig. 12.2).[7] (Note that we have the convention that the engine 'goes forward'—that is, *does* work—during an *expansion*; in other words, net work is done by the engine in a *clockwise* cycle.)

Also in this clockwise direction, and during the isothermal sections, heat will be transferred continuously *from* the heat reservoir A (to the expanding cooling air-in-the-cylinder) and *to* the heat reservoir B (from the compressing warming air-in-the-cylinder). Therefore, when the engine is run forward, there will be a net flow of heat from the higher temperature, T_A, to the lower temperature, T_B.

We thus have an ideal heat-engine that conforms to Sadi's overall specification: work is done, and heat flows from a higher to a lower temperature. Moreover, the engine is reversible; that is, the exact same path can be tracked but in a reverse (anticlockwise) direction. In this *reverse* direction, the engine is a net consumer of work, and there is a net transfer of heat from B to A. But Carnot's directive, that heat flows only from hot to cold, has not been overturned: the heat does not flow *directly* from cold to hot; it trickles from the cold reservoir B to the even cooler, expanding gas; and from the hot, compressed gas to the infinitesimally cooler reservoir A. . . .

The proof of the reversibility of the engine for Sadi was the fact that there was a return to the initial 'state' of the air in the cylinder. For example, at ef (Fig 12.1), the air has exactly the same P and V values however many times this state has been passed through, and from whichever direction it is approached. This idea, of the *state* of a system, uniquely defined by certain macroscopic parameters—say, P and V—is the second great gift of Sadi Carnot to thermodynamics (after the idea of *reversibility*). His third great gift was the idea of a *complete cycle*: for any arbitrary starting point, the ideal engine must run through a complete series of operations (i.e. trace a *closed* path on the indicator diagram) and end up where it began.

The importance of all this was so that different ideal engines could be compared. Only by carrying out a complete cycle, and returning to the exact same starting state, could Carnot be sure that there were no work costs or heat costs that were unaccounted for. One ideal engine could then be compared with another, and so it would be possible to ascertain, first, whether Carnot's amazing proof of the equivalence of ideal heat-engines was correct, and, second, what the ideal efficiency (the theoretical maximum to efficiency) actually was. (We remember that improving the efficiency of engines was Sadi's ultimate goal.)

We must be sure that we know what 'efficiency' means. Surely, an engine using more heat will yield more work. And an engine with a greater drop in temperature will also do more work. Sadi took his clue from the analogy with water-power (like his father, Sadi always had gravity-powered water-engines in mind):

we can compare with sufficient accuracy the motive power of heat to that of a waterfall . . . The motive power of a waterfall depends on its height and on the quantity of the liquid; the motive power of heat depends also on the quantity of caloric used, and on what may be termed, on what in fact we will call, the height of its fall, that is to say, the difference of temperature of the bodies between which the exchange of caloric is made.[8]

In other words, Sadi understood that the heat-engine will do more work for a greater amount of heat used, and for a greater 'fall' in temperature. This answers our earlier question as to the 'size' of engines: only engines using the same quantity and the same fall of heat have the same 'size', and only these engines are to be compared.

Let's assume, therefore, that we are at last comparing engines that are ideal, that are going round a complete cycle, and are the same 'size'. Now, for the first time, comes a real difference between water- and heat-engines. Ideal water-engines, whatever their mechanism, all perform at 100% efficiency.[9] However, we find that heat-engines, even ideal ones, *never* operate with an efficiency of 100%.

Pausing briefly to give an explanation in modern terms, we see that this inefficiency is *sui generis*, built into the very definition of the heat-engine; for Sadi's heat-engine must operate between two temperatures, and heat must always be '*thrown away*' at the lower temperature. Water in a waterfall-engine flows to the bottom level, but it has been completely stripped of its (gravitational potential) energy. Heat also flows to the bottom level (the lower temperature), but it cannot be stripped of its energy: in fact, heat *is* energy.

Perhaps the efficiency of the ideal heat-engine is thus limited to some fixed amount—say, 70%?

We have been careful to compare only engines using the same quantity of heat and the same fall in temperature, but we have not thought to specify the *actual* temperatures. Sadi does not beg any questions:

We do not know . . . whether the fall of caloric from 100 to 50 degrees furnishes more or less motive power than the fall of this same caloric from 50 to zero.

This is quite startling. It is like asking whether a waterfall with a 50 m drop, and for a given quantity of water (say, 100,000 kg), will yield more work at 0 m above sea level than at 1,000 m above sea level.* But this is exactly what Carnot does find in the case of heat-engines:

The fall of caloric produces more motive power at inferior than at superior temperatures.

In other words, a given quantity of heat, falling through a given temperature interval, can do more work the lower the temperature. The efficiency is therefore not only less than 100%; it actually *depends* on the temperature. This is the conclusion that will eventually lead to the Second Law of Thermodynamics—that the 'worth' of heat is temperature-dependent.

Carnot couldn't fully justify this conclusion on his own calculations. Rather, it was his physical intuition that told him that it had to be true. However, this temperature dependence of efficiency can't have come as a

---

* In fact, as the strength of gravity varies inversely with height, there *is* a difference between the waterfalls.

complete surprise to him. After all, he had already (implicitly) conceded a temperature dependence in the very *modus operandi* of his engine (if there was no difference between the work carried out in the low- and the high-temperature sections of the cycle, then the engine would yield no net work).

So, how did the efficiency—let's symbolize it, η—of different ideal heat-engines actually compare? We still have not defined 'efficiency': for a waterfall-engine, it is the work done for a given weight of water falling through a given height; for a heat-engine, it is the work done for a given quantity of heat dropping through a given temperature interval. The 'work done' in each case can be easily determined. But now we come to yet another significant difference between water- and heat-engines. The water can be seen as it flows, but heat ('*calorique*') is invisible, and also weightless; it cannot be seen as it trickles into or out of heat reservoirs, and these reservoirs, being essentially infinite, leave no tally.

The resolution—to check up on the performance of an ideal heat-engine—would have been to build a mock-up near-to-ideal engine, and use special heat reservoirs (say, water-baths) whose take-up or release of heat *could* be monitored. Carnot did not do this[10] but, instead, resorted to a blend of experimental findings (the whole corpus of nineteenth-century work on gases) and 'exact reasoning'.[11] In detail, he used: 'Poisson's result' (that an adiabatic compression of 1/116th leads to a 1 °C temperature rise); Gay-Lussac's expansion coefficient (heating a gas directly by 1 °C at constant pressure leads to an expansion of 1/267th); the (ideal) gas laws of Mariotte and of Gay-Lussac; the measurements of γ; the measurements of specific heats by Dulong and Petit, and by Delaroche and Bérard; the work of Dalton, and of Clement and Desormes; and many other results. (These data have been explained in Chapter 10, but it is necessary for us only to follow the gist.)

At last, for three hypothesized, 'same-sized', ideal gas-in-a-cylinder heat-engines, some very real results were obtained. The hypothesized engines used air, then water vapour, and then alcohol vapour; and Carnot found that using these *different* gases, the engines all had very *similar* efficiencies—in particular, the last two engines had efficiencies agreeing to within 1.5%. His overall conclusions were as follows:

• The ideal efficiency, η, has a maximum, less than 100%, and is independent of the working substance and design of the ideal engine.
• This efficiency depends only on temperature (it depends on pressure only through its dependence on temperature).

- The efficiency is greater the greater the interval of temperature.
- For a given amount of heat used, and a given temperature interval, the ideal efficiency is most probably a continuous function ('Carnot's function') of the starting temperature, T, being larger at smaller temperatures. (He couldn't go as far as to claim $\eta \propto 1/T$, but only that when $1/T$ increased, η always increased.)

These conclusions were momentous; but Carnot had only been able to determine *relative* efficiencies (one ideal engine as compared with another). If he had been able to calculate the absolute efficiencies of these three engines, he may have been less than impressed. Jumping ahead to Kelvin's absolute scale of temperature, and Kelvin's definition of η as $(T_1 - T_2)/T_1$ (explained in Chapter 16; see Equation 16.1), we can make some absolute determinations of the ideal efficiency. For example, for the alcohol-vapour engine running at 78 °C (one of the starting temperatures that Carnot considered), with a one-degree drop in temperature, the ideal (theoretical maximum) efficiency is a mere 0.28%; or, for a more realistic 100-degree drop, it is still only 28%. This low efficiency reinforces the conclusion of Chapter 8: getting work from heat is inherently difficult.

Whether or not Carnot suspected this inherent difficulty is not known. However, it didn't contradict any of his conclusions, it didn't undermine his intention to beat the British steam engineers at their own game, and, most of all, his analysis showed him that his initial proposition (his syllogistic argument) was supported. His confidence in this was such that he elevated his proposition to a law:

Carnot's Law
The motive power of heat is independent of the agents employed to realize it; its quantity is fixed solely by the temperatures of the bodies between which is effected, finally, the transfer of caloric.

Also, in the manuscript version, he writes, triumphantly:

The fundamental law that we proposed to confirm seems to us to have been placed beyond doubt, both by the reasoning which served to establish it, and by the calculations which have just been made.

However, in the *published* version of the *Réflexions*, Carnot's tone is completely different:

The fundamental law that we proposed to confirm seems to us to require, however, in order to be placed beyond doubt, new verifications. It is based upon the theory of heat as it is understood today, and it should be said that this foundation does not appear to be of unquestionable solidity.

What had caused this hesitancy, this change of heart?

As we have argued above, Carnot was not depressed by the low theoretical maximum to efficiency for the ideal heat-engine; rather, his confidence was eroded for a different reason. It appears that, in the short time interval between the manuscript and the published copies, Sadi had the dawning realization that his 'exact reasoning' was flawed in one pivotal respect: he had used the caloric theory of heat, and had adopted without question its central tenet—that heat is always conserved. Water fell, the blades of a water-wheel were set a turning, and all the water then arrived at the bottom level, at the end of its fall. And should not the subtle-fluid, caloric, likewise 'fall', set a heat-engine to do work, and then all be collected at the bottom level; that is, the lower temperature? But Sadi's intuition began to tell him otherwise: did *all* the heat really arrive at the lower temperature? Carnot must have begun to see that this was false, and with this realization came a new idea—heat could be *consumed* and *converted* to or from motive power.

After Carnot's death, a bundle of his papers was found—some 23 loose sheets labelled 'Notes on mathematics, physics and other subjects'[12] and thought to span a time from the writing of his book until his death in 1832. Here is a selection from these posthumous notes, in assumed chronological order:

[A]lways in the collision of bodies there occurs a change of temperature, an elevation of temperature. We cannot, as did Berthollet, attribute the heat set free in this case to the reduction of the volume of the body; for when this reduction has reached its limit the liberation of heat should cease. This does not occur.

. . . how can one imagine the forces acting on the molecules, if these are never in contact with one another . . .? To postulate a subtle fluid in between would only postpone the difficulty, because this fluid would necessarily be composed of molecules.

Is heat the result of a vibratory motion of molecules? If this is so, quantity of heat is simply quantity of motive power.

Can examples be found of the production of motive power without actual consumption of heat? It seems that we may find production of heat with consumption of motive power.

Supposing heat is due to a vibratory movement, how can the passage from the solid or the liquid to the gaseous state be explained?

When motive power is produced by the passage of heat from the body A to the body B, is the quantity of this heat which arrives at B (if it is not the same as that which has been taken from A, if a portion has really been consumed to produce motive power) the same whatever may be the substance employed to realize the motive power?

Is there any way of using more heat in the production of motive power, and of causing less to reach the body B? Could we even utilize it entirely, allowing none to go to the body B? If this were possible, motive power could be created without consumption of fuel, and by mere destruction of the heat of bodies.

If, as mechanics seems to prove, there cannot be any real creation of motive power, then there cannot be any destruction of this power either—for otherwise all the motive power of the universe would end by being destroyed.

At present, light is generally regarded as the result of a vibratory movement of the ethereal fluid. Light produces heat, or at least accompanies the radiant heat . . . Radiant heat is therefore a vibratory movement. It would be ridiculous to suppose that it is an emission of matter while the light which accompanies it could only be a movement.

Could a motion (that of radiant heat) produce matter (caloric)?

Undoubtedly no; it [one motion] can only produce a [another] motion. Heat is then the result of a motion.

Then it is plain that it could be produced by the consumption of motive power and that it could produce this power . . . But it would be difficult to explain why, in the development of motive power by heat, a cold body is necessary; why motion cannot be produced by consuming the heat of a warm body.

Heat is simply motive power, or rather motion which has changed its form. It is a movement among the particles of bodies. Wherever there is destruction of motive power, there is at the same time production of heat in quantity exactly proportional to the quantity of motive power destroyed. Reciprocally, wherever there is destruction of heat, there is production of motive power.

We can establish the general proposition that motive power is, in quantity, invariable in nature; that it is, correctly speaking, never either produced or destroyed. It is true that it changes its form—that is, it produces sometimes one sort of motion, sometimes another—but it is never annihilated.

In summary, we can see that Carnot's thoughts evolved away from the caloric theory and on to the dynamical theory of heat. He came to realize that: heat cannot be material—it must be 'a motion'; all motive power is 'motion' of one sort or another; and interconversions are possible, but not the creation or destruction of 'motion'.

Carnot also suggested various experiments to test and quantify the conversion between heat and work—experiments that foreshadowed those of Mayer and Joule, some 20 years into the future. He estimated the conversion factor between heat and work (what would later come to be known as the 'mechanical equivalent of heat'), and his value is the first to appear in print.

Finally, these posthumous notes show that he wondered whether or not *all* the heat in a body could be *converted* into work ('motive power could be created . . . by mere *destruction* of the heat of bodies' and 'cannot [motive power] be produced by *consuming* [all] the heat of a warm body?'). This was a question that would not be resolved until the Second Law of Thermodynamics had formally arrived, some 25 years later.

However, Carnot's goal was to improve engine efficiency rather than to discover new laws of physics. He concludes the *Reflections* by giving advice on engine improvements to engineers. Overall, he stipulates that: attending to reversibility in the small will ensure efficiency in the large; insulation, and avoidance of contact between parts at different temperatures, will minimize the 'useless flow of heat'; and, finally, the highest starting temperature, and the largest possible fall in temperature, are desirable. For this reason (he speculates), air might be a better working substance than steam.

Despite the contemporary success of high-pressure steam engines, Carnot suspects that the high pressures are actually more of a hindrance than a boon. They can cause cylinders to burst and therefore lessen the maximum temperature that can safely be achieved. Air in a combustion engine would be more suitable than steam in this respect. He had thus anticipated the technology of an age some 40 years into the future (the first internal combustion engine was that of Jean Lenoir in 1859, and by 1885, Karl Benz was driving the first automobile).

Carnot had therefore understood that it is *temperature* rather than pressure that is ultimately the driver of the heat-engine. This is in accordance with our findings in Chapter 10, where we concluded that it was *T* rather than *P* that was the thermodynamic parameter of importance. (This is explained in Chapter 18, 'Which is more fundamental, temperature or pressure?')

The Fate of the Book

Despite Carnot's amazing achievements, he never actively promoted his book. It is true that he wasn't the self-promoting type (by all accounts he was very introverted), but the main reason for Carnot's reticence must lie with his understanding that the very cornerstone of his theory (the conservation of heat) had crumbled away. He was also reticent about this abandonment of the caloric theory, and never disclosed his changing views, even to his friend and associate Nicolas Clement—the prevailing hegemony was too strong, his views too radical.

Carnot would surely have known that his work would stand the test of time, and that it could be reworked to incorporate the dynamic theory of heat—and, of course, this *has* been done (see Chapter 16)—but this was too big a task for one individual to carry out.

The book itself was not a success. It was both too abstract for the engineers and too far from the mainstream for the physicists—in short, it was years ahead of its time. Six hundred copies were printed and cost 3 francs each (1824). One review was given—it was favourable, but then the editor of the journal was Sadi's brother, Hippolyte. Then the book simply disappeared from view.

Apart from a few copies of *The Reflections*, one other research paper, and the posthumous notes already mentioned, all Sadi's work has been lost. He contracted cholera in the great epidemic of 1832 and died within a few hours, aged only 36. All his personal effects (save a picture of a champion boxer[13]) were burned.

The Reflections would have been lost to physics altogether but for one thread. The engineer Emil Clapeyron (1799–1864) found a copy, modified it (chiefly by the addition of the indicator diagram), and applied it to the liquid/vapour boundary, in 1834. This work was discovered by William Thomson (later Lord Kelvin) in 1845, and inspired Thomson to look for Carnot's original work. Thomson scoured the booksellers of Paris, who had never heard of the *Reflections*: 'I went to every book-shop I could think of, asking for the *Puissance motrice du feu*, by Carnot. "Caino? Je ne connais pas cet auteur." With much difficulty I managed to explain that it was "r" not "i" I meant. "Ah! Ca-rrr-not! Oui, voici son ouvrage," producing a volume by Hippolyte Carnot [Sadi's brother]'.[14] Eventually (1848), Thomson did find a copy, and pronounced it 'the most important book in science I have ever read'.[15] It was a turning point of Thomson's career, and of the fortunes of Carnot's work (see Chapter 16).

Overview

Carnot had correctly identified the essential method of a heat-engine: heat flows from a high temperature to a lower temperature while doing work (for example, causing a volume change against an external pressure). Without doubt, Georg Friedrich Reichenbach's* (1771–1826) column-of-water engine was suggestive, but a more subtly suggestive engine was James Watt's steam engine. It is possible that Carnot's crucial link between engine operation and temperature would never have been made were it not for Watt's discovery of the condensing steam engine, with its readily apparent temperature change from the *hot* boiler to the *cold* condenser (see Chapter 8).

Also, Carnot had found that the maximum efficiency of even an ideal heat-engine was inherently limited, and was independent of the method or working substance of the engine (whether gas, vapour, liquid, etc.). He further posited that this maximum theoretical efficiency was probably determined solely by the actual temperatures between which the heat flowed, being greater the higher the starting temperature and the lower the final temperature.

Finally, Carnot had also set out the practical features of how to maximize the work output of real heat-engines: heat leaks must be minimized (and so there should never be direct thermal contact of any engine part with anything else—except in the case of a heat reservoir), the temperature of the hot source should be as high as possible, and the temperature of the cold sink should be as low as possible.

Carnot had therefore accomplished all that he had set out to achieve, and he had unwittingly also set out the groundwork for the whole of thermodynamics. He had introduced the concepts of 'state of a system', 'cycle of operations', and 'reversibility'; defined the ideal heat-engine; recognized that temperature rather than pressure is the primary determinant; and that the maximum amount of work to be obtained from heat probably depended on nothing else but the absolute temperature of that heat. This last would eventually become (a version of) the Second Law of Thermodynamics. However, perhaps the most remarkable aspect of Carnot's work is that so much should be right, and of long-term utility to science, when his fundamental axiom, that total heat is conserved, was wrong.

---

* As far as I know, there is no link between this Reichenbach and the Reichenbach Falls over which Moriarty and Sherlock Holmes plunged to their 'deaths'.

We shall examine this paradox.

In order to determine whether the heat was, or was not, conserved, it was necessary to experimentally track the heat taken up, or discarded, around the cycle: but the heat reservoirs kept their secrets, and the relevant data—for the specific heats of gases—were difficult to measure, and plagued by wrong assumptions. For example, we saw in Chapter 10 the false conclusions of Delaroche and Bérard, who claimed that the specific heat of gases increased with an increase in volume (for a given weight of gas). But Carnot's instinct for what was reliable—and *physical*—stayed with him:

According to the experiments of MM. Delaroche and Bérard, this capacity [the specific heat capacity at the two 'volumes'] varies little—so little even, that differences noticed might strictly have been attributed to errors of observation.

[Although] it is [strictly speaking] only necessary to have observed them [the specific heats] in [just] two particular cases . . . the experiments . . . have been made within too contracted limits for us to expect great exactness in the figures . . .

Experiments have shown that this heat [the specific heat of air] varies little, in spite of the quite considerable changes of volume . . . If we consider it nothing [if we consider the variation to be zero] . . . [then] the motive power produced would be found to be exactly proportional to the fall in caloric.

Apart from their presumption in using just two data points to derive a whole relationship, Délaroche and Berard—along with almost all the physicists of the period—had mistakenly made an equivalence between the *direct* heating of a gas and the *adiabatic* heating that occurs as a gas is compressed or expanded. Laplace was one who had considered adiabatic changes from a theoretical standpoint—the caloric theory, of course. ('Adiabatic'—see Chapter 10—means that no heat enters or leaves the system from the outside). Laplace had famously used these adiabatic effects to derive a correction factor, $\sqrt{\gamma}$, for Newton's erroneous determination of the speed of sound in air (see Chapter 10, Part II). Poisson's results were also derived using Laplace's work. Now Laplace's correction factor turns out to be correct (we hinted in Chapter 10 that Laplace had the advantage of knowing what value he was aiming for). So no contradictions arose for Carnot when he made use of either Laplace's or Poisson's results. Finally, Clément and Desormes had determined γ experimentally. Also, their value does not conflict with modern determinations (within the experimental errors, which are large).

In summary, for all this data available to Carnot, whether theoretical or experimental, there was no disagreement between the caloric theory and the modern dynamic theory of heat, and so Carnot's work was not compromised by his use of these contemporary results. In fact, for adiabatic changes (the short sections in Fig. 12.2), the only point of disagreement between the old and the new theories was over interpretation. In the modern view, heat is *generated* by the work done on the gas during compressions, and *converted* into work during the expansions. In the caloric theory, heat merely changes from latent to 'sensible' (adiabatic compressions) and from 'sensible' to latent (adiabatic expansions). In other words, latent heat totally mops up the effects of work.

That heat should be weightless and subtle (invisible) is bad enough: that it should go latent as well is like adding insult to injury. How would such a fudge-factor ever be detected? If the heat could have been tracked along *all* sections of the cycle, then differences between the caloric theory and the modern theory would have shown up. In particular, it would have been seen that *heat is not conserved*. It is curious that in the *Reflexions*, Carnot, in contrast to his successor, Emil Clapeyron, never stressed the constancy of the heat around the cycle. Rather, he stressed the identicality of a given point on the cycle, a given *state*, however many times it was reached and from whichever direction. As always, his instincts were faultless.

History tells us (as do Carnot's unseen notes) that the caloric theory would only be ousted by looking at the whole of physics, and not just gases. For example, there were problems such as explaining ordinary and radiant heat in one theory; the heats of chemical reaction; and frictional heating, such as occurs in Rumford's experiments. Carnot, in the posthumous notes, suggests repeating Rumford's experiments. Significantly, he also suggests taking the next step that Rumford never contemplated—*quantifying* the work done by the borer. It is an irony that the arena of gases, so long an aid to the study of heat, was now holding back progress.

This is only one of many ironies. Some others include the following:

• Daniel Bernoulli (in 1738; see Chapter 7, Part III) was the first to consider the scenario of the gas in a cylinder with a piston, and so brought forth the first kinetic theory of gases. Carnot used exactly the same scenario but assumed an essentially static gas theory (as implicit in the caloric theory).

- The caloric theory both helped and hindered understanding: the model of heat as a fluid helped to make the analogy with water-engines; however, the caloric theory's axiom of heat conservation held back progress.

- The concept of latent heat was the perfect 'fudge-factor'—it managed to almost completely mask the erroneous axiom of the conservation of heat.

- The abstract construct of the ideal heat-engine managed to further the understanding of real, down-to-earth engines.

- Real, down-to-earth engines were busily and noisily converting heat into work all over Europe even while the natural philosophers were debating whether such conversions were possible.

- The ideal heat-engine was reversible, but it was to bring in the Second Law, which would show that all real processes were irreversible.

- While Carnot's syllogism vetoed a perpetually acting machine, Newton's First Law of Motion required perpetual motion. However, the Second Law of Thermodynamics would eventually show that perpetual motion can never occur *in practice*.

- The very fact of the Second Law would make it harder to oust the caloric theory, and to enable the discovery of the First Law of Thermodynamics (the fact that there are always losses in practice means that the 'energy books' don't always appear to balance; also, conversions from heat to work occur less readily than from work to heat).

- Carnot was despondent because he had adopted a false axiom but, in fact, this made no difference to his conclusions, which were all correct.

Perhaps, wherever there is a proliferation of ironies, this shows that while some premises are fundamentally right, others are fundamentally wrong.

But how can things like engines and their efficiency, which are so anthropocentric, have anything to do with *physics*? This question will be returned to in Chapter 16, where it will be shown how Carnot's ideas were extended from an idealized heat-engine to . . . everything.

In assessing the importance of Carnot's work—and of physics in general—we note that there has been another Sadi Carnot (in fact, he was Carnot's nephew), and this younger Sadi was President of France during the Third Republic—but who has ever heard of him?[16]

Carnot always had an intuition, a hunch, for what was right; but hunches so often lead to falsehood. Perhaps the message to be drawn is that one should always follow one's hunches, but only if one is a genius, like Carnot.

13

Hamilton and Green

What I tell you three times is true.

Lewis Carroll, 'The Hunting of the Snark'

The advancing idea of energy required not just the discovery of new 'blocks' but also an increase in the mathematization of the concept. This occurred due to the work of two men who were contemporaries but who knew nothing of each other.

Green

In this chapter we have yet another case of Thomson rescuing a posthumous book from obscurity. This was Green's *An Essay on the Mathematical Analysis of Electricity and Magnetism.*[1]

George Green (1793–1841) is something of an enigma. He was a baker and miller from Nottingham, England, had less than two years of education (he left school aged nine), and he produced no scholarly work until the publication of his book in 1828, when he was already 35 years old. He had to publish his book privately, after placing an advertisement in the *Nottingham Review* on 14 December 1827:

In the Press, and shortly will be published, by subscription, An Essay on the application of Mathematical Analysis to the Theories of Electricity and Magnetism. By George Green. Dedicated (by permission) to his Grace the Duke of Newcastle, K.G. Price, to Subscribers, 7s. 6d. The Names of Subscribers will be received at the Booksellers, and at the Library, Bromley House.[2]

He received only 51 subscriptions[3] and yet the book turned out to be a major work in mathematical physics.

There are no notebooks or manuscripts (and no portrait)—nothing to indicate what Green's early motivations were. Most puzzling of all is how Green gained access to the source material that he quotes, chiefly the *Mécanique céleste* of Laplace, and Poisson's papers. Such advanced mathematics (and in French) was not being taught in the outdated curricula at British universities, let alone available in Nottingham bookshops. Records show (and see the advertisement) that at the age of 30 Green joined a subscription library. It was here, presumably, that he could read abstracts of the French work and then place an order for the originals (there was a trade in such books, interrupted only during the Napoleonic wars in the 1810s).

What Green did was to mathematically define the potential function, *V*, of a static distribution of electric charges. When multiplied by the electric charge of the test particle, it became identical with the potential energy.[4] Green started from Laplace's potential function for gravity (see Chapter 7, Part II) and adapted it to the case of electrostatic attraction. He then extended it to cases where the test point didn't need to be exterior to but could be within or on the surface of the region containing the source charges. An identical treatment was also given for static magnetic attractions.

The important thing for energy is that the role of forces was lessened while that of 'the potential' was emphasized, becoming more abstract and mathematical. Instead of considering the force between two charged bodies and finding their consequent accelerations, rather, the effect that would be felt by a tiny test charge due to its immediate surroundings was investigated. This effect could vary continuously from point to point throughout all space, and was mapped out—by the 'potential function'. Green thus extended the concept of potential energy, from its eighteenth-century beginnings in the works of Clairaut, Daniel Bernoulli, and Laplace (see Chapter 7), and also the nineteenth-century contribution of Poisson (1781–1840). Green was, in fact, the first mathematician to name and define the 'potential', and express it in functional form, $V(x,y,z)$, where *x*, *y*, and *z* are the position coordinates. He wrote:

if we consider any material point p, the effect, in a given direction, of all the forces acting upon that point, arising from any system of bodies S under consideration, will be expressed by a partial differential of a certain function of [position] coordinates . . . [we] call it the potential function.[5]

Knowing this function ($V(x,y,z)$ or $V(r)$[6]), a test mass, charge, or magnet can 'know how to move' merely from its *local* environment.[7] It doesn't need to know where the 'sources' (of gravitational, electrical, or magnetic force) are actually located. Indeed, Green also proved 'Green's theorem', which shows that a system of sources within a given volume is entirely equivalent to the *net effect* of these sources at a surface enclosing the volume.

Only one person responded to the publication of the book in 1828. This was Sir Edward Bromhead, a Lincolnshire landowner and Cambridge graduate in mathematics. He was so impressed that he wrote to Green, offering to help him publish his work in the memoirs of the Royal Societies of London or Edinburgh. Green's diffidence, and also his domestic preoccupations, prevented him from replying for almost two years (his father had just died, and he had a *de facto* wife, many children, and had to do all his mathematics in the hours 'stolen from my sleep'[8]). However, all turned out well in the end, and Bromhead launched Green's new career, helping him to publish his papers, and to enrol as an undergraduate at Cambridge as a mature student (Green was already 40 years old).

Green published more papers but he hardly ever referred back to his own book, and so it was quickly forgotten. He became a fellow of Gonville and Caius College (despite the requirement of celibacy, Green still qualified as, although he had six children, he had never married), but tragically he died of influenza the very next year (1841).

Four years later (in 1845), Thomson was an undergraduate at Cambridge, and came across a reference to Green's *Essay*. He tried but failed to find it in any bookshops or libraries. After his final exams, and the day before he was due to leave for Paris, Thomson mentioned it to one of his tutors, who promptly gave Thomson his own copy. Upon his arrival in France, the young Thomson showed the book to various members of the Paris Academy: 'I found Liouville at home and showed him Green's Essay, to which he gave great attention. I did not find Sturm at home but I left a card. Late in the evening, when I was sitting . . . at our wood fire in 31, Rue Monsieur le Prince, we heard a knock, and Sturm came along our passage panting . . . he said "*Vous avez un memoire de Green, M Liouville me l'a dit.*" So I handed it to him. He sat down and turned the pages with avidity. He stopped at one place calling out, "*Ah voila mon affaire*".[9]

Green's *Essay* caused 'a sensation' and Thomson arranged for it to be reprinted in *Crelle's Journal*. This was a German publication, and it still took many years before the work became known in Britain. Today, Green is appreciated as being the founder of the concept of the potential function, $V(r)$, and of potential function theory.

Hamilton

William Rowan Hamilton (1805–65) was born in Dublin, Ireland, the only son in a family of five (surviving) children. His name was a permutation of Archibald Hamilton, his father's name, and Archibald Rowan, his godfather's name (the famous Irish patriot). From the age of three he was raised and educated by his uncle, curate of Trim, probably because his father, a Dublin solicitor, had gone bankrupt (the father admitted 'I can manage anything but my own money concerns'.[10]).

The uncle quickly recognized Hamilton's precocity in mathematics and languages and started him on a programme of learning that included Hebrew, Latin, and Greek when he was still an infant. In 1818, Hamilton competed against the famous American Calculating Boy, Zerah Colburn. He lost consistently but later chatted to Zerah and wrote a commentary on why Zerah's methods worked (Zerah himself had no idea).

As a student at Trinity College, Dublin, Hamilton won every prize and passed every examination *cum laude*. When only 22, he was appointed head of the Dunsink Observatory in Dublin. This was not a particularly appropriate appointment, as his talents were not in practical physics. He made the best of it he could (in the early years, at any rate), helped by his sisters, who also lived at the observatory and took turns with the observations.

Hamilton became close friends with the poets Wordsworth and Coleridge. Poetry and mathematics were the wellsprings of his creativity, and he considered them as two equally important aspects of the same underlying reality (Wordsworth tactfully advised Hamilton to stick to mathematics). Under Coleridge's influence, Hamilton followed the idealism of the German philosopher Immanuel Kant. Hamilton's idealism also extended to his love affairs: his first love represented the ideal and was a lifelong passion even while it remained on an abstract plane (she was forced to marry another).

One final anecdote tells how Hamilton was a compulsive jotter and calculator: his children remember him writing notes and computations on anything that was to hand, including his fingernails and the shell of his boiled eggs at breakfast.[11]

Hamilton's Mechanics

Idealist or not, Hamilton liked to formulate every problem in the most general way possible, and to express everything algebraically. This was exactly like Lagrange (see Chapter 7, Part IV), whom Hamilton revered, calling him the 'Shakespeare' of mechanics and comparing the *Analytique mécanique* to a 'scientific poem'.[12] Perhaps this is why Hamilton tackled the same problem area as Lagrange (theoretical mechanics), carrying on where Lagrange had left off. He extended Lagrange's Principle of Least Action to cases where the energy of the system could be allowed to vary in time (this became known as Hamilton's Principle'[13])—but this more rigorous version of Least Action was probably the least of Hamilton's stupendous achievements.

It all started when Hamilton was incredibly young—17 years old. He was considering an optical system (rays of light focused through lenses, bounced off mirrors, and so on), wondering how it could all be explained mathematically, when he came to some profound realizations. First, he realized that just because one doesn't have complete information, it doesn't mean that one can't say anything: in other words, he knew he couldn't follow the path of just one ray but, given a whole bundle of rays, with just a small spread in starting positions and/or starting angles, maybe something useful could be said. Second, Hamilton's big leap forward came from a big leap back to the wisdom of Descartes. Now, in Descartes' coordinate geometry (see Chapter 3), geometric features (a circle, a curve, a plane, etc.) could be described in purely algebraic terms (that is, as functions of coordinates). Perhaps, thought Hamilton, the same could be made to work in optics—the *geometric* features could be described as one overarching *function of coordinates*.

For Hamilton, the crucial geometric feature was that the tip of the light rays defined a specific surface in space. Moreover, the rays pierced this surface at right angles, and would pierce all subsequent 'surfaces of simultaneous arrival time' at right angles.[14] Remembering the near-infinite speed of light, this represents a remarkable insight (this 'ray property' would have been visible to Hamilton's mathematical mind's eye, not to his actual eye). Just as remarkable as the idea, the young Hamilton was able to follow his plan through, and intuit how to model the system (choose the telling coordinates) and find that grand all-determining function. It helped that he was a mathematical prodigy. Finally, using his optical theory, Hamilton made an experimental prediction—that light passing through certain crystals should emerge as a cone—and this 'conical refraction' was, with some difficulty, eventually detected, by Lloyd in 1832.[15]

Now light was thought by some (famously, Newton) to be particulate, but Hamilton's optical theory had demonstrated *wave*-like properties.[16] (Hamilton said at a meeting of the British Association[17] in 1842 that he 'hoped it would not be supposed that the wave men were wavering, or that the undulatory theory was at all undulatory in their minds'.[18]) But if light was particulate and yet showed wave properties, then maybe actual particles (billiard balls, planets, etc.) would display wave properties, and then their motions would also be subject to a 'wave mechanics'. This was the revolutionary idea behind Hamilton's next big advance: his optico-mechanical theory for the motion of material particles.[19] (As head of the Dunsink Observatory, Hamilton was tackling a mechanical problem in the celestial domain—the problem of perturbations in the orbits of the planets.)

In this new optico-mechanical theory, Hamilton again considered a bundle of rays, but now the 'rays' were for particles rather than for light. As before, the geometric 'ray property' was evident, but whereas for light there were 'surfaces of simultaneous arrival time', now there were 'surfaces of equal action'. Also, whereas for light the surfaces were separated by a least time interval, now, for particles, the surfaces were separated by a least action interval. Finally, also as before, the 'ray property' for particles could be described by one overarching algebraic[20] function. Hamilton called this the 'principal function' and it determined the path along which a particle would progress from a starting 'position', q_i, on the starting surface, to an end 'position', Q_i, on the final surface (i runs over all the particles; lowercase refers to positions on the starting surface, and uppercase refers to positions on the final surface).[21] The principal function could also depend on the time, t, but it nevertheless determined only the paths and not the speed of the particles along these paths.

Aside from Hamilton's principal function, another algebraic function was to emerge from the optico-mechanical theory. Now, the principal function determined the 'distance' between action surfaces, and not the arrival time of this or that particle (as mentioned above, the particles did not, in general, arrive at a given action surface simultaneously). But the theory gave birth to another crucially important function, the 'Hamiltonian function', H, which related the 'positions' *and* the 'speeds' for *all* the particles together *at one time*. (Strictly speaking, instead of 'speeds' we have 'momenta', denoted p, but as the mass of each particle is taken as fixed, there is no essential difference between speed and momentum.[22] Also note that we shall be taking the particle subscripts, i, for granted from now on.) Not surprisingly, the principal function and the H function were intimately (mathematically) related to each other.[23]

In summary, Hamilton had set out to find his principal function (dependent on q, Q, and possibly t), but this emergent function, H

(dependent on q, p, and possibly t), would prove to be the most important function in classical mechanics and, later, quantum mechanics.

This is all very interesting, but what does it tell us about energy? It turns out that H, the Hamiltonian function, *is* energy: it has dimensions of energy; in cases with no explicit time-dependence, it is conserved; in some conservative cases, it is the same as $T + V$, the *total energy* of the system;* and, moreover, because it is an algebraic function (a function of the p and q variables but not of the 'speeds', $\dot{q}$, or 'accelerations', $\dot{p}$), it depends just on 'configuration' and is therefore a kind of *potential energy*. One more impressive feature about the Hamiltonian is that (for a given mechanical problem) there is just *one* grand function, H, that does everything—there is no H_1 for particle 1, H_2 for particle 2, and so on. To gain more insight, we need to see how H emerged from the theory, and this will bring out a new emphasis—the theory of mathematical transformations. The quote from *The Hunting of the Snark* at the start of the chapter may aptly be paraphrased to: 'What I transform [at least] three times is [still] true'.

The first transformation, actually carried out by Lagrange, was to switch from everyday 3D space to n-dimensional 'configuration space' (see Chapter 7). The second transformation was due to Hamilton, and was from n-dimensional 'configuration space' to $2n$-dimensional 'phase space'. The third transformation (or rather, infinite set of infinitesimal transformations) was from one state of phase space to the next—and so on, through time.

Let's run through this again more slowly. Lagrange described a mechanical system with, say, n degrees of freedom, by using n generalized position coordinates, q. Hamilton described the same system by using n generalized position coordinates, q, and *also* n generalized momentum coordinates, p (the p were given by $m\dot{q}$).[24] In other words, Hamilton employed twice as many coordinates; that is, twice as many degrees of freedom. Now, on two counts you may be worried: first, aren't the p merely *consequential* upon the q? Second, can H truly be called 'algebraic' if the differential nature of speed has merely been camouflaged by calling it p/m? To begin the answer, we go back to the young Hamilton's wisdom: even if we don't have complete information, this doesn't mean that we can't arrive at some useful results, and instead of *determining* the speed and the acceleration for each q, we tackle scenarios where it is conceded

---

* In this chapter, T and V refer to the kinetic and potential energies, respectively, and not to temperature or volume.

that this cannot be done, but where something useful can nevertheless be said. Consider an analogy from the game of golf. Suppose a golfer is practising a certain shot, and drives 500 golf balls from the tee to hole number 18. Although the golfer is trying to exactly replicate the shot, there will, inevitably, be a small spread in starting speeds and a small spread in starting angles. But rather than predict the actual trajectory of each and every golf ball, we are less ambitious and are satisfied to predict overall trends and symmetries. Given that the balls start with such and such angles (q), and such and such speeds (p), we can determine that all will go through a certain gap in the trees, none will end up in the rough, and—most importantly—we can monitor something called the 'energy of the system'. We can ascertain that this energy is a constant (in the case of an isolated system) or grows or diminishes in time in a certain way (in cases where energy is added to or subtracted from the system—say, a wind is blowing across the golf course).

That this stratagem could be made to work—that the p and the q really are the only parameters, and the clinching parameters, of the entire mechanical problem—was Hamilton's great intuition. This was sheer genius—but he knew he was right, because part and parcel of the transformation (going from 'q' to 'q and p') was that $(T - V)$ was transformed to $(T_{new} - V_{new})$, wherein the kinetic energy, T_{new}, was always standardized to a specific form,[25] T_{new}, and where the 'leftover' non-kinetic energy contributions made up one function, V_{new}. It then turned out that V_{new} *was identical with H*, the function known as 'the Hamiltonian'.

Transformation theory reaped other dividends (apart from the momentous discovery of H). In plotting all the (p,q) values as they changed in time, a very suggestive metaphor appeared—that of a fluid flowing in 'phase space'. This was indeed a subtle sort of fluid, as the streamlines didn't correspond to everyday actual trajectories; they occurred in an imaginary space.[26] Also, by yet another transformation, in which the time, t, was considered as one of the ordinary 'position' coordinates, then the problem of motion could be transformed into a problem of pure geometry: time passing was like a succession of infinitesimal coordinate transformations, mapping the phase space on to itself.

The importance to physics was that this metaphor, this imaginary fluid, brought in great physical insight, such as the highlighting of symmetries and conserved quantities. For example, Liouville's Theorem (for the conservation of the volume of an incompressible fluid) and Helmholtz's circulation theorem (conservation of vorticity) applied to the imaginary phase fluid as well as to real fluids. Also, energy was conserved whenever there was no explicit time-dependence, and this was equivalent to saying that time was homogeneous (one time was not more special than another), and so the system was 'symmetrical' (unchanged) with respect to a shift in time. In short, energy itself sometimes exhibited a symmetry property.

Thus, Hamilton's important contribution to mechanics was that he had generalized the very concept of energy. It no longer needed to be conserved, and it no longer needed to have the form $(T + V)$ (it only had this form if T was 'quadratic', as in '$\frac{1}{2}mv^2$', and if V did not depend on speeds, and if there was no explicit time-dependence). H could be used in more general cases where V did depend on velocity (as turns out to be the case for electric charges moving in a magnetic field, discovered around this time; see Chapter 15), or where V did depend on time (say, the rug is slowly pulled out from under my feet). This didn't mean that the important law of energy conservation was being abandoned, but that the investigation could include a system that wasn't necessarily closed, and where energy could leak in or out in a prescribed fashion (an anchor swinging at the end of its chain but being hauled in at the same time; the 'variable energy cyclotron', in which the magnetic field is steadily increased in order to keep the accelerating protons in their orbits; and so on). In fact, the energy, H, was now more than ever the chief determinant of the properties of the mechanical system.

Contemporary Reception

Although a fundamental advance, Hamilton's mechanics found strangely little application to the physics of his own era. In optics, we have seen his prediction of conical refraction, and in celestial mechanics it was used by the French astronomer Delaunay (1816–72) to calculate lunar motions; but it would be almost 100 years before it would help in the birth of quantum theory. (There are other cases in the history of physics where fundamental advances had few immediate experimental consequences—most famously Einstein's Theory of General Relativity.)

One problem was that Hamilton's mechanics was abstract, and the mathematics very difficult. Also, Hamilton was not a particularly good expositor of his work. For example, he sent a 20-page letter to the British astronomer John Herschel (1792–1871), at the Cape of Good Hope (Herschel was observing the Southern skies), giving an enthusiastic account of his method. Herschel (no slouch at maths) replied: 'Alas! I grieve to say that it is only the general scope of the method which stands dimly shadowed out to my mind amid the gleaming and dazzling lustre of the symbolic expressions . . . I could only look on as a bystander, and mix his plaudits with the smoking of your chariot wheels, and the dust of your triumph'.[27]

However, there was one contemporary mathematician who understood Hamilton's mechanical theory, and appreciated its amazing generality and beauty. This was the German mathematician Carl Jacobi (1804–51). In fact, Jacobi appreciated Hamilton's work so much that he referred to Hamilton's equations as the 'canonical equations', and to the p and q variables as the 'canonical coordinates'. ('Canonical' meant setting the rule, in the same way as canon law governed church procedure—a curious choice of adjective for one of Jewish background. Later, Thomson and Tait complained that they didn't know what was canonical about it.)[28] Jacobi, at the expense of some beauty, turned Hamilton's mechanics into something of practical utility, as he found a way of solving the canonical equations. (Hamilton's theory, while beautiful and abstract, was mostly insoluble as it stood.)

Future Developments

Electromagnetism

Green's potential function, $V = V(r)$, was a more sophisticated mathematical object than, say, $V = mg\Delta h$ (the change in gravitational potential energy between heights, zero, and, Δh). The force on a test mass or electric charge could now be determined at a *point* in space, knowing only the *local* conditions (the partial rate of change of V with position at that point), rather than having to refer to a finite distance, such as an interval in height, Δh.

Green had dealt with a static distribution of electric charges but, contemporaneous with his mathematical work, the experimentalists were uncovering a whole new arena—magnetic effects arising from moving charges (currents). The first such effect was Oersted's observation in 1820 of a compass needle moving in response to a nearby current of electricity. Almost immediately, this was further investigated by Ampère in France and Faraday in England (see Chapter 15). The amazing findings were that the 'induced magnetism' depended on the speed and direction of the charges, and the resulting magnetic force was in yet another direction, at *right angles*—all non-Newtonian results. Faraday, in the 1830s, showed how the magnetism varied from point to point in space, and could be represented by 'field lines' that showed how tiny bar magnets (iron filings) would align themselves. A similar 'field' description applied to electricity. In this new description, each point in space would now have to be flagged with direction and speed as well as with some function of the position coordinates. But Green's potential function theory could easily be adapted to meet the needs of the new *field* theory (see Chapter 17).

Quantum mechanics

If Hamilton's optics had brought out the wave nature of light, then would his mechanics bring out the wave nature of particles? This provocative question was not even asked until the beginning of the twentieth century. First came the various landmark investigations of Planck, Compton, and Einstein, which showed that while light was wave-like, it also came in particles ('quanta') (see Chapter 17 for this, and for the rest of this section). The reverse case, that particles could be wave-like, was finally postulated by de Broglie in 1924, and experimentally shown for electrons by Davisson and Germer in 1927.

De Broglie cited the influence of Maupertuis rather than Hamilton. Only one mathematician was actively promoting Hamilton's great optical-mechanical synthesis—Felix Klein (in the 1890s). Klein bemoaned the fact that his assessment of Hamilton had lain unread in the reading room at Göttingen for 20 years, and that Jacobi's work had 'snatched away'[29] the glory. However, it only takes one person at a time to keep a thread going, and eventually Schrödinger, through Sommerfeld (physicist and clerk of the Göttingen reading room), came across Klein's assessment. Also, it seems that Schrödinger's tutor, Hasenöhrl, who died young in the First World War, promoted Hamilton's mechanics in his lectures at the University of Vienna.[30] Finally, in 1926, Schrödinger formulated his famous wave equation for masses of microscopic size (he considered the hydrogen atom), and he specifically acknowledged Hamilton: 'His [Hamilton's] famous analogy between mechanics and optics virtually anticipated wave mechanics. The central conception of all modern theory in physics is the "Hamiltonian" . . . Thus Hamilton is one of the greatest men of science the world has produced'.[31] The energy function, H, is at the heart of Schrödinger's equation.

In quantum mechanics, we are in a different realm—the particles are miniscule compared to planets or even golf balls. But 'miniscule' can be quantified: whenever the lengths and momenta yield a quantity of action of the order of Planck's constant, $h = 6.63 \times 10^{-34}$ J s, then we are in the quantum-mechanical world. In this world, some totally new features emerge: (1) the probability wave, (2) Heisenberg's Uncertainty Principle, and (3) the fact that order of observations makes a difference. These three features are closely connected with each other, and *all have their origins in Hamilton's theory.* In Hamilton's minimum principle, one 'varied path'[32] is selected in preference to all the others, but in the quantum-mechanical realm we have a new wave feature—even paths 'off the minimum' have a certain *probability* of being followed. Thus, instead of Hamilton's range of *possibilities*, we now have a spread in *probabilities* (for just one quantum-mechanical particle). This multiplicity of paths for one particle leads to a smearing in the values of position and momentum, and to a constraint on the simultaneous determination of position and momentum for that one particle. This is Heisenberg's famous Uncertainty Principle (see also Chapter 17). Now, the French mathematical physicist Poisson (1781–1840) found that the order of 'variations' in p and q could make a difference. (Specifically, his so-called bracket relation had a non-zero value whenever the order was important.) But, over 100 years later, it was realized that what were 'variations' in the macroscopic domain were 'observations' in the quantum domain. The Poisson bracket relations ended up showing that, for any *one* quantum particle, the order of observations of p and q did always make a difference.

There was yet one more spin-off from Hamilton's theory. It was noticed that while Poisson's 'bracket' was sometimes finite, this finite value was invariant—and invariant for just those transformations that were 'canonical'. However, in the quantum domain, the Poisson brackets were more than just a test of the canonicity (!) of the transformations; they were the very cornerstone of quantum mechanics. It was the British theoretical physicist Dirac (1902–84) who had this insight. We are, today, familiar with the idea that classical certainty must be sacrificed in going to quantum mechanics, yet now, going the other way, we are finding that some classical formalism (the Poisson bracket) finds its true relevance only when the quantum realm is reached.

Overview

A brand new emphasis came in with Hamilton. An optimization principle—the Principle of Least Action—was paramount, and even trumped the Principle of the Conservation of Energy. This is not to say that energy conservation was no longer true or important, but that Hamilton's mechanics could analyse systems in which the total energy was not constant and could flow in or out in a controlled fashion.

However, a crucial requirement in Hamilton's mechanics is that the energy must be describable by a mathematical function (the mechanics cannot be used to handle systems that are chaotic). Green's work was relevant to Hamilton's, as having V in functional form allowed H to be in functional form. (It is not known whether either man knew of the other's work, or even of their existence.) H and V are still unspecified functions—the physicist must decide what goes into H or V in any given physical scenario. We are awestruck when we remember that the work of Hamilton and of Green was set against a background where 'energy' had still not been discovered.

But how can defining new mathematical objects, or carrying out lots of coordinate transformations, teach us anything new? In order to understand this, we must appreciate that the advances of Green and of Hamilton represent more than just mathematical techniques and conventions; they correspond to real physical discoveries: for example, things really do warm up in the Sun, even though the energy must be transported vast distances across empty space, and very quickly; and wave–particle duality *has* been experimentally observed. The mathematics shows us a new reality that we wouldn't necessarily uncover otherwise. Still, it is curious that these discoveries often occur in imaginary mathematical landscapes—it is a bit like going to Narnia (configuration space) or Shangri-La (phase space) and bringing back real treasures.[33]

What actually is the Hamiltonian, H? It is the mathematical function that results when the system is modelled in a special way, using 'position' coordinates, p, and 'momentum' coordinates, q.* There is a *doubling* of the number of degrees of freedom, like treating a school as made up of from girls and boys instead of just children. An even better analogy is to consider a class of, say, 20 girls, and follow their exam results from middle school to high school. Now, obviously, there is a connection between the performance of a given pupil at middle school and then high school, but in Hamilton's mechanics we forget this connection and treat all the variables *as if* they were independent. Finally (back in the realm of mechanics), the kinetic energy is standardized to a specific form. All the

---

* The p and q depend on how the system is modelled, but together they must have dimensions of action.

remaining energy terms (the non-kinetic ones) then make up a potential energy function, different for each mechanical problem being investigated. This potential energy function is the Hamiltonian, H, and it is *the* function that determines the entire dynamics of the system. H is a function of p and of q, but, unlike the usual energy function (in Lagrange's mechanics), it can also be explicitly time-dependent, and it can also depend on velocity. Thus, although growing out of an optimization principle (the Principle of Least Action), Hamilton's mechanics ends up vindicating the crucial importance of energy, and making it more general than in Lagrange's or Newton's mechanics.

Hamilton's mechanics was applicable to light and to material particles, and led to an understanding that, at a fundamental level, there was no sharp division between the two.

In the quantum world, it turns out that the p and the q are on the same tiny scale, and this means that a particle can exhibit particle-like and wave-like aspects at the same time ('wave–particle duality'). Hamilton's optical-mechanical theory is therefore ideally suited in this quantum domain. Hamilton's methods are also indispensable in statistical mechanics, fluid mechanics, and continuum mechanics. Here, there may be a system with 10^{30} 'particles' or more (a mixture of gases; a flowing fluid; a material where the stress varies continuously from point to point), and so merely doubling the number of equations in going from Lagrange's to Hamilton's mechanics is no big disadvantage, but the highlighting of symmetries and conserved properties brings in great physical insight.

'Action' (see Chapter 7, Part IV) is still the important quantity but, after Hamilton, it can be sliced up as (p,q) as well as by (E,t). In the (p,q) slicing, both the p and the q have some 'extensive'* aspects; that is, both quantities are both 'fishy' and 'fowly'. On the other hand, in the (E,t) slicing, it seems as if E has pared off all the extensive aspects leaving naked t all to itself.†

---

* Extensive aspects are those that alter as the sample size is altered (e.g. mass, length, volume, and energy); intensive aspects remain unaltered (e.g. density and temperature).

† However, it seems that time can have extensive aspects—for example, a larger egg timer does indeed measure a longer time interval.

14

The Mechanical Equivalent of Heat: Mayer, Joule, and Waterston

It is the 1840s, and the time is at last ripe for the discovery of energy. We have heard of men who were years ahead of their time (for example, Daniel Bernoulli, Sadi Carnot, and William Rowan Hamilton) but, in the 1840s, there was, finally a growing appreciation of *energy*, more often called 'force' ('Kraft' in German), its conservation, and its interconversion between different forms. Men such as Colding (in Denmark), Mayer and Helmholtz (Germany), Seguin (France), and Joule (England) were independently arriving at the same discoveries. We shall cover the work of the two key players, Mayer and Joule.

The time was ripe for discovery, yes, but not for a ready acceptance of the new ideas. Both Mayer and Joule started out on the fringes of the scientific establishment, and it was many years before their work was appreciated. The trouble was that the scientists of the day still found it hard to accept a quantitative link between heat and mechanical energy. It was, by now, generally agreed that heat was a sort of 'motion', also a sort of 'radiation', and also a by-product of collisions, friction, chemical reactions, and so on. But these were all qualitative results. That there could be a *quantitative* link between 'heat' and 'mechanical effects' was hard to contemplate. It was like a category error—like comparing, say, p.s.i. (pounds per square inch) and PSA (pleasant Sunday afternoon[1]).

Mayer

Many of the new thinkers were coming from Germany. Whether a trend towards the unification of Germany (1848 onwards) also ushered in a unification of ideas is an interesting but unanswerable question. What is certain, however, is that there was a German way of thinking even before there was a country called Germany. Specifically, the maxim 'cause equals effect' was invoked and given central prominence in the work of three Germans in energy physics: Leibniz (see Chapter 3), Mayer (this chapter), and Helmholtz (see Chapter 15).

Julius Robert Mayer (1814–78) was the son of an apothecary in Heilbronn, South Germany. He studied medicine at Tübingen and was an average student, but with a fiercely independent spirit (he belonged to a forbidden, secret society, and when he was banned from studying for a year, he protested by going on a six-day hunger strike). He eventually qualified as a doctor and, against parental advice, took employment as a ship's doctor on the Dutch vessel *Java*, bound for the East Indies.

The ship set sail on 22 February 1840, from Rotterdam, and the voyage took three months. There wasn't much to do on board, and Mayer records in his diary that he had little association with the ship's officers, and spent much of his time reading his science books and feeling hungry. Upon arrival at Surabaya, there was at last something medical to be done, and this provoked the incident that was to be Mayer's epiphany. He had to let the blood of some sick sailors, and he noticed that their venous blood was uncommonly bright red—more like arterial blood. Asking advice, he was informed that it was always like this with new arrivals in the Tropics. But then, in a flash of insight, Mayer suddenly saw the whole picture—the redness of the blood was due to the balancing of 'force' in all processes in nature: the air temperature was *higher* in the Tropics, and the body therefore had a *lower* need for oxygen; there was therefore less need to deoxygenate the blood in order to maintain body temperature.

Mayer quickly saw the generality of this idea—the conservation of 'energy'—and he applied it to as many physical processes as he could think of. For example, he learned from a local navigator that the sea was warmer after a storm—so, evidently, the *motion* of the stormy seas had been converted into an equivalent amount of *heat*.

Mayer went into a meditative state, barely exploiting his chances for shore-leave (this was a source of jokes amongst the crew) and took up the train of thought that was to dominate his scientific career—in fact, to dominate the rest of his life from then on.

Upon his return to Heilbronn, Mayer acquired a large medical practice, later was appointed town surgeon, and subsequently married and had seven children (five died in infancy)—but he was a man with a mission: he wanted to understand and promote his new idea of 'force' (energy).

Mayer was a philosopher rather than an experimentalist, and his newly conceived philosophy was: nothing comes from nothing; cause equals effect; and, whenever a 'force' is consumed, then the same amount of 'force'—possibly in another guise—is generated. He gave as his prime example the conversion of 'fall-force' (gravitational potential energy) into 'moving force' (kinetic energy), and he soon generalized this to all other processes in physics: 'motion, heat, light, electricity, and the various chemical reactions, are all one and the same object under differently appearing forms'.[2]

Mayer was also impressed by the fact that total mass was always conserved in any chemical reaction, whatever mutations had occurred. This could be demonstrated by careful weight measurements: 'stoichiometry falls into our lap like ripe fruit'.[3] Now for Mayer, 'force' was to physics what mass was to chemistry. Surely, then, careful *quantitative* determinations of 'force' would make it reveal itself.

While aware of the many possible transformations of 'force', it was particularly the transformation of 'motion into heat' that intrigued Mayer. Motion could disappear, but total force could never be reduced to nothing—the invisible 'force of heat' must be generated, and must exactly make up for the loss of motion. Moreover, there was no need (in Mayer's outlook) to understand what heat actually was at a more fundamental level—its calorimetric measure was the important thing. This was all philosophy, but Mayer was able to assign *numbers* to these *measures* (see later as well): he determined that the motion acquired by a 1 kg mass falling from rest through a height of 365 m (or 1,000 Paris feet), represented exactly the same amount of 'force' (energy) as that implicated in a 1 °C rise in temperature of 1 kg of water. The approach is perfectly captured in Feynman's analogy of Dennis' blocks—the formula's the thing.

While the fundamental nature of heat didn't enter into the calculations, it wasn't likely that heat could be a material substance, as heat had to appear and disappear in strict subservience to the conservation of total 'force'. Mayer was the first to expose the nakedness of the Emperor when he proclaimed: 'the truth—there are no immaterial materials'.[4]

Mayer wrote four papers in quick succession. The first, in 1841, was straight after his return from the East Indies. It was rejected by Poggendorf, and the original manuscript was not even returned. This was disappointing, but gave Mayer time to improve his shaky knowledge of maths and physics. The second paper (1842) was much better, and included the mechanical equivalent of heat (see below): it was published, by Liebig—but in the section on chemistry and pharmacy, and so it sank almost without trace. The third, in 1845, was rejected by Liebig: Mayer had it printed privately, at his own expense, by the bookstore in Heilbronn.

This third paper was almost book-length (112 pages) and was a highly original exposition of energy transformations in living processes. Contemporary physiologists, such as Liebig, did by now (1840s) understand that animal heat arose from the combustion of food (as opposed to earlier ideas, such as attributing the heat to the friction of the circulating blood). Mayer, however, was the first to consider the totality of energy transformations—not just body heat, but also the work done by the animal (in running, lifting, etc.), heat lost in sweating, losses due to friction, and so on. In other words, the calorific value of the food had to account for *all* the energy conversions, and not just for maintaining body temperature. Mayer also considered the cosmic role of the Sun, and the energy transformations occurring in (what we now call) photosynthesis, transpiration, and so on. Sadly, the title of the paper, 'The motions of organisms and their relation to metabolism: an essay in natural science', didn't sufficiently advertise its contents and its originality, and it was largely ignored.

Depressed but undaunted, Mayer printed a fourth paper, again at his own expense, in 1848. This one, on celestial dynamics, was also highly original, and put forward hypotheses for the source of the Sun's energy (meteors falling in), the bright tails of shooting stars (friction in the atmosphere), the effect of the tides in slowing down the rate of rotation of the Earth, the increasing rate of the Earth's rotation as its volume was reduced by cooling, and some other ideas.

But 1848 was a bad year for Mayer. This was the year of revolution in Europe, and Mayer's conservative attitudes led to his being briefly arrested by insurgents, and permanently estranged from his more rebellious brother, Fritz. In addition, two of Mayer's children died, and also a crank named Seyffer ('nothing'?) ridiculed Mayer's heat-to-motion conversion in the newspaper. After years of almost total neglect, and the fact that others were now beginning to take the credit for similar ideas (principally Joule, but also Liebig, Holtzmann, and Helmholtz), this was perhaps the last straw for Mayer. In May 1850 he attempted suicide, jumping from a third-floor window. His feet were badly damaged, but after some time he could walk again. His mental state wasn't so easy to fix, and Mayer voluntarily admitted himself to a private sanatorium. Unfortunately, this led to a loss of autonomy, and to a number of forced admissions to various mental institutions where Mayer was treated rather badly (for example, he was made to wear a strait-jacket). He eventually made a complete recovery, but was out of the scientific scene for almost a decade.

The story has a moderately happy ending. Mayer re-emerged in around 1860, a year that coincided with the time when his work finally began to be recognized. Helmholtz and Clausius had discovered Mayer's papers, and they lauded him as the true founder of the energy principle. Through Clausius, the English physicist Tyndall came to hear of Mayer, and to champion his cause against the chauvinistic claims of Tait (on behalf of Joule). Thomson stayed in the background, but his sympathies clearly lay with Joule. Mayer was full of admiration for Joule, but Joule was less fulsome about Mayer's achievements (see the end of the next section).

Mayer's enduring legacy was his vision, in Java, of a single, fixed quantity of indestructible 'force' in nature, and that heat and motion were but manifestations of it. He carried out no experiments, but he correctly identified $(C_p - C_V)$, the difference in specific *heats* of a gas, as a quantitative measure of the *work* done when a gas expands adiabatically. By using pre-existing data (that of Delaroche and Bérard, and also Dulong; see Chapter 10) he was able to determine the first value for the conversion between heat and mechanical work, the so-called 'mechanical equivalent of heat'. His value was 365 kg-m kcal^{-1}.[5] That is, the 'force' in a mass of 1 kg falling from a height of 365 m was equal to the heat required to raise the temperature of 1 kg of water by 1 °C. (This corresponds to 3,580 J kcal^{-1}, and compares well to the modern value of 4,186 J kcal^{-1}.) Mayer was even able to respond to a criticism of Joule's: Joule wondered how Mayer could justify his assumption that *all* the heat had been converted into work—hadn't some of it gone just to cause the expansion of the gas? Mayer replied that he knew of Gay-Lussac's twin-flasks experiments (see Chapter 10, Part II), which showed that no heat is consumed when a gas expands 'freely' (changes its volume worklessly).

All this is impressive: Mayer had conquered the 'category error'; considered a variety of scenarios (free fall, expansion of a gas, shaking of water, physiology, and the solar system), and had realized that quantification was crucial.

Nevertheless, looking at Mayer's original papers, one finds parts that are almost incomprehensible to the modern reader. Mayer writes (in 1841): 'The falling of a weight is a real decrease in the volume of the earth, and must therefore stand in some relation to the heat produced'.[6]

In the 1842 paper, this is explained: 'If we assume that the whole earth's crust could be raised on suitably placed pillars around its surface, [then] the raising of this immeasurable load would require the transformation of an enormous amount of heat . . . But whatever holds for the earth's crust as a whole, must also apply to every fraction thereof . . . [and therefore] by the falling of [even a small] weight to the earth's surface, the same quantity of heat must be set free'.[7] In other words, Mayer had likened the heat produced by a falling weight to the heat from adiabatic compression of a gas . . .

This example jolts one into an appreciation of the difficulties in applying a new philosophy.

Joule

James Prescott Joule (1818–89) was born in Salford, near Manchester, into a family of successful brewers. Like Mayer, Joule was outside the scientific profession and, like Mayer, it took many years before his work was noticed or accepted. But in other respects, Joule was altogether more fortunate than Mayer. His father employed a private tutor (for James and his elder brother Benjamin)—not any old tutor, but the illustrious John Dalton, founder of the Atomic Theory (see Chapter 10)—and Dalton was not only a famous scientist, but a born teacher. Joule's father also equipped James with a laboratory of his own in their house in Salford, and freed him from the obligation of earning a living—although he did have some duties regarding the brewery.

Initially, Joule was so far removed from any ideas about a finite totality to Nature's resources that he was, rather, being pulled in the opposite direction, researching an 'electro-magnetic machine' that could outperform the steam engine, and possibly even yield perpetual motion. But how could the genie of perpetual motion surface again when it had been dismissed at the end of the eighteenth century?

Moritz Jacobi (brother of Carl Jacobi; see Chapter 13) was to blame. He proclaimed (in 1835) that electromagnetic engines could provide a source of power that was very likely unlimited: the rotating electromagnets should keep on accelerating perpetually, as their magnetic poles were attracted to the fixed poles on approach, and repelled from the fixed poles upon receding. Now perpetual mechanical engines and perpetual heat engines were clearly seen as a no-no—when the wind or water stopped flowing, or the coal stopped burning, then the engine came to a halt. However, the case wasn't so obvious for the new electromagnetic engine. The battery components got used up, certainly, but this was gradual and somewhat mysterious—it wasn't clear what exactly was powering the engine.

In synopsis, the history of the electric motor is as follows. In 1820, the Danish physicist Oersted had made a serendipitous discovery—'galvanic electricity' in a wire could cause the needle of a magnetic compass to be deflected. This link between electricity and magnetism was immediately explored by the French school (Biot, Savart, Arago, and especially Ampère) and also by Davy's young assistant at the Royal Institution, Faraday (see Chapter 15). In the 1820s, Arago discovered the solenoid, Sturgeon the electromagnet, and Faraday the phenomenon of electromagnetic induction. These discoveries opened up the possibilities of electromagnetic machines to generate electricity (the 'magneto') or to generate motion (the electric motor). The prototypes were made, respectively, by Hippolyte Pixie in 1830 and Salvatore dal Negro in 1831.

The electromagnetic machine had thus been proposed by professional scientists, but thereafter its perfection was entirely in the hands of amateurs. Men from all walks of life—physicians, priests, surgeons, lawyers, teachers, bankers, and one brewer—enthusiastically tried to match Jacobi's promise. This resulted, in the 1830s, in an 'electric euphoria' that swept across Europe and the United States.

So it was that Joule, a teenager, became enthused. He started by giving himself and his friends electric shocks, and by subjecting the servant girl to a steadily increasing voltage until she became unconscious (at which point the experiment was stopped). He took up the challenge of perfecting the 'electro-magnetic machine', and submitted his first paper on this, a letter to Sturgeon's *Annals of Electricity*, in 1837, aged 19.

Joule took encouragement from the fact that the 'power of the engine' was proportional to the square of the current, I^2, while the zinc consumption in the battery was proportional to just the current, I. Thus 'the cost of working the engine may be reduced *ad infinitum*'.[8] (We shall use the modern symbols W, V, I, and R, to denote power, voltage, current, and resistance, respectively, although no standards or units had been developed at this time.) Further research showed that, in fact, the 'duty' of the engine *decreased* as the current was increased, and Joule's hopes of perpetual electric power were dashed. However, he presciently recognized that electricity might be a useful alternative to steam in special cases—it was cleaner, safer, and could easily provide rotative power—and so he persisted with his investigations.

Joule quickly established the first of two relationships—what we now call the laws of electrical *energy*—that the 'power' of the engine (the strength of the magnetic attraction) is proportional to both the number of batteries and to the strength of the current. (In modern notation, $W = VI$.) He then went on to identify heating in the coils as a waste of power. This wasn't particularly surprising as—after all, heating was also a loss in mechanical engines—but it led Joule into a systematic examination of heating in circuits. This eventually led to his second electrical energy law: that the amount of heat lost in an electric circuit was proportional to I^2R.

This is easy to state but conceals an enormous amount of physics—it wasn't as if a package of 'I^2R' lay waiting to be recognized. In condensed form, Joule's achievements were as follows:

- Establishing standards for a quantity of static electricity, current electricity, resistance, and voltage ('electromotive force' or 'emf').
- Devising measuring instruments (the galvanometer and voltmeter).
- For metallic conductors, measuring the relation of heating to: the type of metal (copper, iron, or mercury), the length and thickness of the wire, and the shape of the circuit.
- For the battery itself, determining the resistance of the battery, the heat lost to the surroundings, and the heat capacities of the various liquid and solid battery components; and estimating the heat evolved due to the solution of zinc oxide in sulphuric acid.
- For electrolytic cells, estimating, again, the 'heats of solution', and also the 'heats of dissociation' and the 'heat of gas formation'. He also estimated the 'back emf' and how this depended on the material of the electrode.

All this work culminated in Joule's appreciation that, for any voltaic[9] circuit, the heating was proportional to I^2R. Now Joule knew of Ohm's Law (although maybe not with that attribution), that $V = IR$, and so he recognized straight away that the measures I^2R and VI were equivalent. He had therefore shown, in two separate series of researches, that: (1) the *mechanical* power of an electric machine was proportional to VI, and (2) the *heating* effect was also proportional to VI. In other words, he had shown that *the mechanical and heating powers were proportional to each other*.

We are on the very brink of Joule's discovery of the interconvertibility between heat and mechanical power. But Joule went cautiously, step by step. He remarked: 'Electricity may be regarded as a grand agent for carrying, arranging and converting chemical heat'. But how much of the heat was, perhaps, merely transported from the battery, and how much was truly *generated* from work? Joule knew exactly what experiments to carry out to clinch the matter: he would investigate electric currents arising from mechanical work alone, in a 'magneto' (a generator, in today's language)—in other words, he had to examine circuits with no *chemical* sources, such as batteries or cells, and he had to set the armature spinning mechanically (he used falling weights for this).

These experiments, carried out in 1843, ended with Joule's landmark paper: 'On the calorific effects of magneto-electricity and on the mechanical value of heat'.[10] Once again, extraneous sources of heating and cooling had to be accounted for (heating due to eddy currents in the metallic cores, cooling due to the spinning movement of the armature, and so on). When these effects had been separately measured, and corrected for, the 'I^2R law' could shine through.

Joule experimented further, and found that when a battery *was* included in the circuit, the induced 'magneto' currents could be made to enhance, cancel out, or even reverse, the battery current (and the resultant heating was still always proportional to the net current squared). All this argued against the transference of an actual material heat-fluid, as how could a fluid be cancelled out? Joule summed it up by saying that in magneto electricity we have 'an agent capable by simple means of destroying or generating heat'.[11]

There was still the question of the work done, and whether this was proportional to the heat generated or destroyed. Now the armature was rotated by cords attached to weights via a pulley. First, the weights required just to overcome friction and air resistance were found—by running everything with no current through the electromagnets. Then, with the currents switched back on, the heat generated was measured, and all the corrections made, even that due to heat loss because of sparks at the commutators (it is not known how Joule accounted for this). After many careful experiments, Joule found that, yes, the 'mechanical work' did maintain a 'constant ratio'[12] to the heat generated.

Finally, as well as measuring the heat generated from work put *in* (running the electromagnetic machine as a 'magneto'), the machine was run in reverse, as a motor, and Joule determined the work put *out*—from 'heat consumed' (see the discussion at the end of this section).

Fig. 8. Scale $\frac{1}{16}$.

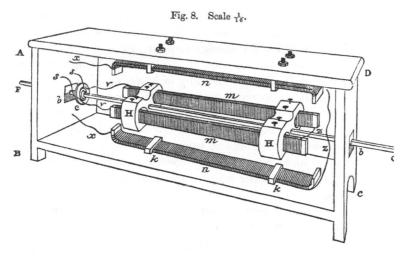

Fig. 14.1 Joule's electromagnetic engine, from *The Scientific Papers of James Prescott Joule*, volume 1 (with the permission of the Institute of Physics).

Joule gathered together the results from all these various experiments and determined the 'mechanical equivalent of heat'; in other words, 'the work done in order to generate the heat that could raise the temperature of 1lb of water by 1 °F'. He found values varying between 587 and 1,040 ft-lb. Then, finally, taking the average of all the data, his mechanical equivalent of heat was 838 ft-lb:

The quantity of heat capable of raising the temperature of a pound of water by one degree of Fahrenheit's scale is equal to, and may be converted into, a mechanical force capable of raising 838 pounds to the perpendicular height of one foot.[13]

Joule then went on to consider a totally different scenario: the work that was done in forcing water through fine, capillary tubes, and the consequent rise in temperature. The work was determined,[14] and the heat also, and the 'mechanical equivalent of heat' came out at 770 ft-lb. This was just a 'look–see' experiment, but it was strongly confirmatory of the earlier results with the electromagnetic machine. Joule had no doubts:

the grand agents of nature are, by the Creator's fiat, indestructible; . . . wherever mechanical force is expended, an exact equivalent of heat is always obtained.[15]

He presented these results at the British Association (BA) meeting in Cork, Ireland, in 1843. Very little notice was taken of them. Joule was disappointed but not discouraged—he knew he had uncovered a new principle, and one that would, surely, eventually command the attention of science.

Joule then looked for yet another arena in which to test his grand new principle. This arena was gases; in particular, the compression and expansion of gases (Joule knew nothing of Mayer's work at this stage). He monitored the temperature changes as air was compressed to, or expanded from, a very high pressure (22 atmospheres, or 'atm'). There were large errors due to the high take-up of heat by the copper vessels and the water-bath, but Joule also wondered about an error of a completely different sort: were the heat changes due entirely to work done on/by the piston, or did the gas consume heat merely by virtue of its volume change?

Joule tested this by repeating (unknowingly) Gay-Lussac's experiment of 40 years before (we have mentioned both men's experiments in Chapter 10, Part II; 'Free Expansions'). He employed twin copper vessels, linked by a special stopcock, and with the whole apparatus surrounded by a water-bath. Air pumped up to 22 atm was fed into the first vessel, and there was a near-vacuum in the second vessel. The stopcock was then opened, and the high-pressure air allowed to expand from one vessel to the next. After waiting a long time for equilibrium conditions to be reached, Joule observed no rise or fall in temperature. From this 'null result', he drew the important conclusion that a volume change, *per se*, caused no heat change.[16] The compression or expansion of a gas, while doing external work on a piston, *could* therefore yield a value for the 'mechanical equivalent of heat'. The values so obtained were consistent with the ones determined already from the electric motor experiments, and the capillary experiments—but the Royal Society, as before, declined to publish.

Joule continued with his now-famous paddle-wheel experiments (1844–45), in which falling weights drive a paddle-wheel that causes water to heat up (Fig. 14.2). As always, there were experimental challenges. The heating effect was tiny; and Joule used his own extra high-precision thermometers, and could measure changes in temperature as small* as 1/200 °F. Baffles were used to break up the motion of the water and prevent its rotation *en masse*. It was crucial to minimize or account for all heat losses, such as the heat lost to the surrounding air, and the heat gained from friction at the bearings. After much painstaking attention to all these details, the 'mechanical equivalent of heat' was found to be 890 ft-lb.

---

* Amazing—note that degrees Fahrenheit are even smaller than degrees Celsius. (Joule always used the Fahrenheit temperature scale.)

Fig. 14.2 Joule's paddle-wheel calorimeter, 1845 (Science Museum, London/ SSPL).

Joule presented these findings to the BA meeting in Cambridge in 1845. He evidently tried to make his talk entertaining, and suggested that those who 'reside amid the romantic scenery of Wales or Scotland'[17] might like to measure the temperature difference between the top and bottom of a waterfall (the temperature would be only around one tenth of a degree higher than at the top). He also sent letters to two leading men of English science, Faraday and John Herschel. (In all of Joule's career, there was surprisingly little communication between himself and Faraday.) However, Faraday was not in sympathy with such quantitative determinations (see the discussion in Chapter 15); while Herschel was not a fan of the dynamical theory of heat, and wrote (in 1846): 'I confess I entertain little hope of the success of any simply mechanical explanation of the phenomena of heat . . . you will excuse me if I say that I have no time for the subject'.[18]

By 1847, after no response from his talks, papers, and correspondence, Joule, presumably in some desperation, gave a public lecture at a little church in his home territory of Manchester. He explained that 'living force' and 'attraction through space' were interchangeable, as when fingers wind up a watch-spring.[19] Also, shooting stars glowed because they moved very fast through the atmosphere, and their 'living force' was ultimately all converted into heat of sufficient intensity to make them burn away. The lecture was covered by the *Manchester Guardian* newspaper, competing with news items such as the 'shocking murder' at the 'rural village of Chorlton-cum-Hardy, a sweet, quiet spot'. The public were enthusiastic, but Joule admitted that people who were not 'scientific folk'[20] found it hard to accept that water could be heated simply by being agitated; also, his theory of shooting stars was contrary to common experience—objects are usually *cooled* when travelling through cold air.

However, 1847 was to be a good year for Joule. It was the year when he got married and when he also presented his work at the BA meeting in Oxford. Amongst many well-known figures of science attending (Airy (Astronomer-Royal), Herschel, Hamilton, Le Verrier, Adams, Baden-Powell, Wheatstone, Stokes, and other notables), there was one very junior physicist, William Thomson. It was Thomson alone who took an interest in Joule's work, and at last brought it to the attention of science. (We shall cover the work of Thomson, later known as Lord Kelvin, in Chapter 16.)

However, it still wasn't plain sailing for Joule: straight after the meeting, Thomson wrote to his brother: 'Joule is, I am sure, wrong in many of his ideas',[21] and it was almost three years before Thomson came round to accepting Joule's findings. Thomson's objections were that Joule's work clashed with Sadi Carnot's theory (which required heat to be *conserved*), and that Joule had shown the conversion of work into heat but not the conversion of heat into work. (These objections were finally resolved by the famous paper of Clausius in 1850, and the subsequent researches of both Clausius and Thomson; see Chapter 16.)

Joule and Thomson later (1850s onwards) went on to form a famous collaboration, and discovered 'Joule–Thomson cooling' and the 'Joule–Kelvin effect' (see Chapter 10, Part II; 'Free Expansions, Again'), amongst other things. Thomson was also the source of a famous anecdote about Joule. Soon after the BA meeting, Thomson was walking in the Chamonix district in Switzerland when he had a chance encounter with Joule and his young wife on their honeymoon. Joule apparently had with him a long thermometer, and was measuring the temperature of a waterfall, while his wife was in a carriage, coming up the hill. The encounter really did take place but the bit about the thermometer is probably too good to be true—Thomson knew how to improve a story.[22]

Joule had shown that heat and work are interconvertible, and that the 'exchange rate' for the conversion is 772.24 ft-lb for a temperature change of 1 °F in 1 lb of water. This is the same as 4,155 J kcal^{-1}, and is only 0.75% lower than the modern value of 4,186 J kcal^{-1}. Significantly, Joule had shown this in so many different domains (weights falling, electrical heating, the motor and the 'magneto', water forced through capillaries, expanding gases, and so on) that his work would open the door to a new abstract quantity—energy.

It wasn't quite fair for Thomson to complain that Joule hadn't shown the conversion of 'heat' into work—Joule *had* demonstrated it, and in two ways: the cooling of a gas expanding against a resisting pressure; and the generation of motion in an electric motor. However, it *is* true that the conversion in this direction was less evident and somewhat counterintuitive. Take the case of the electric motor. Joule had compared the running of the motor, first with it stalled by the addition of weights, and then running faster and faster as progressively more and more weights were removed. He noted that the faster the motor spun round, the less current was drawn, and the less was the consequent heating. His conclusion?—that heat had been converted into the work of the motor.

Nowadays, we find Joule's reasoning almost as alien as Mayer's compressive heating from a falling weight. Instead, we argue the case as follows: as the motor is more and more loaded with weights, so it has to work harder and harder, and so it draws more and more current from the battery. The heating therefore increases (as I^2R) until eventually the stalled motor burns out. The heating represents a *dissipation* of energy, and in no way goes to increase the work of the motor. It is not heat but, rather, 'electrical *energy*' that has been converted into work. Joule's reasoning is correct[23] provided that we take the term 'heat' as an alias for 'electrical energy'.

Incidentally, the fallacy in Jacobi's argument (for a perpetually acting electric motor) can now be explained in the following way. As the motor spins around, it acts as a 'magneto', and induces currents in itself and in the fixed circuitry. These currents are always in a direction such as to *counter* the pre-existing magnetic fields, and so they always reduce the rate of the motor. Jacobi, and also Lenz, soon discovered these counter-currents, and the effect became known as 'Lenz's Law'.

In a historical account such as this, we are following the trail of the winners, but consider, for example, the tale of the unflagging Professor Wartmann. He investigated (in the 1840s and 1850s): the effect of high mechanical pressure, and of coloured lights, on an electric current; the influence of an electric current on the diffraction and polarization of light; and also, whether there was a difference in the rate of cooling of electrified and non-electrified bodies. All gave null results. (Faraday finally did show a link between electricity and the polarization of light; see Chapter 15.) Part of Joule's skill lay in having a hunch for what phenomena to follow up (or, you might conclude that he was just lucky).

Waterston

An exact contemporary of Joule was John Waterston (1811–83), an engineer from Edinburgh, and yet another scientist in the tradition of British lone researchers who had uncannily correct intuitions, but who struggled for recognition by the scientific establishment. However, Waterston's scientific career was altogether less fortunate than Joule's.

The kinetic theory, as we have mentioned before, had a beleaguered start. Daniel Bernoulli, Cavendish, and Herapath were all independent discoverers (see Chapters 7, 8, and 11), and now Waterston joined the ranks of co-discoverers whose work barely saw the light of day.

Waterston had two biographical details in common with Joule and Mayer: his family were involved in the liquor trade (Sandeman's Port), and he took a job with the East India Company. While in Bombay, Waterston wrote a short book—published anonymously in Edinburgh in 1843—in which the basic principles of his kinetic theory were included. He likened the gas atoms to a swarm of gnats in the sunshine.[24] The book drew little attention, perhaps because its title gave no clue as to its contents: *Thoughts on the Mental Functions*.

Still in India, Waterston followed the book up with a more detailed paper, submitted to the Royal Society in 1845. In it, he foreshadowed many aspects of the modern kinetic theory: equal average kinetic energies even for molecules of different masses; a constant ratio of the specific heats for a gas, C_p/C_V; and rotational as well as translational modes of motion.

It was evidently too advanced for the referees, one of whom wrote: 'This paper is nothing but nonsense'.[25] To make matters worse, the rules of the Royal Society meant that the manuscript could not be returned. As Waterston had not kept a copy, he could not try to publish it elsewhere. (It's hard for us to imagine today, in our computer age, that an author would not have a backup copy.) A brief abstract was published in 1846.

Waterston went on to other scientific work (he estimated the size of atoms to be around 10^{-8} cm—in agreement with modern knowledge, and the temperature of the Sun's surface as 13 million degrees[26]). However, after a further rejection, in 1878, he shunned all scientific societies, and scientific contacts.

Waterston's manuscript finally came to light in 1891, eight years after his death. The physicist Lord Rayleigh was looking at a subsequent paper of Waterston's, on sound, when he saw the earlier work mentioned in a reference. As Rayleigh was then secretary of the Royal Society, he had no trouble locating the original submission in the archives. He recognized its great worth, and published it in 1892—but adding a caution: 'a young author who believes himself capable of great things would usually do well to [first] secure the favourable recognition of the scientific world by work whose scope is limited, and whose value is easily judged, before embarking on greater flights'.[27]

Waterston's death was mysterious—he drowned after falling into a canal in Edinburgh, possibly due to a dizzy spell brought on by heat stroke.

Overview

Although it took some time for the British to accept it, there can be no doubt that Mayer got to energy first. He was awarded the Prix Ponçelet of the Paris Academy of Sciences in 1870, the Copley medal of the Royal Society in 1871, and was lionized in his home town of Heilbronn (there is a statue of him there today).

Joule eventually became a grandee of science, was awarded the Copley medal of the Royal Society in 1872, and has been honoured by having his name used to denote the modern unit of energy: 1 joule (J) is the work done by a force of 1 newton acting through a distance of 1 metre. But his enduring image is that of someone who was merely a brilliant experimenter, and accurate evaluator of the 'mechanical equivalent of heat'. Having now heard all about his work, we can appreciate that he was more than just exceptionally painstaking. He showed great physical intuition in his identification of obscuring effects, and in his close reasoning. Quantities such as 'I^2R', or the 'mechanical equivalent of heat', were not just waiting there, to be dusted down and discovered. As Newton had said, with reference to his own '*experimentum crucis*' on colour, Nature has to be coaxed into revealing her secrets.

If the time was ripe, then why did it still take so long for Mayer's and Joule's work to be recognized? There are many possibilities: both were outsiders to the scientific establishment; Joule further noted a North/South divide (in provincial Manchester, we 'have dinner at noon'[28]); Joule was not a charismatic speaker (he was shy, perhaps due to a congenitally slightly hunched back); and Mayer made some egregious errors in his early work. We must also realize that we have hindsight—we know what energy is, and that it is important. In the 1840s there were other exciting trails to blaze: there was the discovery of Neptune by Adams and by Le Verrier, Hamilton's prediction of conical refraction, and Armstrong's discovery that a jet of high-pressure steam was electrified[29]—could electricity be generated this way? Heat physics was a bit passé, and Fourier had already said the last word on it.

Crucially, there was also the 'category error'—heat and work were radically different sorts of things; and the heat-to-work conversion was particularly hard to demonstrate. Also, it seemed that in order to bring out the laws of *nature*, some very *unnatural* contrivances had to be employed: Joule's 'electro-magnetic engine' (Fig. 14.1) was a very complicated device, and bore about as much relation to dal Negro's prototype (a pendulum swinging near magnets) as Watt's steam engine did to Hero's 'kettle'. (The reason for this increase in complexity becomes apparent only after the arrival of yet another new concept—'entropy'; see Chapters 16 and 18.)

Mayer and Joule always referred everything back to heat⇄work conversions, but there is no doubt that they had inklings of the more abstract and comprehensive concept of *energy*. They insisted that the principle of the conservation of total 'force' (energy) would apply within the diverse domains of physiology, light, electricity, magnetism, 'living force' (kinetic energy), and 'force of attraction' (potential energy), as well as heat and mechanical work. (For example, some researches of Joule's that we have not had space to include are his 28 experiments on the 'heat-equivalent' of light generation in combustion.[30])

Mayer, however, took a positivist approach: when 'heat' is consumed, it simply ceases to exist and is replaced by a 'mechanical equivalent'. The numbers all come out right, and there is no need to ask what heat actually is. Joule, on the other hand, took Dalton's atoms, and the dynamical theory of heat, as the bridge between the very different realms of work and heat. As work was motion, heat must be a 'state of vibration' and not a substance. Joule formulated a proto-kinetic theory (he had already come across Herapath's work; see Chapter 11) and calculated that the speed of atoms in a gas, and also the speed of 'water atoms', was just over a mile per second. (This figure is realistic; and also consistent with the speed of Joule's incandescent meteors.)

Mayer's first published value for the 'mechanical equivalent of heat' appeared in 1842, and Joule's appeared in 1843. In 1848, Mayer quickly sent a letter to the *Comptes Rendues*, fearing that his work was being overlooked. There followed a vigorous priority dispute, but this was conducted entirely by third parties (Tyndall, Clausius, and Helmholtz for Mayer; Tait, Thomson, and Rankine for Joule). Throughout their lives, Mayer and Joule never met or even exchanged letters. Mayer was admiring and respectful of Joule's work; and Joule's private verdict was that Mayer had predicted but not established the 'law of equivalence'. In a letter to Tyndall in 1862, he expanded on this: 'I think that in a case like that of the Equivalent of Heat, the experimental worker rather than the mere logical reasoner (however valuable the latter) must be held as the establisher of a theory. I have determined the mechanical equivalent in nearly a dozen ways, and the figure I arrived at in 1849 has not yet been altered or corrected . . . Believe me, Dear Tyndall, Yours always truly, J.P. Joule'.[31]

The neglect of Waterston's work highlights the enormous contemporary difficulty in conceptualizing a gas as a 'swarm' of a near-infinity of miniscule molecules moving at very high speeds, and—the greatest difficulty of all—in a random or chaotic fashion. The story is continued in the work of Clausius, and, especially, the work of Maxwell, and of Boltzmann, in Chapter 17, in the section on 'Kinetic Theory'.

15

Faraday and Helmholtz

The time was indeed ripe for the discovery of energy, but this didn't mean that the path to it was obvious or that there was only one such path. Already (in the previous chapter), we have met those researchers whose main goal was to determine the exact equivalence value between heat and mechanical energy. Now (in the late 1840s) there were those who, rather, looked for qualitative unifying features in all the 'forces' or 'powers' of nature. One such was Faraday. Finally, there was one researcher, Helmholtz, who sought to set out the entire theoretical framework, deriving the actual formulae for energy in all its various forms, and coming up with the first definitive statement of the conservation of energy.

Faraday

Michael Faraday (1791–1867) was born into a poor family in South London (he was once given a loaf of bread to last a week) and had only a very rudimentary education (reading, writing, and ciphering). At 13, he had a job delivering newspapers. His employer, a French émigré, Monsieur Ribeau, not only let out newspapers but also sold and bound books, and so, at 14, Faraday became an apprentice bookbinder—and, thereby, an avid reader. Two books, in particular, aroused his interest in science: an article on 'Electricity' in a copy of the *Encyclopaedia Britannica* that he was rebinding; and Mrs Marcet's *Conversations on Chemistry*. Jane Marcet had written it after attending Sir Humphry Davy's lectures at the Royal Institution.

One day, a customer at Ribeau's offered Faraday some tickets for Davy's lectures. He eagerly accepted the tickets, attended the lectures, took careful notes, and bound them in a special volume. However, by October 1812, his apprenticeship had finished and he wrote: 'I must resign philosophy entirely to those who are more fortunate in the posses-sion of time and means . . . I am at present in very low spirits'.[1]

But then an accident occurred that had great promise in it, for Fara-day at any rate. Davy was temporarily blinded by an explosion with nitrogen trichloride (the same substance that had injured Dulong's eye and finger; see Chapter 10) and was recommended Faraday as amanu-ensis. This came to pass, but the work only lasted for a few days. In late December, a desperate Faraday wrote to Sir Humphry begging for em-ployment, and sending along the bound volume of carefully written out lecture notes. Davy was flattered but still couldn't help. Then, fortune smiled on Faraday for a second time. The 'fag and scrub' at the Royal Institution, a Mr Payne, lived up to his name, and became involved in a brawl. He was summarily discharged. 'That evening . . . Faraday . . . was startled by a thundering knock at the door. On the street below he saw a carriage from which a footman alighted and left a note for him in which Sir Humphry requested him to call the next morning . . .'[2] The rest, as they say, is history.

Faraday came from a small religious sect, the Sandemanians, and he followed their beliefs for the whole of his life. To understand God's Uni-verse—this was the driving spirit behind all his ambitions in science. (His friend, the physicist Tyndall (see Chapter 14), observed perplexedly, 'he [Faraday] drinks from a fount on Sunday which refreshes his soul for [the whole] week'.[3]) For Faraday, as for Joule, God's 'powers' could not be created or destroyed. As Faraday was later to write: 'The highest law in physical sciences which our faculties permit us to perceive—[is] the Conservation of Force'.[4] But in addition, Faraday was convinced that there had to be an underlying *unity* in all the powers or forces: 'the various forms under which the forces of matter are made manifest have one common origin'.[5] To demonstrate this—by experiment—was the common thread behind the whole of Faraday's long career.

The scientific world was agog after Oersted's linkage between elec-tricity and magnetism, in 1820 (see Chapters 9 and 14). The young Faraday was one of many who immediately began to investigate this further, but Faraday brought his philosophy of the 'unity of force' to bear on it: if electricity caused a magnet to rotate, then surely magnet-ism would cause electric 'rotations'. His experiment to demonstrate this (in 1821) was one of the most simple yet ingenious experiments ever devised (Fig. 15.1) and brought out the symmetry between magnetism and electricity, as well as the 'unity of force'. Leaving aside the sorry tale of Davy's accusations of plagiarism, Faradays predictions were borne

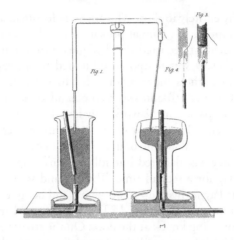

Fig. 15.1 Faraday's electric and magnetic rotations, from *Experimental Researches in Electricity*, volume 2, 1844 (on the left, a magnet is free to rotate through mercury around a fixed conductor; on the right, the magnet is fixed and the conductor rotates freely).

out—the free wire (with cork attached) rotated around the fixed magnet when the battery was connected.

André-Marie Ampère (1775–1836) (sometimes dubbed the 'French Faraday'), had recently shown that electric currents exhibited magnetism. Faraday therefore predicted that magnetic 'currents' should exhibit (produce) electricity. This led to the most famous of all Faraday's discoveries, his law of electromagnetic induction. But he showed this in *two* completely different ways: first, by a varying 'magnetic field' (as we should say now) and, second, by moving a permanent magnet (moving it up or down the axis of a helix of coiled wire, a wire in which electric currents were flowing). Feynman and Einstein were struck by this, and Feynman wrote: 'We know of no other place in physics where such a simple and accurate general principle [the induction of electric currents] requires for its real understanding an analysis in terms of *two [utterly] different phenomena*'.

Less memorable but also important were Faraday's researches (around 1832) that showed that the various kinds of electricity (whether from electrostatic generators, voltaic cells, thermocouples, dynamos, or electric fishes and eels) were all identical. In the Fifteenth Series of Experimental

Researches, on the 'character and direction of the electric force in the *Gymnotus* [electric eel]', Faraday explicitly stated that the convertibility of force included *all* manifestations:

Seebeck taught us how to commute heat into electricity; and Peltier . . . how to convert the electricity into heat. Oersted showed how we were to convert electric into magnetic forces, and I had the delight of adding the other member of the full relation . . . converting magnetic into electric forces. [Now I have further shown that the electric eel can] convert nervous into electric force.

This was followed by Faraday's researches into electrolysis, in 1833, which showed the links between electricity and chemical affinity: the amounts of different substances deposited or dissolved by the same quantity of electricity were proportional to their chemical equivalent weights. This law of electrochemistry was Faraday's only *quantitative* law. Surprisingly, it didn't lead Faraday on to accept atoms: 'if we adopt the atomic theory . . . then the atoms of bodies which are equivalents to each other . . . have equal quantities of electricity . . . But I must confess I am jealous of the term *atom*; for though it is very easy to talk of atoms, it is very difficult to form a clear idea of their nature'. While his law of electrochemistry didn't lead him on to atomism, it did cast doubt on the contact theory of electricity.

The contact theory had been promulgated by Volta (1745–1827), inventor of the Voltaic pile (see Chapter 9, 'Connections'). He had argued that electromotive force was generated by the mere contact of dissimilar metals—no chemical action was required. This smacked of perpetual motion: a metal could be brought up to another metal, any number of times, generating electric power, but without consumption of anything. Faraday, in 1839, carried out a series of experiments demonstrating unequivocally that chemical action was *always* present, although one might have to look hard to find it. But Faraday's objections to the contact theory were metaphysical as well as experimental:

By the great argument that no power can ever be evolved without the consumption of an equal amount of the same or some other power, there is *no creation of power*; but contact would be such a creation.

This was as close as Faraday ever came to a statement of the conservation of energy.

Still searching for evidence of the 'unity of force', in August 1845 Faraday resumed experiments that he had started some 25 years earlier. Polarized light was passed through an electrolytic cell, and the plane of polarization was checked for rotation after various alterations to the external conditions. The variations investigated were as follows: different electrolytes (distilled water, sulphate of soda solution, sulphuric acid, and copper sulphate) and different currents (constant, intermittent, beginning, ceasing, and rapidly recurring secondary currents)—but the light was not rotated. He later wrote to Sir John Herschel, 'It was only the very strongest conviction that Light, Mag[netism], and Electricity must be connected [must be somehow equivalent] that could have led me to resume the subject and persevere through much labour before I found the key'.

In September, Faraday tried even more variations: electromagnets instead of galvanic currents, and transparent materials instead of electrolytes. The polarized light was passed through flint glass, rock crystal, and calcareous spar, and the magnetic 'field' strength was varied—but still no rotation was observed. Then Faraday tried using some glass that he had made for the Royal Society back in 1830—a 'silico borate of lead' (a glass with an extremely high refractive index). At last, '*there was an effect produced on the ray*, and thus magnetic force and light were proved to have relation to each other'. (Faraday went on even further, and showed that he had discovered a new kind of magnetism, which he named 'diamagnetism', and which, with characteristic thoroughness, he tested on everything—from glass to foolscap paper, from litharge to raw meat.)

We have already stated Faraday's misgivings about atoms. Influenced by his former patron, Davy, influenced in his turn by Coleridge and by Kant, Faraday veered rather towards point atoms and 'force fields', similar to those advocated by Boscovich (see Chapter 7, Part II). (These views were not popular—Gay-Lussac and Thenard even threatened police action if one of Davy's papers was published.) The force fields had the advantages of being continuous, and extendable to infinity. More germane was the fact that they ruled out the need for imponderable subtle fluids. This could only encourage the emerging concept of energy—for how could energy be conserved if heat-, electric- and magnetic fluids were all *separately* conserved?

By the late 1840s and 1850s, Faraday had further evolved his worldview in a unique and revolutionary way—'powers' were disseminated by three-dimensional 'lines of force'. These lines of force filled all space, and accounted for the 'unity of force' and the 'harmony of the cosmos'. It was all part of the 'Grand Design', where 'nothing is superfluous'.

There was only one effect left to consider, and that was gravity: surely it had to be linked to electricity. Faraday's own powers were waning, and this was to be his last piece of research. (Faraday had a breakdown in 1838–40, probably due to overwork, and after 1860 he could do no more research.)

Gravity troubled him greatly: a body at a height of several thousand feet weighed less than one on the surface of the Earth, but where (Faraday asked) was the compensatory force accounting for the loss of weight? 'Surely this force must be capable of an experimental relation to Electricity, Magnetism and the other forces, so as to bind it up with them in reciprocal action and equivalent effect', he wrote in 1849. Experiment would, as ever, be the guide, and these investigations would perhaps be the most important of Faraday's whole career, putting his cherished principle of the unity of force to the test: 'It was almost with a feeling of awe that I went to work'.

Cylinders of copper, bismuth, iron, gutta percha, and so on fell from the ceiling on to a 'soft cushion' on the floor in the lecture room at the Royal Institution. The cylinders were surrounded by a helix of copper wire, 350 feet long, connected to a sensitive galvanometer—but no effect was found. In the next experiments, the cylinders were vibrated rapidly up and down—still no effect, but 'They do not shake my strong feeling of the existence of a relation between gravity and electricity'. Finally, the whole experiment was scaled up. A 280-lb lead weight was raised and lowered through 165 feet within the Shot Tower on the River Thames near Waterloo Bridge. Again, the results were negative—but Faraday insisted 'I cannot accept them as conclusive'. He wanted to repeat them with more sensitive instruments, but this was not to be. This was Faraday's last paper submitted for publication (and rejected), in the spring of 1860.

Faraday recognized that gravity was strange in at least two respects: it had no neutral state (whenever there was mass, there was gravity),[6] and it was much weaker than the other forces, quite different to electricity: 'a grain [about 0.002 oz] of water is known to have electric effects equivalent to a very powerful flash of lightning'.[7] He continued: 'many considerations urge my mind toward the idea of a cause of gravity which is not resident in the particles of matter merely, but constantly in them *and all space*'[8] (my italics). With hindsight, it seems as if Faraday was on the trail of a *field* theory of gravity . . .

This would be consistent with Faraday's overall outlook. It was the intuitive concept of force rather than the more abstract concept, energy, that fired his scientific imagination. He was the discoverer of lines of force in electricity and magnetism, and was a tireless researcher into the 'unity of force'. Even though his discoveries had led directly to the electric motor and the dynamo (bringing them to an advanced state, beyond Joule's earlier researches), the final linkage—that between the various forces and mechanical *energy*—doesn't appear to have commanded his attention. The measures of energy, such as 'mechanical equivalent', 'work', and 'gravitational potential', are scalar (have no direction); and while they may be tracked by lines (like the contour lines on an ordnance survey map), these lines give only the height; they don't show *which way* a pebble will actually roll.

All along, Faraday seems to have been questing for a unified field theory, a quest that continues to this day, and in which gravity still refuses to join the party. Thus Faraday, like Newton, missed 'energy'. We now turn to our last scientific personage, Helmholtz, who, when he wrote 'force', really did mean 'energy' (most of the time).

Helmholtz

Herman von Helmholtz (1821–94) (the 'von' was added by Kaiser Wilhelm I in 1882) was what one might call a physicists' physicist: he made outstanding contributions in both experimental and theoretical physics, was 'exceptionally calm and reserved',[9] and a physics professor and patriarch of German science from 1871 until his death in 1894. It comes as something of a surprise, therefore, to find out that his early researches were in physiology and, worse, his first and most influential paper in physics started out with a long introduction on philosophy.

Helmholtz was born in Potsdam, Germany, the eldest of four children. His father was a teacher at the Potsdam Gymnasium and a highly cultured man. (He was a personal friend of the philosopher Immanuel Fichte, son of *the* philosopher, Gottlieb Fichte.) The young Hermann was very accomplished in the arts as well as the sciences. In one of his first letters home when a student in Berlin, he writes: 'Any spare time I have during the day is devoted to music . . . I play sonatas of Mozart and Beethoven . . . In the evenings I have been reading Goethe and Byron . . . and sometimes for a change the integral calculus'.[10] His father advised him: 'don't let your taste for the solid inspiration of German and classical music be vitiated by the sparkle and dash of the new Italian extravagances—these are only a distraction, the other is an education'.[11]

Helmholtz's first love was physics, but his father's salary didn't extend to university fees and, besides, 'physics was not considered a profession at which one could make a living'.[12] He readily accepted medicine as an alternative, as the fees were paid in return for working for some years as a Prussian army doctor after graduation.

Within medicine, Helmholtz veered towards physiology, but he always saw every problem through physicist's eyes. (He taught himself mathematics and physics through private study in his 'spare time'.[13]) The vitalist philosophy of Stahl (also inventor of the phlogiston theory; see Chapters 5, 6, and 8) and the '*Naturphilosophie*' of Hegel held sway in Germany at this time. The vitalists attributed life in organisms to the presence of a 'life force' in addition to food, air, and water. Helmholtz felt sure that this was 'contrary to nature',[14] but he was unable to state this in the form of a proposition that could be put to the test.

Finally, in his last year as a medical student, Helmholtz 'realized that Stahl's theory treated every living body as a *perpetuum mobile*'.[15] But he had known, ever since his teenage years, that perpetual motion was supposed to be impossible, and this was reinforced by his recent reading of the works of Daniel Bernoulli, Euler, d'Alembert, and Lagrange (as usual, in 'spare moments'). He now had an aim—to rid physiology of vitalism. He was joined in this quest with three other young physiologists in Berlin, especially his friend and fellow student du Bois-Reymond.

Thus Helmholtz's first researches, between 1845 and 1848 (having recently qualified as a doctor), were all intended to rebut the vitalists' claims. Specifically, he carried out experiments to try to show that the 'mechanical force and the heat produced in an organism could result entirely from its own metabolism'.[16]

This was easier to state than to demonstrate—we all know, for example, that some people can eat like a horse and others like a sparrow. Some of the problems that Helmholtz had to contend with were: he had no value for the 'mechanical equivalent of heat'; heat could be generated in the blood and the muscles as well as in the lungs and the stomach; heat of excreta had to be accounted for; and, also, what were the heats of combustion of different foods, and had the reactions gone to completion? Helmholtz estimated that 2.6% of heat was lost to excrement, 2.6% to the heating of expired air, 14.7% by evaporation from the lungs, and 80.1% by evaporation of sweat, and radiation and conduction from the skin.[17] He realized that the previous studies (of Dulong and Despretz) were in error in using the heats of combustion of hydrogen and of carbon rather than the heats of combustion of the complicated molecules of fat or sugar. (He also invoked Hess's Law to show that the order of the decomposition reactions didn't matter.)

Helmholtz still had to show that food accounted for the mechanical work done by the animal (as well as heating), and so he continued on with experiments on the thigh of a frog (he 'waited impatiently for the spring and the frogs'[18]): he determined that a single muscular contraction caused a temperature rise of 0.001–0.005 °C. He improved or invented instrumentation as he went along: for example, the myograph, whereby a contracting muscle leaves its trace on the blackened surface of a revolving cylinder, or on a moving glass plate; the 'moist chamber' for keeping the muscle in a good condition; and electrical apparatus to apply shocks of known duration and known intensity. (His most famous invention was the opthalmoscope in 1851, and others include the Helmholtz resonator and the Helmholtz coils.) All this research did show that vital forces were redundant, but the experiments could not be considered as absolutely conclusive.

Then, in 1847, while still a Prussian army surgeon with the Royal Hussars, Helmholtz had a change of tack. The motivation was still to show that perpetual motion was impossible, but he now widened the scope to cover all physical processes, with physiology just a subset, and also moved from an experimental to a theoretical attack. He had a grand ambition: to base his new principle of the impossibility of perpetual motion on the securest possible foundation—on philosophical bedrock as it were—and to apply it to the whole of physics.

With remarkable assurance (he was only 26, and this was his first work in physics, a subject in which he was entirely self-taught), he composed a memoir, some 60 pages long, entitled 'Über der Erhaltung der Kraft' ('On the conservation of force').[19] It started off: 'I have [formed] a physical hypothesis . . . and then developed the consequences of this hypothesis in the several branches of physics and, finally, have compared these consequences with the empirical laws'.[20]

For the justification of his ideas, Helmholtz looked to Kant's transcendental idealism. There he found that there were two kinds of law in science: empirical laws (a summary of observations—for example, the laws of refraction of light, and Boyle's Law) and theoretical laws (laws of the hidden causes behind the observations). But which kind of law was the prohibition against perpetual motion? We have seen that perpetual motion was vetoed because of the *experimental* findings—people kept trying, but the machines always failed (see Chapter 2). However, Helmholtz, like Mayer, Carnot, and others, thought that the principle was rather of the second kind—it just *had* to be true.

From Kant again, science had to be comprehensible, and its comprehensibility lay in the law of causality. Some things might happen spontaneously—that is, without cause—but they were outside the remit of science. Helmholtz took the impossibility of perpetual motion as an example of a causal law and, in fact, equivalent to the old nostrum of 'cause equals effect'. But did the impossibility of perpetual motion mean that something was being conserved? Kant's philosophy was helpful yet again. It asserted that the physical world was comprehensible because of two 'intuitions'—space and time; and two abstractions, matter and force. Matter was inert and could only be experienced through its effects; but—and this was the crucial link—force was the *cause* of all the effects. Thus, Helmholtz was able to associate 'cause' with 'force'; and 'cause equals effect' with 'conservation of force'. The grand finale—and, at last, a resolution of the question posed at the end of Chapter 2: the impossibility of perpetual motion *necessitated* the conservation of 'force'.

Now, what if several different causes could all explain the observations equally well? Helmholtz argued that only the *simplest* (i.e. those satisfying the principle of sufficient reason) could be considered as the 'objective truth'. Kant's 'matter' had only two properties—mass and position—and so, for Helmholtz, the simplest type of force had to be 'central'; that is, depending only on the masses and the positions of bodies, and acting in the direction of the line joining the bodies.

Helmholtz's '*Über der Erhaltung der Kraft*' started with a lengthy philosophical introduction, in which all these antecedents were explained. Then he proceeded to apply his new principle, 'the conservation of force', across all branches of physics. We give brief excerpts from all the major disciplines, as follows.

First, Helmholtz turned to that most well-established discipline—mechanics. He had been so impressed by Carnot's argument against perpetual motion, applied in the arena of heat engines, that he immediately set about applying it in the case of mechanics. He imported not only Carnot's argument but also Carnot's new constructs of 'states' and 'cycles'. For a system of bodies, acted upon only by internal central forces, F, he defined work as $\int F dr$, and saw that a state of the system was completely defined by the positions and the velocities of all the bodies at a given time. The work released in changing the system from an initial state, A, to a final state, B, had to be *equal* to the work required to take the system from B back to A, or 'we should have built a *perpetuum mobile*'. From this start, Helmholtz was able to prove all the standard results in mechanics: 'conservation of *vis viva*'; the maximum quantity of work in

going from A to B was definite and fixed (shades of Carnot); the principle of 'virtual velocities'; a system of bodies can never be set in motion
by the action of its internal forces (cf. Torricelli); the final speed on free
fall depends only on the vertical distance travelled; this speed is just sufficient to return the body to its starting height (assuming no friction);
the work done in a simple machine is inversely proportional to the speed
of its moving parts; and the laws of elastic impact (when combined with
the constancy of the motion of the centre of gravity).

Next, Helmholtz considered waves and wave motion. Fresnel had
shown that all the laws of light (reflection, refraction, and polarization)
could be deduced from two premises: the conservation of *vis viva* and
the principle of continuity. Helmholtz took Fresnel's demonstrations
as confirmation of his principle, 'the conservation of force', and added
something of his own: the intensity of a wave had to decrease in accordance with the inverse square law. Furthermore, as 'interference' between
two wave trains caused only a redistribution of intensity, the amount of
total 'force' was unaffected. But Helmholtz warned that experimental
confirmation was still required, especially to show that, in a system of
just two bodies, A and B, the total heat radiated by body A was equal to
the total heat absorbed by body B.

There were two arenas, collisions and friction, in which (Helmholtz
wrote): 'an absolute loss of force has until now been taken for granted'.
However, for inelastic collisions, he noted that the deformation forces
were increased, and heat and sound were generated. Also, where friction
occurred, then there was always an equivalent amount of thermal and
electrical change.

In the conversions from mechanical work to heat, Helmholtz mentioned Joule as being the only investigator (he had not yet come across
Mayer), but he was not particularly complimentary about Joule's methods: 'His [Joule's] methods of measurement . . . meet the difficulties of
the investigation so imperfectly that the results can lay little claim to accuracy . . . a quantity of heat might readily have escaped . . . loss of mechanical force in other parts of the machine was not taken into account'.
Helmholtz again mentioned Joule's work when it came to the generation
of heat from 'magneto' electricity, endorsing all of Joule's careful deductions against the materiality of heat (see Chapter 14) but neglecting to
credit Joule with them. Regarding the conversion of heat to work, Helmholtz commented that 'nobody has yet bothered', but then went on to
cite the very experiments where Joule had done just that (Joule's experiments on the adiabatic expansion of gases). At least this time Helmholtz

admitted that the experiments had been 'rather carefully made'. All in all, one gets the impression of a young man (Helmholtz was 26) impatient to have his ideas experimentally corroborated, rather than of any serious criticism of Joule. (In the revised edition of the memoir in 1881, Helmholtz was more generous towards Joule: 'His [Joule's] later investigations, [were] carried out with complete professional knowledge and indefatigable energy, [and] merit the highest praise'.)

Helmholtz was convinced of the wrongness of the caloric theory of heat. He attributed heat instead to the *vis viva* of the constituent particles ('free heat'), and to any tensional forces between these particles ('latent heat'). He also acknowledged Ampère's theories in which the *vis viva* could manifest itself as rotational as well as translational molecular motion. If the dynamic theory of heat was adopted, then the 'principle of the conservation of force holds good wherever the conservation of caloric was previously assumed'.

Electricity and electromagnetism took up a third of Helmholtz's memoir. This was an area in which the Germans were especially active, and Helmholtz was fully up to date with the researches of Gauss, Weber, Neumann (father, Franz, and son, Carl), Lenz, Ohm, Kirchoff, and others.

In static electricity, the forces were central, and balanced the 'living forces'—so the 'principle of the conservation of force' was guaranteed. Helmholtz defined the equipotential surfaces of an isolated conductor and outlined operations for establishing the unit of electrical potential. He also gave the (now) standard expression for the energy of a capacitor, $\frac{1}{2}Q^2/C$, and compared it with experiments measuring the heat generated by the discharge of a Leyden jar.

In 'galvanic' electricity (that is, electrical currents), Helmholtz straightaway dismissed Volta's contact theory: 'the principle which we are presenting here directly contradicts the earlier idea of a contact force'. Instead, the contact force had to be replaced by (central) forces of attraction and repulsion, 'which the metallic particles at the place of contact exert upon the electricities at that point'.

Helmholtz went on to use his conservation principle to deduce the 'electromotive force' of a battery by balancing this against the heat generated chemically in the cell, and the resistive heating in the wire. He brought to bear all the contemporary, quantitative laws: Ohm's Law, Lenz's heating law, Joule's more general heating law (I^2R), Kirchoff's circuitry laws, and Faraday's laws of electrolysis.[21] He then derived a new heat-balance equation for the case of thermoelectric currents (those arising out of junctions held at different temperatures, as in the Peltier effect; see the end of Chapter 9) and bemoaned the fact that there were no quantitative experiments with which to compare his predictions.

Electrodynamic and induction phenomena were concerned with the motion of magnets near currents in wires. Lenz's Law, and Franz Neumann's extension of it, showed that 'the force of the induction current . . . acts always in opposition to the force which moves the magnet'.[22] This was encouraging—a perpetual motion was thereby prevented. Carl Neumann had defined an 'electromotive force of induction', and had an expression for the work done by this force in moving the magnet. Helmholtz used Neumann's result to show that 'if a magnet moves under the influence of a current, the *vis viva* which it gains must be supplied by the tensional forces consumed in the current'.[23] (The heating in the wires, I^2R, had also to be remembered.) He was further able to show that Neumann's undetermined, empirical constant of proportionality was the reciprocal of the 'mechanical equivalent of heat'. However, Helmholtz did not sufficiently acknowledge Neumann's work, and the wording in the memoir suggested that the deductions had followed from Helmholtz's conservation principle alone. Neumann, and also Clausius, protested, and Helmholtz eventually capitulated. For one thing, Helmholtz had not taken into account self-induction; for another, he had implicitly assumed that the 'energy' of the circuit-magnet system didn't depend on the position of the magnet (in fact, it didn't, but this was not self-evident[24]).

These examples show that the case of electromagnetism was tricky. In fact, it was worse than tricky; it was the first instance in which Helmholtz's programme failed—the forces were no longer all of the simple, central type that Helmholtz had assumed. Weber had traced back electrodynamic and induction phenomena to 'the forces of attraction and repulsion of the electric fluids themselves',[25] but the intensity of these forces was found to depend upon 'the [magnet's] velocity of approach or recession, and upon the increase in this velocity'. Helmholtz commented dryly that 'Up to the present time no hypothesis has been found by which these phenomena can be reduced to constant central forces'. However, the major part of his project, the conservation of force, *was* still upheld.

Finally, Helmholtz returned, very briefly, to the initial impetus for all his work—the absence of vital forces in organic processes. He said that for plants 'there is a vast quantity of chemical tensional forces stored up . . . the equivalent of which we obtain as heat when they are burned'. Also, 'the only *vis viva* which we know to be absorbed during the growth of plants is that of the chemical rays of sunlight; we are totally at a loss, however, for means of comparing the force equivalents which are thereby lost and gained'.

For animals, he said that 'Animals take in oxygen and the complicated oxidizable compounds [food] which are generated in plants, and give them back partly burned as carbonic acid and water, partly reduced to simpler compounds. Thus they use up a certain quantity of chemical tensional forces and . . . generate heat and mechanical force'. From contemporary data (the calorimetric experiments of Dulong and Despretz), he noted that the validity of the conservation of force can 'at least approximately, be answered in the affirmative'.

Helmholtz wrote in conclusion:

I believe, by what I have presented in the preceding pages, that I have proved that the law under consideration [the conservation of force] does not contradict any known fact within the natural sciences; on the contrary, in a great many cases it is corroborated in a striking manner . . . I have attempted to guard against purely hypothetical considerations, [and] to lay before physicists as fully as possible the theoretical, practical, and heuristic importance of this law, the complete corroboration of which must be regarded as one of the principle problems of physics in the immediate future.

Reception of the Memoir

This was a bravura performance: Helmholtz had founded the 'conservation of force' on the securest of foundations, had shown an extraordinary knowledge of contemporary theory and experiment, and of the work of the ancients, and had identified, or had a good stab at identifying, the appropriate formulae—the 'blocks of energy'—in fields as disparate as heat, electricity, magnetism, physiology, chemistry, light, and mechanics. In outlining a future programme for physics, he had shown enormous self-confidence (not to say, chutzpah). But what did his contemporaries make of it?

The older physiologists (still followers of Stahl) were not to be persuaded by an exposition that was so mathematical, theoretical, and, above all, in a foreign discipline.

Helmholtz had high hopes of impressing the physicists, at least, but to his 'astonishment' the physicists 'were inclined to deny the correctness of the law [of the conservation of force] . . . because of the heated fight in which they were engaged against Hegel's philosophy of nature, [and] to treat my essay as a fantastic piece of speculation'.[26] What irony, that Helmholtz, in trying to banish Hegel's *Naturphilosophie* from science, should have come over as too Hegelian. The scientist and publisher Poggendorf (who had rejected one of Mayer's papers) declined to publish Helmholtz's work as it was too theoretical, too speculative (meaning, too philosophical), and, most damning of all (wrote Poggendorf), had no new experimental findings. As regards theoreticians, only one, Carl Jacobi (see Chapter 13), was enthusiastic, while Franz Neumann was mildly supportive, Wilhelm Weber indifferent, and Clausius not receptive at all[27] (see below).

It was really only the British school—in particular, Thomson (see Chapter 16)—who gave wholehearted support to Helmholtz's memoir. Thomson came upon Helmholtz's 'admirable treatise'[28] in 1852, and wrote that had he 'been acquainted with it in time',[29] he would have used its results in many of his papers. Helmholtz and Thomson first met in 1855, and became close friends for life. They evolved to have somewhat similar outlooks, both considering the law of the conservation of energy as the heart of physics. They also grew to occupy similar positions as establishment figures of science in their respective countries. (In the priority dispute over Mayer and Joule, Helmholtz, Thomson, and all the other physicists except for Tyndall split along national lines—but this didn't sour international physics relations for long.)

Maxwell (see Chapter 17) was also appreciative of Helmholtz's memoir, writing (in 1877):

To appreciate the full scientific value of Helmholtz's little essay on this subject [the conservation of energy], we should have to ask those to whom we owe the greatest discoveries in thermodynamics and other branches of modern physics, how many times they have read it over and over, and how often during their researches they felt the weighty statements of Helmholtz acting on their minds like an irresistible driving-power.[30]

(Helmholtz was the first to promote Maxwell's electromagnetic theory on the continent. He subsequently developed his own theory, which, apparently, included Maxwell's as a limiting case.)

It is fascinating to learn about contemporary life, and the interrelations between these energy physicists. Helmholtz met Faraday on his first trip to England in 1853. He wrote to his wife: 'I succeeded in finding the first physicist in England and Europe . . . Those were splendid moments. He is as simple, charming, and unaffected as a child; I have never seen a man with such winning ways. He was, moreover, extremely kind, and showed me all there was to see. That, indeed, was little enough, for a few wires and some old bits of wood and iron seem to serve him for the greatest discoveries'.[31]

Helmholtz met Thomson in Kreuznach, a German resort, in 1855: 'I expected to find the man, who is one of the first mathematical physicists of Europe, somewhat older than myself, and was not a little astonished when a very juvenile and exceedingly fair youth, who looked quite girlish, came forward. He is at Kreuznach for his wife's health . . . she is a charming and intellectual lady, but in very bad health. He far exceeds all the great men of science with whom I have made personal acquaintance, in intelligence and lucidity and mobility of thought, so that I felt quite wooden beside him sometimes'. He met Thomson again in Glasgow, in 1864, and Thomson's brother, a professor in engineering, was also there: 'It is really comic to see how both brothers talk at one another, and neither listens, and each holds forth about quite different matters'.

On this trip, he also met Joule at a dinner party: 'Mr Joule, a brewer, and the chief discoverer of the conservation of energy, and [another guest] were both very pleasant lively individuals, so we spent a most interesting evening'. Finally, he also met Maxwell: 'I went with an old Berlin friend to Kensington, to see Professor Maxwell, the physicist at King's College, a keen mathematician, who showed me some fine apparatus for the Theory of Colours which I used to work at; he had invited a colour-blind colleague, on whom we experimented'.

Of London, Helmholtz writes to his wife: 'And now you shall hear about this great Babylon. Berlin, both in size and civilization, is a village compared to London'; and, in humourous vein, (after talking at the BA meeting at Hull) 'Here in England the ladies seem to be very well up in science . . . they are attentive and don't go to sleep, even under provocation'.

Overview

First, a comment on words. In the birth of a new concept, as with the birth of a child, there is usually an interim period during which an appropriate name must be found. In the memoir, Helmholtz used only the term 'force', sometimes in the old Newtonian way, but usually in the new way, meaning 'energy'. It is invariably clear from the context which meaning is intended; thus, 'forces of attraction and repulsion' (force means force), and 'the principle of the conservation of force' and 'force equivalent' (force means energy). The term 'energy' was introduced by Thomson in 1852, and Helmholtz very much approved.

History can sometimes miss the point: Joule is regarded as having 'merely' carried out extremely precise measurements, and Helmholtz as determining lots of formulae, as if working back from the correct dimensions of energy. So, what new did Helmholtz's memoir bring in? Maxwell, giant of physics in the nineteenth century, says it all in his appraisal of Helmholtz:

[Helmholtz] is not a philosopher in the exclusive sense, as Kant, Hegel . . . are philosophers, but one who prosecutes physics and physiology, and acquires therein not only skill in discovering any desideratum, but wisdom to know what are the desiderata.[32] (1862)

and

the scientific importance of [Helmholtz's] principle of the conservation of energy does not depend merely on its accuracy as a statement of fact, . . . it [further] gives us a scheme by which we may arrange the facts of any physical science as instances of the transformation of energy from one form to another.[33] (1877)

Helmholtz's memoir marked the beginning of a new era, the 'epoch of energy'.[34] However, there was a big difference in the philosophical underpinning of the German and British schools (this will become evident in Chapters 16 and 17). Thomson, Joule, and Faraday, and to a lesser extent Maxwell, invoked the permanence of 'God's Creations', while Mayer, Clausius, and Helmholtz argued from the rationality of science.

There can be no doubt that it was Helmholtz's expertise across disciplines (the three 'phs' of physics, physiology, and philosophy, and also mathematics, the 'science of aesthetics', and the 'moral sciences') that gave him such an extraordinarily broad outlook. Each discipline helped the others along, as we shall see.

From his work in physiological optics and physiological acoustics, Helmholtz learned that there was no such thing as raw sense data. Fichte's view was that we only experience a succession of conscious states—but this doesn't permit us to know if there is really anything 'out there'. Helmholtz cleverly adapted Fichte: our free will permits us to contrive experiments, and these lead to correlations between experimental results and our conscious states—thus there *is* something out there. (Helmholtz waited until after the death of his father—a Kantian Fichtean idealist—before expounding these views.) According to Helmholtz, the correlations are captured in the laws of physics, and these laws grew to have more significance than causes for him. (As Helmholtz's friend, du Bois-Reymond, commented, nothing is to be gained by introducing causes such as the hand 'that shoves the inert matter silently before itself'.[35]) There were also other ways in which Helmholtz progressed from Kant. For example, his work on the origins of geometry convinced him that space doesn't have to be Euclidean. But in the final analysis, and by Helmholtz's own reckoning, he always remained a Kantian, and never departed from Kant's dichotomy of 'existence' and 'activity', from which Helmholtz derived 'matter' and 'energy'. This is still the prevailing conceptual divide in physics today (cf. the physicist explaining his subject at the start of this book).

One can take philosophy too far. Helmholtz had wanted to prove not only that total energy was conserved, but also that the elemental force was *necessarily* central. Clausius, in 1853, disagreed—perhaps nature isn't so simple, and the force between two mass-points *does* depend on factors such as speed, acceleration, or direction? In the case of electrodynamics, nature decided in favour of Clausius. The disposition of whole bodies (such as magnets, and other electrical components) was found to be important (i.e. the whole is more than the sum of its mass-points), and, worse still, Weber showed that the force depended on speed and acceleration (e.g. on *how fast* a magnet was moved). Helmholtz conceded defeat, but took hope from the fact that Weber's force law led to instabilities, and sometimes to perpetual motion, and therefore it had to be flawed. But Clausius, always able to see through to the bare logical bones of an argument (see especially Chapter 16), found that if Newton's Third Law was sacrificed, then perpetual motion was avoided, and the conservation of energy upheld even *with* velocity-dependent forces. Helmholtz, with reluctance, had to agree. He added an appendix to the 1881 edition of his memoir, and stated that the forces were central only in those cases where the Principle of the Equality of Action and Reaction also applied.[36]

There was yet one more problem with Helmholtz's philosophy. Helmholtz, still influenced by Kant, could never sanction forces or energy in empty space. His pupil, Heinrich Hertz (1857–94) (famous for detecting radio waves), commented that Helmholtz always considered that matter was the seat of force, and that, without matter, there was no force. But Faraday's lines of force were curved, even where there were *no* mass-points along the way. Even stranger, sometimes the lines of force were completely closed loops, neither starting nor ending on a massy source. Finally, Maxwell's equations (1873) were to show that the electromagnetic field had energy, even in empty space (see Chapter 16).

Helmholtz struggled to accommodate the new physics into his worldview. He developed an electrodynamics that was a bridge between the Continental action-at-a-distance theories and Maxwell's theory (in fact, as mentioned earlier, Maxwell's theory was a limiting case of Helmholtz's theory). He went on to apply thermodynamics to chemical processes, and defined the 'free energy' of a chemical system as the energy that is available for conversion into work. He then used the Second Law of Thermodynamics to derive an equation (the famous Helmholtz–Gibbs equation) showing that it is sometimes useful to consider the 'free energy' as minimized, rather than the entropy as maximized (see Chapter 17, 'Chemical Energy'). (This explained the seemingly anomalous occurrence of spontaneous endothermic reactions—those chemical reactions in which the products are cooler than the reactants.) While never abandoning his cherished law of the conservation of energy, Helmholtz moved away from seeing it as the pre-eminent guiding principle of physics, and pinned his hopes instead on the principle of least action. He spent the last few years of his life working on this, but with no very great success.

After the work of Helmholtz, 'vitalism' disappeared, and the Germans, reluctant at first, came to adopt the energy principle with such enthusiasm that some even advocated energy as *the* primary quantity, in place of mass or force. These 'energeticists' even went so far as to deny the existence of atoms—remarkable in view of the fact that most were chemists.

16

The Laws of Thermodynamics: Thomson and Clausius

The final players in our history of energy are Thomson and Clausius. Their seminal contributions were written in and around 1850, but neither made any reference to Helmholtz's memoir of 1847. Instead, they looked back some 25 years to Sadi Carnot's *Réflexions* (see Chapter 12). They initially had access only to Clapeyron's revision and extension of this work, but they both peered through Clapeyron to the genius of Carnot.

Thomson

William Thomson (1824–1907) was the most gifted in a gifted family—his father and elder brother were professors of mathematics and engineering in their time. Born in Belfast, William's mother died when he was six years old and the family moved to Glasgow, where his father became professor of mathematics at the university. Taught initially at home, William's mathematical prowess showed up early; aged only 16, he had mastered Fourier's *Analytical Theory of Heat* and shown Professor Kelland of Edinburgh University to be in error regarding the Fourier series. (Fourier's work was to be an influence for the whole of Thomson's life—see Chapters 11 and 17.) Thomson studied mathematics at Cambridge (1841–45) and was reputedly so confident of success that he asked who came second in the mathematics tripos exams, only to find that *he* had come second. He revised better for the next lot of exams and came first as Smith's Prizeman in 1845. We have already heard—in Chapter 13—how he tracked down Green's work at this time. After Cambridge, Thomson went to Paris for a year to work in Regnault's laboratory and learn some practical skills. (Regnault (1810–78) was a master experimentalist, and was researching the thermal properties of gases, especially steam, the gas laws, and precision thermometry.) Thomson then returned to the University of Glasgow as professor, a position he held from the young age of 22 until his retirement.

While in Paris, Thomson came across Clapeyron's 1834 paper, 'On the motive power of heat',[1] a reworking of Sadi Carnot's *Reflections on the Motive Power of Fire*. It inspired Thomson to find Carnot's original work, which he managed after some effort (see Chapter 12). Thomson was greatly impressed by the *Reflections* ('the most important book in science'[2]) and saw what neither Clapeyron nor anyone else had seen—that Carnot's theory opened the door to the possibility of another scale of temperature—apart from, say, the height of a column of mercury, or the volume or pressure of a gas. According to Carnot's extraordinary proof, the maximum efficiency of a heat-engine didn't depend on the engine design or working substance, but depended only on the actual operating temperatures (see Chapter 12). Therefore (Thomson realized), Carnot's Law could yield a temperature scale that was *absolute*—independent of any device or material. This was the holy grail of temperature scales, once and for all removing questions regarding the uniformity, and the cross-calibration, of different types of thermometer.

After a first paper in 1848 and then another in 1854, Thomson finally arrived at *the* temperature scale that is still in use today and called the Kelvin scale (Thomson was elevated to the peerage in 1892, becoming Lord Kelvin, Baron of Largs).

The Kelvin Scale of Temperature

The scale can now be explained as follows (where Q refers to heat taken from or added to a reservoir within a Carnot cycle—see Chapter 12—and Q_T refers specifically to the heat taken up or put out at temperature reservoir T, within a Carnot cycle). Imagine a complete Carnot cycle (an ideal heat-engine) in which heat, $Q_{T_{high}}$, is taken in at a high temperature, T_{high}; and heat, $Q_{T_{low}}$, is given out at a lower temperature, T_{low} (in other words, the engine is driven in the forward direction, and net work is done). Now, it is evident that $Q_{T_{high}}$ must be greater than $Q_{T_{low}}$, as $Q_{T_{low}}$ is the heat *left over* after work has been done. (We remember from Chapter 12 that this asymmetry is endemic in Carnot's engines.) Also, T_{high} is defined as being higher than T_{low} from Carnot's directive that the heat-engine goes in the forward direction—*produces* motive power—when transferring heat from *higher* to *lower* temperatures.

Now, to set up a scale across all temperatures, we need to imagine an infinity of idealized heat engines, operating between any two temperatures (heat reservoirs). We have the further requirement that all these engines must be of the same 'size'—they must be hypothetically scaled to each other before an overall scale can be set up. This scaling can be achieved (hypothetically, at any rate) by making sure that all the engines give out the same amount of heat at some reference reservoir. These conventions (that, in the 'forward' direction, $T_{high} > T_{low}$, and $Q_{high} > Q_{low}$), and the proviso that all the engines are 'scaled' to one another, are all that is required to arrive at the absolute scale of temperature (granted that we adopt the definition of an ideal heat engine, as given by Carnot). What results is remarkably simple: T is proportional to Q_T. In other words, the absolute temperature of a heat reservoir, T, is determined by the heat, Q_T, that would be taken in or put out at that reservoir by a pre-scaled ideal heat-engine. It is not necessary to ask what has happened, or what will happen, to that heat—whether it shall be conducted, radiated, converted to/from work, or whatever.[3] In honour of Thomson's (Kelvin's) achievements, the degrees of the absolute temperature scale are called 'kelvin', and symbolized K (not °K). Also, the absolute scale has been cross-calibrated with the Celsius scale, such that an interval of 1 °C equals an interval of 1 K, and 0.00 °C equals 273.15 K.

Now the efficiency, η, of an ideal heat-engine operating between T_1 and T_2 is defined as W/Q_1, where W is the work done for an input of heat, Q_1. From the First Law of Thermodynamics, we know that $W = Q_1 - Q_2$, and therefore we know that the ideal efficiency is $(Q_1 - Q_2)/Q_1$. Using the absolute temperature scale, in which T is proportional to Q_T (with the same constant of proportionality for all temperatures), we can now rewrite this as follows:

$$\text{Ideal efficiency, } \eta = (T_1 - T_2)/T_1 \qquad (16.1)$$

As this efficiency can never be greater than unity, so *the temperature can never be less than zero.* There is therefore an absolute zero to the temperature.

With this absolute scale, we can then calculate the ideal efficiency (in other words, the theoretical maximum to efficiency) of any engine operating between any specific absolute temperatures. For example, Carnot's test case of an alcohol-vapour engine running between 78.7 °C and 77.7 °C (in other words, 351.85 K and 350.85 K) has an ideal efficiency of (351.85 − 350.85)/(351.85), or a mere 0.28%. This just goes to confirm our earlier suspicion (see Chapter 8) that getting work from heat is inherently difficult. However, for a more realistic case of an engine running between, say, 110 °C and 10 °C, the ideal efficiency is 26%. The greater the starting temperature, and the greater the drop in temperature, the greater is the ideal efficiency.

Conflicting Theories

While developing the absolute scale of temperature, Thomson still (in 1849) hadn't come round to accepting that heat could be *converted* into work. He thought that Carnot's conclusions would come crashing down if the axiom of the conservation of heat was abandoned. But he couldn't reconcile this axiom with Joule's experiments, which showed that heat and work were interconvertible, and that heat was actually consumed or generated in the process (see Chapter 14).

One problem in particular that troubled Thomson was the phenomenon of heat conduction. In this process, heat was transferred from a high temperature—say, at one end of an iron bar—to a lower temperature at the other end of the bar (he may have been thinking of Fourier's conduction equation, yet again)—but where was the work that, according to Carnot, should have been developed by this 'fall' of heat? Thomson wrote:

When thermal agency is thus spent in conducting heat through a solid, what becomes of the mechanical effect [work] which it might produce? Nothing can be lost in the operations of nature—no energy can be destroyed.[4]

This is the first recorded use (in 1849) of Thomson employing the word *energy*.

Another similar problem was the case of a swinging plumb line that loses all its motion when submerged in water—what has become of the 'mechanical effect'? Thomson, independently of and later than Joule, carried out his own experiments with a paddle-wheel: he found that, like Rumford, he could bring water to the boil merely by friction. He had thus amply confirmed Joule's findings—but this experiment still didn't suggest to Thomson the solution to the conflicting theories.

Yet another problem concerned the hypothetical case of an engine based on the freezing of water: as the water froze and expanded, it could do work (e.g. cause a pipe to burst). But this all occurred at a *constant* temperature (32 °F) and so contradicted Carnot's requirement for a fall of temperature. Thomson's brother James came up with an ingenious resolution: as the pressure on the ice slowly increased, then perhaps the temperature of melting might be *lowered*. William immediately set about testing this suggestion with some very careful experiments; he managed to confirm James' prediction, and thereby give precise quantitative support to Carnot's theory.

It seemed as if both Carnot's and Joule's opposing theories were becoming more and more confirmed and entrenched. Thomson couldn't resolve this conflict: nor could he explain the conundrum of workless heat conduction; or the fact that Joule appeared to have shown conversions of work to heat, but not of heat into work. (Thomson was a bit unfair to Joule over this, as Joule *had* shown the generation of work from heat by the adiabatic expansion of a gas against an external pressure; see Chapter 14).

Salvation came from Clausius, with his amazing clear-sightedness and cool logic. He saw that Joule's interconversions had to be correct, and that Carnot's conservation of heat had to be jettisoned. But he found a way of achieving this without putting 'Carnot's Law' into jeopardy.

Clausius

Rudolf Clausius (1822–88) was born in Köslin, Pomerania (now Poland). His father was a Lutheran pastor and principal of a small private school at which Rudolf and his many older brothers had their early education. He then went to a Gymnasium in Stettin, and later to the University of Berlin (1840–44), and then presented his doctoral dissertation in 1847 at the University of Halle.

There is a three-year gap (1847–50) during which Clausius disappears from the record; and then comes the publication of his famous paper, 'On the moving force of heat and the laws of heat that may be deduced therefrom',[5] in 1850. The importance of this work was soon recognized in Germany and in England (thanks to Tyndall's translations), and Clausius' academic career was launched—he was successively a professor in Berlin, Zurich, Würzburg, and finally Bonn, from 1869 until his death in 1888. He became one of the leading physicists (the others were Helmholtz and Kirchoff) in the newly unified Germany, a nation that was to become pre-eminent in science and industry during the second half of the nineteenth century.

There is scant background information about Clausius. He married Adelheid Rimpam in 1862, and volunteered to lead an ambulance corps in the Franco-Prussian war of 1870–71. He was wounded in the leg, and awarded the Iron Cross for bravery, but suffered severe pain and disability for the rest of his life. (In this same war, Regnault's laboratory was destroyed and his son killed.) Clausius' wife died in childbirth in 1875, and Clausius was left to look after their six children, a task which he apparently took on with dedication and kindness. In 1886, he married again and had one more child. After this chapter, we make further brief mention of Clausius in connection with the kinetic theory (see Chapter 17).

The First Law of Thermodynamics

We often hear it said that the Second Law was discovered before the First, but this isn't exactly true. Clausius was convinced by Joule's heat⇄work conversions (the reversible arrow means both heat-to-work and work-to-heat) and saw that these were ultimately explained by the dynamical theory of heat—that heat is a motion of the microscopic constituents. He was nevertheless conscientious in developing the First Law (his 'First Principle of Heat') without recourse to assumptions about microscopic structure. How he did this was as follows.

Clausius tackled the usual scenario of a gas in a cylinder, expanding against a piston (see Chapter 12).[6] Carnot's analysis had been hampered by the fact that he couldn't track the amounts of heat transferred during individual sections of the cycle. Clausius got round this difficulty by considering an infinitesimal cycle—a cycle with sides approximating to straight lines, and therefore resembling a tiny parallelogram (imagine the cycle in Fig. 12.2 scaled right down). He could then work out the heat taken up along the isothermal sections (by using Mariotte's Law), and the heat taken up along the (approximately vertical constant-volume) adiabatic sections (by using data on C_V).[7] Also, the work done by the cycle, W, was given by the area of the parallelogram.[8] Armed with all this data, Clausius found—as he had expected—that Carnot's heat-conservation axiom was not upheld: but he found another conserved quantity.

Before this is explained, we must understand that Clausius' investigation differed from Carnot's in two ways: he considered an infinitesimal cycle (as we have just described); but he also considered cycles from one 'state of the system' to *another* 'state of the system' (the cycle was not necessarily a complete cycle). The two states of the system (say, positions ik and gh in Fig. 12.2) could be connected by compounding lots of tiny cycles together. Moreover, these tiny cycles could be added together in different ways, but still connecting the same two states (for example, one could start at ik, expand halfway along the top isotherm, then drop down (adiabatically) to the bottom isotherm, and continue the expansion on to gh). With infinitesimal cycles, there was evidently an infinity of combinations leading between the same two states.

However, what Clausius found was that between any pair of start and final states, the total heat taken up, ΔQ, plus the total work done, ΔW, *together* summed to a *constant* quantity, ΔU:

$$\Delta U = \Delta Q + \Delta W \qquad (16.2)$$

(Note that, in this equation, Clausius assumes that ΔQ and ΔW are both in new 'mechanical units'—he later called them 'ergons'.) This was to become the First Law of Thermodynamics—but what was U?

It was clear to Clausius that ΔQ and ΔW were not separately conserved and, even worse, they were dependent on the actual route taken from the start to the final state. But Clausius saw that the *change* in U, that is to say, ΔU, *was* route-independent, and was constant between any two end-states. Moreover, ΔU could be expressed as a function of the parameters that defined those end-states. Also, if it so happened that the start and final states were the same (the system returned to its start state), then Equation 16.2 still applied,* and ΔU was a function of the parameters that defined this one state. Clausius' big conclusion: U was 'a function of state', and was determined solely by the properties intrinsic to that state. U was, in fact, what would later be called the 'internal energy' of the state. It was given this designation soon afterwards, by Thomson, but Clausius was the first to truly understand that more than just interconversions were at stake—a new abstraction, *energy*, was required.[9]

To summarize, Clausius' 'First Principle of Heat', or the First Law of Thermodynamics, went beyond Mayer's and Joule's heat⇄work conversions, and posited a new entity—the total internal energy of a system. For an isolated system, or for a state returned to by whatever route, the total internal energy was a constant and, unlike ΔQ or ΔW, could be expressed as a function of the state parameters, V and T, or P and T, and so on.[10]

---

* ΔU was then constant and equal to zero.

The Second Law of Thermodynamics

After this establishment of the First Law came Clausius' great resolution of Thomson's problem, the conflict between Carnot and Joule. For work to be generated, Joule required the *consumption* of heat and Carnot required the *transfer* of heat—but Clausius saw that it wasn't a case of either/or—both requirements could be met. Clausius realized that the amount of work depended on the *proportional split* between these two processes (the heat consumed and the heat transferred). Also, the value of this proportional split depended only on the initial and final temperatures.[11] He wrote, in his famous work of 1850, 'On the moving force of heat and the laws regarding the nature of heat itself which are deducible therefrom':[12]

it may very well be the case that at the same time [as work is produced] a certain quantity of heat is consumed and another quantity [of heat is] transferred from a hotter to a colder body, and [that] both [these] quantities of heat stand in a definite relation to the work that is done.

In order to determine quantitatively what this 'definite relation' was, the extremum case had to be examined; in other words, Clausius had to look at the *maximum* work done by an *ideal* heat-engine. He therefore had to consider Carnot's ideal engines, and so he again analysed infinitesimal Carnot cycles, now incorporating the First Law: he then compared his findings with the calculations of Clapeyron and of Thomson, and using the extensive data of Regnault. His results agreed with these earlier analyses tolerably well,[13] and they corroborated Carnot's and Joule's suspicion that the ideal efficiency was temperature-dependent, and was an increasing function of $1/T$ (see Chapter 12).

But how did the familiar form of the Second Law come out of this? It is a testimony to Carnot that the Second Law arrived, not from any calculations or comparisons with data, but from another run-through of Carnot's original syllogistic argument (see Chapter 12 again). This time, the argument (Clausius' version) employed a subtly different starting premise from Carnot.

Clausius imagined one heat-engine (let's call it 'superior') that could do more work than an ideal heat-engine, or, 'what comes to the same thing',[14] that it could do the same amount of work for a smaller amount of *transferred* heat (the heat *transferred* to the lower temperature, as opposed to heat *converted* into work). Running the ideal engine in reverse, and following it with the 'superior' engine going forward, the work consumed and the work done would exactly balance out, but there would also be some residual heat—heat transferred to the higher temperature. The net result would be a 'combined engine' that did no work, but transmitted heat from the cold reservoir to the hot reservoir.

Clausius didn't like this result—and so he ruled out the possibility of a 'superior' engine. But why didn't he like this result? It didn't contravene the First Law, as the net work done was zero, and all the heat was conserved; all that had happened was that the heat was *distributed* differently, transferred from the lower to the higher temperature. Clausius' objection was simply that it didn't tie in with common experience—heat *doesn't* flow from cold to hot of its own accord, in other words, without the input of work.

In the same way that Carnot had ruled out the 'superior' engine as it would permit a perpetual source of motive power, Clausius vetoed his version of the 'superior' engine as it would allow heat to flow from cold to hot, unaided. In other words, both men advanced physics by appealing to age-old wisdom and common experience: perpetually acting machines, and heat flowing 'uphill', don't happen—the world just isn't like that. This statement, of Clausius', that heat cannot flow from a low temperature to a higher temperature unless aided by work, was the first appearance of the Second Law of Thermodynamics (Clausius' 'Second Principle of Heat'[15]).

Thomson immediately saw the logic of Clausius' (beautiful) argument, and realized that his own conflict had been resolved. Also, he had at last (during 1850) come to accept Joule's results and, crucially, to accept the dynamical theory of heat. In 1851, Thomson published a paper swiftly after Clausius' paper of 1850, acknowledging Clausius' priority but putting Clausius' statement into a more precise form.

Statements of the Second Law

First, here is Clausius' version:

It is impossible for a self-acting machine, unaided by external agency, to convey heat from one body to another at a higher temperature.[16]

Thomson's version is as follows:

It is impossible, by means of inanimate material agency [in other words, with no man-made device employed to do work], to derive mechanical effect from any portion of matter by cooling it below the temperature of the coldest of the surrounding objects.[17]

Thomson had unknowingly addressed the query in Carnot's posthumous notes (see Chapter 12) as to whether 'motive power could be created . . . by mere destruction of the heat of bodies'.[18] According to Thomson, this doesn't happen: left to themselves (i.e. without an input of work), it never happens that if the ocean cools, or if the atmosphere cools, then the heat so liberated is converted into work. Either of these statements—that due to Clausius or that due to Thomson—constitutes a statement of the Second Law of Thermodynamics. Thomson stated (and it can easily be verified) that the two versions are completely equivalent to each other: the violation of one implies the violation of the other.

Finally, in 1854, Thomson put the Second Law into a more quantitative form.[19] From his absolute temperature scale (defined in this same paper—a paper mostly on thermoelectricity), he already knew that Q_T was proportional to T (referring, as always, to the operations of ideal heat-engines). Then, in a moment of profound perception, Thomson realized that this was tantamount to a mathematical expression of the Second Law. Simplifying the notation by writing Q_T as Q, and adopting the convention that heat taken out of a reservoir is positive, and heat put into a reservoir is negative, then however complex the cycle (however many different temperature reservoirs are involved), the sum of Q/T for a complete ideal cycle (that is, a complete reversible cycle) must always come to zero:

Thomson's mathematical version is:

$$\sum(Q/T) = 0 \text{ (complete reversible cycle)} \qquad (16.3)$$

Cosmic Generalizations

So far, the Second Law applied only to heat-engines. It was Thomson who, seemingly out of the blue, extended the canvas of the Second Law to . . . all of nature.

How this happened is more or less as follows (there is some conjecture in this). Implicit in the Second Law is the assumption that heat *does* flow spontaneously from hot to cold. Thomson saw that such allowed heat-flows not only occurred readily in nature, but gave an overall direction to natural processes: hot things cool down; heat flows from the hot end of a bar to the other end; heat flows from the Sun to the Earth; and so on. Thomson then saw a commonality between these examples and other examples that had nothing overtly to do with heat but were nevertheless directional: the swinging plumb line slowing down in water, planets slowing down over aeons, frictional losses in a mechanical device, and so on. The unifying feature for all these multifarious phenomena, on both terrestrial and planetary scales, was a loss of useful 'mechanical effect', whether actual (the plumb line stops swinging) or potential (the work that didn't happen during the conduction of heat). Another unifying feature was that all these processes gave a cosmic direction to time—they were all irreversible. Now, what were the links between heat flowing, loss of 'mechanical effect', and irreversibility?

By 1851 Thomson had understood it all. In the first place, Carnot's Law only applied to ideal *reversible* processes, but heat conduction was clearly *ir*reversible. Second, Carnot only imposed a maximum and not a minimum to the amount of work generated (the minimum might be zero, as in the case of heat conduction). Finally, Thomson saw that energy conservation—which he had at last accepted—was not contradicted by irreversible processes (as, according to the dynamic theory of heat, bulk motion lost was made up for by an increase in the microscopic motions). Some 'mechanical effect' (a bulk phenomenon) *was* lost to mankind, but the total energy (including bulk effects *and* microscopic effects) was not lost.

Some common irreversible processes were heat conduction, friction, and the absorption of light. All were 'dissipative'. This was a word from the Victorian lexicon (Thomson's father had apparently told him cautionary tales about idle young men dissipating their talents[20]). It was *dissipation* that seemed to Thomson to lie behind all the disparate phenomena. Then, in 1852, all these ideas came together in three bold declarations, stated abruptly at the end of a brief paper entitled 'On a universal tendency in nature to the dissipation of mechanical energy':[21]

There is at present in the material world a universal tendency to the dissipation of mechanical energy . . .

Any *restoration* of mechanical energy, without more than an equivalent of dissipation, is impossible . . .

Within a finite period of time past, the earth must have been, and within a finite period of time to come the earth must again be, unfit for the habitation of man . . .

It is hard to be sufficiently impressed by the boldness and sweep of these assertions. We cannot easily un-know what now seems so 'obvious' and well-attested—that cosmic time has a direction. We shall discuss at the end of the chapter the impact of Thomson's words, and the validity of such generalizations. We shall find that dissipation wasn't the end of the story—it was still not the clinching characteristic that dictated the direction of time. The next leap forward in understanding came with Clausius.

Entropy

Clausius' achievements were outstanding: he had realized the necessity of a new abstraction (energy), resolved Thomson's conflict between Joule and Carnot, discovered the Second Law, checked up that Carnot's ideal engine efficiency did indeed vary as $1/T$, and made great strides into the kinetic theory (in 1857; see Chapter 17). All this would have been enough to guarantee his place as a founder of thermodynamics—but he pushed back the frontiers even further (in the 1850s and 1860s), striving to get to the very heart of the Second Law.

Energy conservation ensured that the 'energy books' balanced—that for every bit of heat that was consumed an equivalent amount of work was generated, and vice versa. But, Clausius wondered, did the proportional split between the heat transferred and the heat consumed imply the existence of *another* abstract quantity, and *another* conservation law?

As usual, enlightenment would only come from quantification, and quantification would only come by looking at the extremal case—the case of the ideal heat-engine. We have already heard how Clausius saw that there were basically just two kinds of 'transformations': (1) the conversion of heat into work; and (2) the transfer of heat. Clausius wondered whether there was an *equivalence* between these two transformations—after all, they both concerned heat and they always occurred together (in the Carnot cycle). Heat did not flow spontaneously from cold to hot, but when this process occurred *non*-spontaneously—say, in a reversed ideal engine—it was because it was always *compensated* for by a simultaneous conversion of work into heat. Presumably, there was some 'equivalence value' for each type of transformation that guaranteed this compensation.

What could the quantitative form of the 'equivalence value' be? Clausius had a remarkable ability to see past distractions and through to the interior logic, to see the general in the particular. First, he saw that Carnot's quintessential heat-engine involved, most generally, *three* temperatures and not just two. In the most general case,[22] heat was transferred between temperatures T_x and T_y, and work was generated from heat at *yet another* temperature, T_z. Second, he saw that the transfer of heat was really the same as two other processes: (1) a conversion of heat into work at the higher temperature, followed by (2) a conversion from work back into heat at the lower temperature. In other words, *all* transformations were of just *one* type, heat⇄work, and it was not necessary to ask whether the work was of the 'shifting weights' or the 'shifting heat' variety. Lastly, even this one type of transformation, heat⇄work, was really only an example of heat added or subtracted from a heat reservoir, *never mind about the work*. Moreover, because heat and work were quantitatively connected (by the First Law), then all the heat magnitudes were taken care of—they would all come out in the wash, so to speak.

The money-laundering analogy can be followed a little way. We have seen in Chapter 14 that the 'currency exchange rate' (Joule's mechanical equivalent of heat) is fixed even though the proportions may change (the *bureau de change* may see a long queue and, to avoid running out of notes, change only 20% of all the money proffered). Now, we are further finding that the money (the heat) bears no trace of its origins, whether stolen or clean, whether obtained from heat transferred or from work. But now we must ditch this analogy. In money transactions, the money that can be withdrawn from a bank is not related systematically to that bank. However, in physics, the 'value' of the heat *is* dependent on the temperature of the heat-bank: the higher the temperature, the less 'valuable' is a given quantity of heat.

Carnot had defined the worth of heat in terms of how much work it could do. Now, Clausius was seeking to generalize this notion: in essence, his 'equivalence value' was another, more abstract, measure of the 'worth of heat'. The 'equivalence value' that Clausius was looking for had to have the mathematical form $Q/f(T)$, as it depended only on the heat, Q, and on some increasing function,[23] f, of the absolute temperature, T. In fact, as $f(T)$ is always increasing, it might just as well be *defined* to be *equal to* T. Thus, Clausius finally arrived at the following: for every addition or subtraction of heat, Q, at temperature, T, the equivalence value is Q/T. (Clausius was undoubtedly influenced by Thomson's mathematical formulation of the Second Law; see Equation 16.3) There was only one thing left to do, and that was to specify a convention for the sign of Q. Clausius decided that wherever a process was natural—that is to say, spontaneous—then it should be given a positive sign. Therefore, *adding* heat *to* 'the surroundings' (*to* a heat reservoir) should be positive, and *withdrawing* heat *from* a reservoir must be negative.

Clausius' original aim had been to show that something is being conserved. In going around a complete[24] Carnot cycle, with an arbitrary number of different-temperature heat reservoirs, the system is brought back to its initial state, and *nothing has changed*. Clausius interpreted this as meaning that the total equivalence-value was conserved; or, in other words, the sum of the equivalence-values at each temperature reservoir, T, around the whole cycle, must sum to zero. Generalizing to infinitesimal transfers of heat, dQ, Clausius finally arrived at[25]

$$\int dQ/T = 0 \text{ (complete cycle, ideal transfers)} \quad (16.4)$$

This sum, or integral, was plainly the conserved quantity that Clausius had been searching for. He labelled it S (some say, to commemorate Sadi) and, in his paper of 1865, gave it the name entropy:

$$S = \int dQ/T = 0 \text{ (complete cycle, ideal transfers)} \quad (16.5)$$

He especially constructed a word that sounded similar to 'energy', and was likewise rooted in classicism:

I hold it better to borrow terms for important magnitudes from the ancient languages so that they may be adopted unchanged in all modern languages, I propose to call the magnitude S, the *entropy* of the body, from the Greek word τροπη, transformation.[26]

To recapitulate, Clausius was claiming that, rather than caloric, it was energy that was conserved, and entropy that was conserved. This was an outstanding advance, but an advance that was somewhat barren—it only applied to ideal engines, not real engines. It was Clausius alone among his contemporaries who saw how to make the result universal and applicable to everyday life. For real engines, a quantitative result could still be teased out: ideal engines used the ideal—that is to say, the minimum—amount of heat for a given amount of work done; real engines always used *more* than the minimum amount of heat—so, for real engines, the summation over dQ/T was always *greater* than zero. This was not something that needed to be mathematically proven; it was just what was evident by looking around, observing everyday phenomena: therefore for real processes, the entropy changes did not scatter equally on both sides of zero—they were always clumped together on *one* side of zero:

$$S = \int dQ/T > 0 \text{ (complete cycle, real transfers)} \quad (16.6)$$

For complete cycles with some mixture of ideal and real processes, (in other words, combining Equations 16.5 and 16.6), Clausius finally arrived at:

$$S = \int dQ/T \geq 0 \text{ (complete cycle, real and ideal)} \qquad (16.7)$$

But what actually was entropy? Consideration of this question caused Clausius such difficulties that he delayed publication of his work for over a decade (the relevant papers came out in 1854, then 1862, and 1865). The definition of entropy had been arrived at by a paring away of elements until there was nothing left but Q and T. Seemingly by 'sleight of physics', but actually by profound thinking, the quantitative relations between heat, temperature, and entropy had emerged. Nonetheless, Clausius realized that in order to physically understand entropy, he had to put the specificity back in. *What* work had been done? Was the volume greater? Had weights been raised? And *what* were the microscopic changes? Had atoms rearranged[†] themselves? Were molecules moving faster, on average? Clausius saw that any or all of the following might happen when heat was added to a system: an increase in 'internal heat'; an overall change in physical dimensions (usually an expansion); some other changes in structure (melting, evaporation, dissolving, chemical reactions, etc.). An increase in 'internal heat' corresponded to an increase in temperature, while all the other effects amounted to changes in molecular arrangement ('disgregation'[27]). These changes in molecular arrangement were all examples of 'internal work', which sometimes showed up as real (macroscopic) work—the raising of weights and so on. 'Disgregation' implied that entropy was an *extensive* quantity—it had extension, such as length, breadth, mass, or volume.

One final discovery was that S, like U, was a 'function of state': the change in entropy between specific states was route-independent (the net ΔS was the same for any ideal processes connecting the same end-states). Neither U nor S could be determined absolutely, but the *differences*, ΔU or ΔS, *could* be determined.[28]

Clausius finished off this paper (in 1865), like Thomson's paper of 1852, with some sweeping cosmic assertions:[29]

The energy of the universe is a constant; E = constant (16.8a)

The entropy of the universe tends to a maximum; $\Delta S \geq 0$ (16.8b)

[†] For example, due to chemicals reacting, ice melting, water boiling, and so on.

Overview

Two new concepts have emerged, energy and entropy; and two universal laws, the first governing the conservation of energy in all its forms, and the second concerning the distribution of energy, specifically, thermal energy. (That Equation 16.8b has something to do with the *distribution* of energy will be explained later in Chapter 18.)

It is impossible to say who contributed more out of Thomson and Clausius. Thomson was more influential in introducing 'energy' to the whole of physics, whereas Clausius' legacy was in the foundations of thermodynamics. To Clausius goes the honour of the discovery of entropy, and of the realization that the subtle fluid, caloric, had to be replaced by a subtle concept, energy.

From the late 1840s onwards, the torch of discovery was passed back and forth between Thomson and Clausius, the work of each one crucially shaping the work of the other. This happened through their publications—there is no record that they ever met. They each had one other physicist closer to hand (Rankine and Helmholtz, respectively), but Clausius barely gave Helmholtz's work a mention—it was on energy in all its forms rather than on thermodynamics—and Rankine, although he made important contributions, was criticized for not keeping separate his new theories in thermodynamics and his speculative molecular theories.

There was, nevertheless, a difference in style between the British (Thomson, Rankine, and Joule) and the Germans (Clausius, Helmholtz, and Mayer). The former appealed to religious convictions (there were many references to 'the Creator' and quotes from the Bible), while the latter invariably couched their theories in neutral terms. However, for all, the theories themselves were ultimately judged by purely *physical* aesthetic criteria.

It is something of a puzzle as to why Thomson was the last to finally come around to accepting energy, the more so as Joule was his close friend and associate. The following must all have had something to do with it. The enormous success of the caloric theory, and the fudge-factor of latent heat, continued to operate right into the middle of the nineteenth century. The works of Clapeyron, Thomson, and Regnault, within the caloric theory, produced a wealth of data—data that was not immediately overturned by Clausius' almost identical analysis using Carnot cycles. (Clausius' analysis, of course, had the signal difference that he invoked energy, and the First Law.) Also, the very fact of the Second Law made the First Law harder to spot (the energy 'books' didn't always exactly balance in actual experiments, and heat→work conversions were not so evident as work→heat conversions). Thomson had been won over by Fourier at an early age. Thomson's biographers[30] make the interesting point that Fourier's heat-conduction theory and Thomson's theories of electricity bore many parallels. During a reversed Carnot cycle, heat could be 'raised to a higher state': in the same way, work could be done to move electricity to a 'higher state' (a higher electrical potential); electricity was conserved in this process and so, Thomson presumed, heat was conserved in like process.

Although Thomson was the last to accept energy, he then proceeded apace, with all the fervour of the recent convert. It was he who introduced the terms 'energy' and 'thermodynamics' into physics (in 1851), and then expanded 'energy' to cover all applications, not merely the interaction of heat and work. In 1852, he adopted Rankine's terms 'potential' energy and 'actual' energy, and subsequently (1862), substituted the term 'kinetic' for 'actual' energy. With fellow Scot, Peter Tait (1831–1901), he wrote what was to become the standard textbook in physics, *Treatise in Natural Philosophy*,[31] and so launched what his biographers refer to as the 'epoch of energy'.[32] Despite this, Thomson stuck with an essentially force-based Newtonian outlook, as opposed to endorsing Hamilton's mechanics, where energy takes centre stage.[33] By contrast, Thomson (and Tait) started a tradition in which it was crucial to determine the 'work function'—all the ways in which work could be done in the given system.

Thomson was also emerging as an establishment figure. He amassed a personal fortune as a result of his work developing the first transatlantic telegraphic cable, and was a rising star in the Victorian scientific firmament. Clausius was more reticent. It had been completely out of character for him to put forward his cosmic generalizations (Equations 16.8a and b)—one wonders whether he felt compelled to compete or was simply inspired by Thomson's forthright style and public persona. Their relationship seems to have been one of mutual respect but no particular warmth: Clausius thoroughly approved of Thomson's introduction of the word 'energy' into physics,[34] but Clausius and Thomson were in opposing camps as regards the priority dispute between Mayer and Joule. Clausius also had to contend with his concept of entropy being (wilfully?) misunderstood by Tait (see Chapter 17, 'Interactions between the Physicists').

The expression $S \geq 0$ is the first time that an inequality comes into physics (some later examples are Heisenberg's Uncertainty Principle, and the limiting speed of light; see Chapter 17). This, along with Thomson's assertions on dissipation, introduces an overall direction to time. It is hard to overstate the impact of this on physics. Up until Thomson and Clausius, the legacy of Newton's 'clockwork mechanics' was that the universe was cyclical, and might continue on forever in the same form. (The question of whether it had always existed in the past was, in Europe, answered in the negative, for religious reasons.) Even when Laplace, in his *Celestial Mechanics*, found that frictional and tidal forces would make the planets and their moons gradually acquire circular orbits, he nevertheless assumed that once this had been achieved, the orbits would continue unchanged forever. The cooling of bodies was, as we have seen (in Chapters 6 and 11), well known to Black and to Fourier, but their emphasis was on achieving equilibrium rather than on direction. Water found its own level, but who noticed the connection between this and the fact that rivers always flow *down*hill? Fourier's heat-conduction equation did have a direction for time built into it, but it seems that Fourier never noticed this. The waste and inefficiency of engines was known about and lamented by engineers the world over. Also, it was known that there was a loss of motion in inelastic collisions, and the flight of projectiles. There was also the conversion of wood into ash and gases in burning; the disappearance of salt and sugar in dissolving; the heating of electrical wires in a circuit; the ageing of people and of things; the cooling of bodies; and so on. But, before the work of Thomson and Clausius (and, of course, Carnot), these various phenomena were totally disparate phenomena, not connected with each other or to any other body of theory.

One puzzle with thermodynamics is: what have engines and their efficiency—such anthropic things—got to do with physics? The extension of the empirical base from heat-engines to *everything* probably happened as follows. First, Thomson and Clausius realized that any real mechanism is like a heat-engine of sorts—some heat is always thrown out; there are always heat losses. Subsequently, it was understood that the losses are not always *heat* losses (Dennis' blocks don't just cool down; they wear down as well, and they also get dispersed—cf. Feynman's allegory). This loss or dissipation is always asymmetric in time (for example, the blocks wear down, not up). In summary, it is 'dissipation' rather than 'cooling' that is the more general process (see also 2, Appendix III).

But nature is subtle, and if one looks even more closely, one can find processes that do have a direction in time but don't involve cooling or even dissipation (in the sense of things getting degraded or wasted). One such is that puzzling phenomenon looked at in Chapter 10, the 'free expansion' of a gas. Here, 'free' means that the ideal gas does no work as it expands. The gas follows a given isothermal curve (Fig. 10.4a), but always in the direction towards larger volumes and lower pressures. No work is done, no heat transfers take place, and *nothing is lost, or getting more disorganized*. What is driving this progression? Clausius was able to show that as the expansion proceeds, the average molecular separation increases, and so the 'disgregation' and hence the entropy increases. But why does this happen; why must entropy increase? This can't be answered until a fully microscopic and, above all, a statistical approach is adopted. All is explained in Chapters 17 ('Kinetic Theory') and 18 ('Difficult Things').

Microscopic explanations were also crucial to the acceptance of energy and the First Law of Thermodynamics. For example, they explained where the energy had gone to when the paddle wheel, plumb line, or projectile had slowed down and come to rest. (Incidentally, if Thomson could have looked even more closely at his 'paradoxical' plumb line, he would have found that it never comes completely to rest, but has a tiny residual *thermal* motion; see Chapter 18.)

The laws of thermodynamics stand in relation to Regnault's experimental work in the same way as Kepler's theories of planetary motion stand in relation to Tycho Brahe's measurements. However, in saying that the Second Law is empirically based, we mean that it stemmed from the wealth of 'common experience' as much as from detailed measurements. Newton's Second Law of Motion, by contrast, is not primarily empirical but arises by the introduction and definition of new concepts—mass and force. The law of the conservation of energy appears as, perhaps, an intermediary case, based on both common experience and ideology—perpetually acting machines don't exist and can't exist.

The Second Law, in the original statements of Clausius and Thomson, was strangely wordy, not like usual laws of physics (the Second Law was put in a mathematical form by Carathéodory in the twentieth century; see Chapter 18). What did contemporary physicists, and society in general, make of it? Thomson's ideas of dissipation, and of a future Earth not fit for the habitation of man, could have engendered a deep despondency, or been as shocking as Copernicus' moving Earth, Kepler's non-circular orbits, or Galileo's discovery of a pitted lunar surface. We mention the opinions of the nineteenth-century physicists in Chapter 17, but it seems that there was very little response from the public[35]— the ideas were simply too big.

At this stage (the first decade or so after mid-century), energy had really 'come in' to physics. By the end of the century, there were even those, the German 'energeticists'—Mach (see Chapter 3, 'Newton') and the physical chemist Ostwald (1853–1932)—who advocated that energy was *the* important concept, superseding force, mass, and even atoms. This therefore ends our historical chronology of the discovery of energy. In the next chapter, we try to condense the whole of the rest of physics having special relevance to energy. We also look at some interactions between the nineteenth-century physicists, and end with a very brief history of how 'energy' came into the public domain.

17

A Forward Look

Electromagnetism and Other Kinds of Energy-in-the-Field

Apart from 'energy', there was one other outstanding development in physics in the nineteenth century. This was Maxwell's theory of electromagnetism, the first field theory (James Clerk Maxwell, 1831–79). The theory emerged in the late 1860s, a decade or so after 'energy'.

The idea of the field was first put forward by that preternaturally gifted experimentalist, Michael Faraday (see Chapter 15). Faraday noticed that tiny iron filings near a bar magnet moved at *right angles* to the force, and aligned themselves along *curved* lines in space, lines that sometimes went round closed loops, not starting or ending on a source—all shockingly un-Newtonian. Faraday's first thought was that the 'atoms' making up the material of the bar magnet must be orienting themselves in special ways (note that atoms were still conjecture), but he dismissed this idea when he found that rotating the magnet about its axis left the pattern of 'field lines' unaltered. Faraday's daring conclusion: there was a 'field' right there in the void.

Maxwell was also daring: he took Faraday's suggestion at face value, and tried to make his mathematical theory fit in with Faraday's detailed and numerous experimental findings. Thus was born Maxwell's towering achievement, the theory of electromagnetism, which combined the previously disparate effects known as 'electricity' and 'magnetism'. Central to the theory was the idea that these effects existed in the field.

What is a field? It is a way of describing a quantity that, instead of being in a clump of such and such dimensions, is seeded throughout space, including both empty space and the interior of objects. If you consider a field of grass, then you will find a blade here and a blade there, with earth in between. But for a field in physics there is no in between, and the 'grassiness' is found everywhere. This doesn't mean that the amount of 'grassiness' is a constant—it can depend on position and on time, and is therefore to be specified at each point in space and time, (x, y, z, t).

What does all this have to do with energy? The connection is through the fact that the electromagnetic field contains energy. As Maxwell wrote:

In speaking of the Energy of the field . . . I wish to be understood literally . . . The only question is, Where does it reside? On the old theories it resides in the electrified bodies, conducting circuits, and magnets, in the form of an unknown quality called potential energy, or the power of producing certain effects at a distance. On our theory it resides in the electromagnetic field, in the space surrounding the electrified and magnetic bodies, as well as in those bodies themselves.[1]

The field solved the controversy of action-at-a-distance ('action'* was transmitted from the source to a neighbouring bit of field, and then to the next neighbouring bit of field, and so on) but the price was that empty space was now filled with a mysterious invisible mathematical construction—the field itself. However, there is no longer any controversy: the field does exist. For example, things warm up in the sunshine (even though the Sun is separated from Earth by 150 million kilometres of mostly empty space); and fluorescent tubes glow when placed near power lines, even when the tubes are not plugged in.

As extra confirmation, Maxwell's theory predicted that an oscillating electrical disturbance (or field) generates an oscillating magnetic disturbance, and vice versa. The result is an electromagnetic wave that can move through empty space (cf. the aforementioned sunshine), leaving the source charges far behind. Such waves have been (and are continually being) detected; for example, radio waves, microwaves, visible light, UV, and so on. Now, as each kind of disturbance generates the other kind, it is crucial that they don't get out of step, or else the wave might be cancelled out, or grow infinitely large. It turns out (as Maxwell discovered to his surprise) that in order for the electric and magnetic disturbances to keep in step, they must both travel at the same, extremely fast, constant speed—the 'speed of light'. So it came about that different predictions from totally different areas of physics beautifully confirmed one another in one consistent whole body of theory—and there is therefore even less doubt that electric and magnetic fields exist, electromagnetic waves exist, and all contain energy.[2]

---

* A loose term, not the same as the 'action' in Chapters 7 and 13.

There was still a problem with exactly where the energy resides (cf. Maxwell's quote, above). For example, there is the paradox that the energy associated with a point charge is infinite;[3] and in certain experimental configurations (e.g. a bar magnet with an electric charge nearby) the energy must perpetually circulate (in order for angular momentum to be conserved). Also, it turns out that in order to make Maxwell's theory of electromagnetism consistent with Einstein's theory of Special Relativity (see below), the conservation of energy must be qualified by the more stringent requirement that energy is conserved *locally*. The electromagnetic energy in a given small volume (the energy density) must then be balanced against the energy flowing into or out of this volume (the flux of energy). This energy flux was defined by Poynting (1852–1914) in 1884, and explained the transfer of electromagnetic energy from one place to another through space. (One way to finally resolve the location problem for electromagnetic field energy might be to locate the gravitational pull of this energy[4]—but experiments of such fineness have never been done.)

When it came to circuit electricity, Maxwell's field theory brought in an interesting new feature: mere topology was not enough—the actual configuration in space of wires, magnets, capacitors, and so on made a difference. (Readers will have experience of draping wires in different ways, or changing the angle of a radio, in order to improve reception.) Not just space, but the 'configuration in *time*' made a difference as well (for example, the speed of moving a magnet, the speed of opening a switch, playing the theremin, and so on).

Obviously there has to be consistency between the abstract field energies and the less abstract circuit energies (quantities such as VIt, $\frac{1}{2}CV^2$, $\frac{1}{2}LI^2$, etc.*—some of these formulae were discovered by Joule and by Faraday). The field energies are given at each and every point in space and time, whereas the circuit energies are tallies—measures of how much work is done in moving electric charge 'uphill' through the circuit. However, there is consistency between the two approaches, as the circuit components (a wire, a capacitor, an electromagnet, etc.) contain spatial information implicitly (for example, the C might refer to an arrangement of parallel plates, an electromagnet is made from a wire that coils in space, and so on). As an analogy, consider an irrigation system where we want to know the total amount of water. We can employ a

---

* V, I, C, L, and t are the voltage, current, capacitance, inductance, and time, respectively.

moisture-detector and track down every tiny volume-element of water throughout space; or we can add the number of standard components of known volume: 'length of pipe', 'holding tank', and 'cistern'.

There were other field theories apart from electromagnetism. In the early nineteenth century there was Green's potential function, $V(r)$ (see Chapter 13), a proto–field theory approach for 'force fields' such as gravity, electrostatics, and magnetostatics. It was used mostly by mathematicians (Green, Poisson, Gauss, and others), whereas 'work [F.dr] was for the workers'[5] (engineers such as Coriolis, Navier, and Poncelet). Hooke's Law of elasticity applied to a spring extended by a weight—but what about forces and materials that were extensive and continuous? Engineers realized that energy could be stored in a 'continuum' (throughout the extended volume of a material), and so the energy density had to specified point by point, like a field property, and depended on the type of material, and whether it was stretched, squashed, or twisted, and whether electric or other external forces were applied. Augustin-Louis Cauchy (1789–1857) made a crucial advance in constructing a mathematical object,[6] the stress energy tensor, that could store the information* in just the right way, with mathematical transformation properties built in (a bit like a modern-day computer spreadsheet). Finally, in Einstein's Theory of General Relativity (1915) (see below), it was proposed that gravity exists as a field. (This also needed a stress energy tensor, showing a family connection between the theory of continuum mechanics and Einstein's Theory of Gravitation.) Even more profound was Einstein's realization that matter could *emerge* from the field, and that there was essentially no qualitative difference between matter and field: 'Matter is where the concentration of energy is great, field where the concentration of energy is small'.[7] Today, there are modern field theories (for example, quantum field theory) and, as always, there is energy in the field.

Magnetic Energy

In breaking open an electrical circuit, the magnetic field surges and there may even be a spark—so there is unquestionably energy in magnetism. Other examples include the energy needed to separate two strong magnets, and the energy needed to power an electromagnet.

---

* The energy density at each point.

In Maxwell's equations, it is shown that magnetic effects arise from electric charges in motion. This is easy to demonstrate in the case of circuits (just watch a compass needle as it is brought near a current-carrying wire), but what of permanent magnets such as a static lump of 'lodestone' (magnetic iron oxide)? Here, also, modern theories and careful experiments show that the magnetism derives from the ever-changing distribution of electrical charge at the atomic and nuclear levels.

Now, if we have two, parallel, current-carrying wires, with the currents in the same direction, then the magnetic field associated with each is such that the wires attract each other (an effect discovered by Ampère; see Chapters 13 and 15). But if we could 'sit' on an electric charge in the current, there would be no motion, and the magnetic fields would disappear—but would the wires still be attracted to each other? Yes, but now because of electric rather than magnetic attraction (a fuller explanation of a similar scenario is given in Chapter 18). It turns out that in order to have equivalence between moving and stationary frames of reference, magnetism must be brought in as a 'correction' to electricity (a correction needed to satisfy the axioms of Einstein's theory of Special Relativity). So, all those magnetic phenomena, such as the properties of lodestone, well-known to the Ancients, are to be considered as 'relativistic corrections'. This is true even where the speeds involved are very small, far less than the speed of light (e.g. a typical drift speed for electrons in a current is 0.0001 m s^{-1}).

From Oersted's discovery of a link between electricity and magnetism (see Chapter 9) to Faraday's electrical rotations (that wonderful experiment shown in Fig. 15.1), there was a quest to find the parallels between electricity and magnetism. However, what the experiments and Maxwell's theory all show is that electricity and magnetism, while connected in various ways, are not completely symmetrical: magnetism is a much weaker effect—for example, we get electric shocks rather than magnetic shocks; and there are no isolated magnetic 'charges'. That is why we say that magnetism, rather than electricity, is a 'correction'. The reason why we see any magnetic effects at all is that occasionally the very strong electric effects are completely cancelled out, leaving the weak magnetic effects to be noticeable.

There are some 'paradoxes' associated with magnetic energy. One is the fact that no work is done by a magnetic force acting on a moving electric charge—it takes no energy for an electric charge to circle around a magnetic field line. This is because the displacement of the charge as it moves is perpendicular to the force (and both are perpendicular to the magnetic field). Another 'paradox' arises in circuits. Suppose we have a circuit with two inductances (solenoids) in it, carrying currents I_1 and I_2, respectively, and with a mutual inductance, M_{12}. It turns out that the work required to change the separation in space of the inductances is $-I_1 I_2 \Delta M_{12}$ in one analysis, and $+I_1 I_2 \Delta M_{12}$ in another. However, in the first analysis, the energy needed to keep the currents at their constant values, I_1 and I_2, has not been included, whereas in the second analysis it has been included. The lesson of this is that, where energy is concerned, it is crucial to make sure that the boundary of the system is unambiguously defined. One other curiosity is Lenz's Law (see also the discussion of Joule's electromagnetic engine in Chapter 14): induced currents are always in a direction such as to produce magnetic fields that *oppose* the pre-existing magnetic field. This is impossible to explain except by appeal to the deeper principle, the Principle of Least Action (see Chapter 7, Part IV). Finally, we can ask whether magnetic energy is kinetic or potential. As it arises from *moving* electrical charges, we are inclined to call it kinetic, but the *arrangement* of tiny 'bar magnets' (e.g. the dipole moments of nuclei) in a magnetic field is a kind of potential energy: so, as we have mentioned before (see Chapter 7), and shall mention again (see Chapter 18), these two forms of energy are not necessarily antithetical to each other.

Kinetic Theory and Statistical Energy

The kinetic theory explains the gross properties of matter from the motions of the individual atoms.* Feynman says the theory is 'extremely difficult',[8] and this must be part of the reason why it had such a chequered early history, being discovered and lost, rediscovered and lost again, many times over: the mathematics was too hard, and the arguments too subtle, for it to win a ready acceptance—and also, atoms hadn't been discovered yet.

The kinetic theory is very important for energy, as it explains what heat is at the microscopic level, the link between heat and work (see the discussion on Thomson's struggles in Chapter 16), and the way in which energy is distributed when it is shared between a very large number of tiny moving components. We shall try to side-step the maths but still give you the gist of the arguments, which are too good to miss.

---

* Or molecules, or ions, or other microscopic particle in the given model.

So far, the discoverers we have discussed are Hermann (1716), Daniel Bernoulli (1738), Cavendish (around 1787), Herapath (1816), and Waterston (1845). None of them hesitated to apply Newton's Laws in the atomic regime. Daniel Bernoulli considered a gas made from a swarm of 'corpuscles' (atoms or molecules) residing in a container fitted with a piston (Fig. 5.2). He saw that an elastic atomic collision with a wall of the container resulted in the atom reversing its direction, thereby pushing the wall with a certain tiny force. Summing over all the atom–wall collisions that would occur in a given time, he managed to prove a macroscopic outcome: Boyle's Law, $P \propto 1/V$ (where P and V are the gas pressure and gas volume, respectively; and the temperature, T, is held constant). He further speculated that if the gas was heated, then the atomic speed would increase, and P would be proportional to the *square* of the speed.* But what was the magnitude of this speed? Herapath was the first to estimate it by considering the speed of sound in a gas; and Joule, later, estimated the speed of hydrogen molecules at 60 °F as 6,225 ft s^{-1} (around 2 km s^{-1}; in excellent agreement with today's value). Waterston made some outstanding advances, but unfortunately none were picked up in his lifetime (see Chapter 14). (Note that, right up to the mid-nineteenth century, the prevailing view of a gas was still one of *static* repulsion between molecules.)

Then Clausius, in 1857, and with no knowledge of these precursors, derived his own kinetic theory (we have skipped over Krönig in 1856). Clausius tried to account for the collisions between the gas molecules themselves, and not just collisions against the container walls. The chief innovation he introduced was the concept of the 'mean free path'. This is the average distance that a molecule travels in a straight line in between collisions. In air at standard pressure and temperature this is miniscule, of the order of 10^{-7} m. With the aid of this concept, Clausius could answer the criticism made by the Dutch physicist Buys-Ballot as to why gas molecules moving at the rapid† rate of, say, 0.5 km s^{-1}, nevertheless cause only a slow diffusion of the smell of dinner across the dining room—the smell molecules travel fast but *not very far*.

Now, all the early kinetic theorists had described macroscopic phenomena (such as the Ideal Gas Laws) by working their way up from the

---

* Bernoulli saw that this was because a speed increase causes both (1) an increase in the intensity of an impact and (2) an increase in the number of impacts (in a given time interval).

† Faster than a supersonic airplane.

actual atomic or molecular collisions. Could this bottom-up approach be continued indefinitely; in other words, could they track the molecules individually, work out what collisions they would have, what velocities they would end up with, what collisions they would have next, and so on? Absolutely not, and for reasons of the scale of the undertaking. For example, there are around 10^{27} gas molecules in my study, and it would take more ink, paper, and time (or more computer power, computer memory, and time) than exist in the whole universe just to write down the starting conditions. Why didn't this hinder the early kinetic theorists? Because, in their 'simplistic' analysis (for example, that of Daniel Bernoulli), all the gas particles had the *same velocity*. Therefore, whether dealing with $n = 6$, 60, or 6.024×10^{23} (where n is the number of particles per unit volume), it was only necessary to rescale n. It is almost certain that Daniel Bernoulli and the others did appreciate that the particles would not all have the same velocity (Bernoulli describes the 'corpuscles' as moving 'hither and thither', so he realized that the directions, at least, were all over the place), but how could these theorists proceed? They were stymied by the impossibly large amount of input data. It was Maxwell, that towering genius of physics, who resolved the impasse.

Maxwell understood (in his kinetic theory of gases, 1867) that the mechanism for the smearing of molecular velocities arose from the near-infinite number of collisions, each involving very tiny changes in speed and direction. These redistributions weren't random at the level of two or three molecules but, en masse, they could be treated *as if* they were random. This meant that the spread ('distribution') of velocities was *probabilistic*, and so the 'error law' could be used. Gillispie[9] suggests that Maxwell was influenced by Quetelet's *Treatise on Man*[10] via the work of John Herschel. Adolphe Quetelet (1796–1874) was a Belgian astronomer and social scientist (in today's terms), who applied the 'law of errors' to human characteristics (he looked at the distribution of height, weight, etc., within a population). Prior to this, the error law had only ever been applied to measurement errors and to gambling games; Quetelet's and then Maxwell's application of it to the actual physical things themselves constituted an enormous advance. As to the use of probability, Maxwell felt that he had to justify this depravity:

[probability calculus], of which we usually assume that it refers only to gambling, dicing and betting, and should therefore be wholly immoral, is the only *mathematics for practical people*, which we should be.[11]

But this 'top-down' approach meant that it was essential for Maxwell to make various starting assumptions. The main assumption was the one of randomness, later known as the assumption of 'molecular chaos'[12]: that the starting speeds were independent of each other,* and that the individual miniscule changes in velocity were all *equally probable*. An unusually fast (or slow) molecule would arise from a succession of collisions, each causing a tiny increment (decrement) in speed; and the probability of this high final speed would be the same as the product of the probabilities for each tiny increment. This is exactly the same rule as for errors of measurement (the probability of a succession of independent tiny errors is the product of the probabilities for each tiny error); and this is why Maxwell's distribution of speeds in a gas is qualitatively the same as the distribution of random errors (a bell-shaped curve or 'normal distribution'). Not just the overall velocity, but each of the three velocity components *independently* had to follow the error law, an assumption that Maxwell conceded 'may appear precarious'.[13]

This is all very well, but is any of it true? Nowadays, the distribution of velocities can be checked by experiment; for example, by injecting a puff of gas through a collimating slit and into a drum-shaped chamber, with a detecting screen on its inner surface. The drum is made to rotate at a steady pace, and so its horizontal axis is an indication of the time of flight of the gas molecules. Maxwell's distribution has been amply confirmed.† However, the new probabilistic methods often led to very counterintuitive outcomes. For example, the kinetic theory predicted that the viscosity of a gas was independent of its density. Maxwell was so surprised that, aided by his wife, he performed experiments to check up on this prediction— it was corroborated.

The Austrian physicist Ludwig Boltzmann (1844–1906) took Maxwell's distribution as his starting point[14] (1872), but he took several important steps beyond Maxwell.[15] Boltzmann contemplated:

(1) a distribution of *energy* rather than velocity;

(2) inelastic collisions; and

(3) external potentials.

---

* But still satisfying the conservation of total energy.
† When Maxwell's assumptions apply: we are talking of an 'ideal' gas at one temperature.

We'll take each of these in turn:

(1) Maxwell's choice of 'velocity' as the parameter was perfectly correct—*if* the gas molecules were all of one type. However, as soon as the possibility of mixtures was allowed—say, O_2 and N_2 molecules mixed together—then it was essential to consider a distribution of *energy* rather than velocity. Now, in Chapter 10, Part I we stated that the average kinetic energy is the same irrespective of the molecule type. This assertion is supported by an ingenious argument in Feynman's Lectures on Physics.[16] Feynman shows that the multitude of elastic, two-body, molecular collisions that occur in a gas serve no other 'purpose' than to randomize the molecules' directions, and, furthermore, this randomizing can *only* happen if there is an equality between the average kinetic *energies* of the colliding molecules.

(2) Boltzmann not only switched from velocity to energy, but to energy of a more generalized kind: translations through space, yes (as occur in purely elastic collisions), but also internal vibrations and rotations of molecules. Consider as an analogy for a molecule, an ideal spring.[17] It is well known that an oscillating spring will, on average through time, have exactly the same amount of kinetic* and potential energy. Also, the spring doesn't care whether it is squashed up because of interconversion between its internal energies or because of a collision with another spring. Boltzmann had the profound intuition that after a near-infinite number of collisions, all the *rates* of energy transfer would equalize,[18] whatever the kind of energy transfer involved (translational or internal), and therefore the actual individual *amount* of each energy transfer would also (eventually) *equalize*. This would mean that, at equilibrium, the actual amount of energy tied up in each 'thermal degree of freedom'† would be equal—the famous 'equipartition theorem'. (This is explained again more fully below.)

(3) The equipartition theorem applies even when there is a mixture of types of molecules—say, O_2 and N_2. Employing the spring analogy once more, changing the mass (by going to a different type of molecule) is equivalent to changing the stiffness of the spring; but this is

* We are talking here about its *internal* kinetic energy, not its kinetic energy due to the whole spring being translated from one place to another in space.

† Independent mode of motion: translation in one of three independent directions, internal vibrational kinetic energy, and internal rotational kinetic energy.

equivalent to introducing an external potential (e.g. a spring may be more stretchy because of a change in stiffness *or* because of a change in the stretching force). Now, the crucial fact is that the stiffness of a spring does not affect the *rate* of energy transfer, and so it will not affect the inexorable process of 'thermalization'. The upshot is that the equipartition theorem must apply to different types of molecule, and to external potentials, as well as to the cases discussed in (1) and (2). In short, it applies to *all* kinds of thermalized energy, whatever the molecule types, and whether the energy is internal or external in origin.[19]

The resulting spread in molecular energies is known as the Maxwell–Boltzmann distribution, and is shown at three representative temperatures in Fig. 17.1. There is a peak near the middle of the curve, at the mean energy.[20] A large molecular energy, far from the mean, is very unlikely, as the molecule would have to be pushed from the left—say, by a fast molecule—then pushed from the left again, then pushed from the left again, in quick succession. Likewise, an extremely sluggish molecule is very unlikely to remain so for long. Therefore, all extremes of energy are uncommon—but by the same token, there is also the opposite trend, a trend *towards* uniformity. This is because the rate of energy transfer is optimized when the donating and receiving energies are very similar[21]—therefore, as equilibrium is approached, any extreme—that is, any excess (or lack)—in a given molecule's energy tends to be reduced (or increased). This is at the heart of the equipartition of energy. We could put it yet another way: the more probable a given energy transfer is, the more frequently it will occur; and the more frequently it occurs, the more probable it becomes.

Boltzmann was at first (1866) motivated to explain the Second Law of Thermodynamics and Clausius' entropy principle purely mechanically (i.e. without need of probability theory). He was able to show that introducing an external potential (that is, doing work on the system) resulted in a reshuffling of the microscopic thermal energies. (This was encouraging: Clausius had specifically invented entropy in order to explain the redistribution of heat consequent upon the performance of work; see Chapter 16.) Then, in 1872, in his celebrated '*H* theorem', Boltzmann found a functional form, *H*, of Maxwell's distribution, and noticed that *H* was always minimized after any change to the system.* Boltzmann immediately identified '–*H*' with Clausius' entropy, *S*—both were maximized—but not everyone was convinced, and the question of *S* was still controversial.

---

* This *H* has nothing to do with the Hamiltonian function.

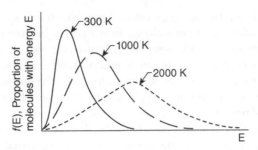

Fig. 17.1 The Maxwell–Boltzmann distribution of energies in a gas.

Then, in his long paper of 1877, Boltzmann took two more important steps. (We are already impressed by Boltzmann's advances (1), (2),
and (3), but the next two innovations were to guarantee Boltzmann im-
mortality in the history of science.) First, 'as an aid to calculation',[22]
he considered the 'fiction' of *discrete* energy *levels*, each with the same
probability of being occupied. Second, Boltzmann introduced a radically new approach, although it slipped in without fanfare: he switched
from looking at *one molecule at all times* to looking at *all the molecules
at one time*. In this snapshot view (the second approach), he considered
the fraction of the total number of molecules that occupied each energy
level. In other words, the solution of the dynamical equations of motion
for each of, say, 10^{24} molecules was assumed to be completely equivalent to the *non*-dynamical question 'how many permutations of 10^{24}
molecule energy-assignments will yield a certain energy profile?' (This is
exactly the same switch in strategy that we adopt when we try to predict
the number of heads after tossing coins: considering, say, a million coins,
we can either solve a million equations of motion, or we can assume that
each coin has an equal chance of landing heads or tails, and then use
Pascal's laws of probability.)

In this way, Boltzmann solved the problem of the distribution of molecular energies—by reducing it to a problem of combinatorial analysis
(assuming that the 'architecture' of the energy levels for the given system
was already known). The counting of combinations is a standard procedure in statistics and leads to each energy profile having a certain 'statistical weight', log W. This is how Boltzmann's famous equation entered
physics:

$$S = k \log W \qquad (17.1)$$

where S is the entropy, W is the statistical weight, k is a constant of proportionality (Boltzmann's constant), and 'log' is the natural logarithm.
This is the equation on Boltzmann's tombstone, but he didn't write it
exactly like this, and he never had a constant called k—that was introduced later by his student, Max Planck (see the next section).

Boltzmann had no hesitation in identifying his microscopic formulation with Clausius' S—both tended towards a maximum value for real
processes, and both resulted in a final condition in which the energy
was distributed—'dissipated'—as uniformly as possible. This was momentous; Boltzmann's statistical approach supplied an *explanation* of the
Second Law of Thermodynamics.

Physicists of today may reel and gasp in appreciation, but Boltzmann's
contemporaries were slower to understand what he had accomplished.
First, we must remember that molecules had still not been detected.
Second, that macroscopic irreversibility—that is, a preferred direction
in time—could come out of *reversible* molecular collisions was, and still
is, deeply puzzling. It was hotly debated by Boltzmann and his contemporaries (see below, 'Interactions between the Physicists') and we will
discuss it further in Chapter 18. For now, we return to energy.

The relevance of Boltzmann's work to the new concept of energy was
likewise profound. He recognized that, in statistical physics, energy was
a more telling parameter than velocity. In fact, his equipartition theorem showed that energy is IT, *the* paramount parameter. It is a curious
fact, but sometimes microscopic processes have aspects that are *easier* to
understand than processes on everyday scales—and Boltzmann's use of
discrete energies is one such example. At equilibrium, tiny discrete dollops of energy are constantly being redistributed by a hive of random
activity (billions of collisions per second for the 'air' molecules in my
study, for example), blind as to whether the energy is doled out as rotational, translational, or vibrational; blind also to different types of gas;
blind even to the huge mass of a macroscopic object such as a 'grain of
pollen' or 'a whole piston'.[23] The crucial thing is that, at equilibrium, the
size of the average tiny dollop of energy is *exactly the same*, whatever the
degree of freedom, and depends on nothing except the temperature—in
fact, it *is* the temperature (explained further in Chapter 18).

Boltzmann realized that his statistical approach could apply to more
than just gases, but it was an American, J. Willard Gibbs (1839–1903),
who vastly extended it to other thermodynamical systems (a substance
undergoing a phase change, heterogeneous mixtures, and other chemical, elastic, surface, electromagnetic, and electrochemical phenomena).

Gibbs is therefore rightly called the founder of statistical mechanics and of physical chemistry. In his monumental work of 1878, *On the Equilibrium of Heterogeneous Substances*,[24] Gibbs applied three special techniques, including two adapted from earlier eras: (1) the idea of an 'ensemble'; (2) the use of 'phase space' (growing out of Hamilton's (p,q) space; see Chapter 13); and (3) the 'variational calculus'—as, at equilibrium, infinitesimal variations in energy (with respect to the overall energy distribution) had to sum to zero—reminiscent of Johann Bernoulli's principle of virtual work from over 150 years earlier (see Chapter 7). As ever, the new concept of energy was indispensable.

Radiant Energy

Problems with the classical theory of heat occurred at both high frequencies (the so-called 'ultraviolet catastrophe') and low temperatures (the anomalous specific heats). The first problem concerned radiation from 'black bodies', and was resolved by Boltzmann's student Planck (1858–1947) in 1900. In 'an act of desperation',[25] he was forced to adopt a radical heuristic—the energy of radiation must come in discrete chunks or quanta. It was almost certainly Boltzmann's discovery of discrete energies in a gas that gave Planck the permission to take this unpalatable step—a bit like allowing running only at certain speeds, or stair treads only at certain heights.

Planck's new theory had the radiation energy given by

$$E = h\nu \qquad (17.2)$$

where ν is the frequency of the radiation and h is a constant, now known as Planck's constant. This corrected the anomalies—but it is remarkable to think that it made any difference at all, and to everyday phenomena, considering that the spacing of the energy intervals was miniscule (h is around 6.6×10^{-34} Js). Planck had, of course, started off the quantum revolution. 'Small', 'smaller', and 'smallest' are all relative terms, but when the relevant physical quantities are on the same tiny scale as h, then we know that this is *absolutely* small, the domain of a new mechanics called quantum mechanics.

When light is shone on a metal surface, then electrons may be ejected. This phenomenon, called the photoelectric effect, was explained by Einstein in 1905 (it is the work that gained him his Nobel prize, in 1921). Einstein noticed that the energy of the ejected electrons depended not on the intensity but, rather, on the frequency (the 'colour') of the light. If the light had too low a frequency, no electrons were emitted, no matter how intense the light was. Einstein realized that a certain minimum amount of energy was required to liberate an electron from the metal's surface, and no slow accumulation of low-frequency radiation would ever be enough.[26] He concluded that the light's energy must be arriving in discrete packages or quanta, and nothing would happen (no 'photoelectrons' would be ejected) unless the quantum had at least as much energy as the requisite energy threshold for the given metal.

But what was the quantitative relationship between the quantum of radiation energy and the frequency of that radiation? For radiation with a frequency above the threshold, experiments showed that the *frequency* (of the light) and the *energy* (of the ejected electron) increased *in proportion* to each other. Einstein immediately recognized Planck's relation, $E = h\nu$, occurring again, and saw that in this new setting it now applied to absorbed and transported radiation, as well as to 'black body' radiation. In other words, the radiation, or light itself, was quantized—that is, particulate. (The light particle was later called the photon.)

If light waves could be considered as particles, then maybe other particles could be considered as waves. This deep idea was suggested by de Broglie in 1924 and experimentally confirmed for electrons by Davisson and Germer in 1927 (cf. Chapter 13). It was found that the wavelength, λ (of the wave), was inversely proportional to the momentum, p (of the particle), and satisfied de Broglie's relationship, $\lambda = h/p$. Here is Planck's constant cropping up again. Now, can a 1 kg bowling ball travelling at 1 ms^{-1} be considered as a wave? Yes, but its wavelength would be a mere 6.6×10^{-34} m long. (It is therefore not surprising that most quantum effects do not show up on everyday scales.)

Anomalous Specific Heats

The trouble with the specific heats (the capacity of a given substance to store energy at a given temperature) was also resolved by the newly emerging quantum theory. Dulong and Petit had found that the specific heat per atom was a constant for different solids (around 25 J per mole of atoms per degree; see Chapter 10). However for diamond, and some other hard solids, the specific heat was found to decrease continuously with temperature rather than remain constant. Einstein resolved this, in 1907, by adopting Planck's quanta of energy yet again (the resolution

involves quantum statistics and cannot be explained simply). Quantiza-
tion also explained other anomalies such as the discontinuous decrease
in the specific heat of liquefied gases: at very low temperatures, certain
levels were 'frozen out'—the quantum of energy wasn't enough for these
levels to be reached. Finally, certain polyatomic gases (later dubbed
'greenhouse gases') were found to have a much larger specific heat capacity than monatomic gases—due to the much greater number of thermal degrees of freedom (avenues through which heat could be absorbed,
such as internal rotations and vibrations) in these polyatomic cases. So,
as commented upon in Chapter 10, the specific heat was proving to be
a macroscopic window into microscopic structure.

The Quantum World

We have commented earlier that quantum mechanics is not just ordinary mechanics scaled down (atoms instead of marbles), but is *absolutely*
different. Here is an example. If we wish to analyse the motion of
marbles, then we'll look at many collisions and may even take some
video footage: essentially, we are bouncing photons off marbles and
into our eyes or cameras. But in the case of atoms and fundamental
particles, how do we 'see' them? Can we watch a photon by bouncing
another photon off it? We find that, in the quantum domain, the very
act of observing or measuring something affects that observation or
measurement. The German physicist Werner Heisenberg (1901–76)
discovered that there was a fundamental limitation to the precision
with which we could simultaneously measure certain properties relating to *one* quantum-mechanical particle. He found that this limitation
is inherent, a matter of principle: there is no technological change,
now or in the future, that can ever improve the joint precision of two
specific parameters. What are these two parameters? They are the pair-
ings (p,q), or, equivalently, (E,t), for the given particle (p and q are,
respectively, the particle's momentum and position; E and t are, respectively, the particle's energy and time). The limit of joint precision
is tiny, around a twelfth of Planck's constant (where Planck's constant,
as we remember, is extremely small). There is nothing uncertain about
it, but the relationship came to be known as Heisenberg's Uncertainty
Principle (1927):

$$\Delta p\Delta q \geq \hbar/2 \quad \text{or, equivalently,} \quad \Delta E\Delta t \geq \hbar/2 \qquad (17.3)$$

(where $\hbar$ is $h/(2\pi)$). Energy conservation could thus be contravened to the extent ΔE as long as this loan was paid back within time Δt. For example, one Joule could be 'borrowed' for around 0.5×10^{-34} seconds.[27]

Erwin Schrödinger (1887–1961) saw that Hamilton's mechanics (see Chapter 13), with its 'optico-mechanical' analogy, would be ideally suited to coping with de Broglie's wave–particle duality. He therefore adapted Hamilton's mechanics for use in the quantum world, and so derived the famous Schrödinger equation. One important difference between the two methods, however, was that Hamilton's mechanics led to a range of 'possibilities' for many particles, whereas Schrödinger's mechanics led to a spread of 'probabilities' for just one particle. Hamilton's 'range of possibilities' (streamlines in phase space) was very useful in that overall properties (symmetries and conservation theorems) provided an essential physical handle (see Chapter 13). Likewise, Schrödinger's new wave mechanics (and the other versions of wave mechanics that succeeded it) yielded conservation and symmetry relations for the given problem.

Hamilton's methods were especially useful in quantum mechanics, as almost all the problems are too hard to solve exactly. The best line of attack is to delimit the problem, to make sure that the relevant conservation and symmetry relations are satisfied, and then to use some appropriate method of approximation. As ever, a guiding principle is the conservation of energy. (One earlier theory—due to 'BKS'[28]—had to be discarded, in part because energy was only conserved *on average*.) Also, the 'Hamiltonian' (in effect, a potential energy function) is as crucial in quantum mechanics as it was in classical mechanics.

Richard Feynman, in 1948,[29] developed Hamilton's methods even further: he calculated the 'action'—the integral over the entire path in time (or 'history')—for all possible 'histories' of the quantum-mechanical particle. This eventually led him on to the theory known as 'qed' (quantum electrodynamics), which is the most precisely confirmed theory in all of physics, and amply demonstrates the applicability of the concept of energy on the tiniest of scales.

Special and General Relativity

The laws of thermodynamics were very influential in all of Einstein's thinking. He wrote 'classical thermodynamics . . . is the only physical theory . . . [that] will never be overthrown'.[30] In the same way as thermodynamics was a 'theory of principle',[31] so were Einstein's Theories of Special Relativity (SR) and General Relativity (GR).

A founding axiom of SR was that the laws of physics had to be cast in a way that ensured that different, valid, viewpoints ('frames of reference' or 'observers') were equivalent, and ultimately indistinguishable. This was Einstein's Principle of Relativity.[32] The other founding axiom was that the speed of light in a vacuum was a constant for all such observers, whether they were moving towards the light, away from the light, or staying still.

Obviously, this would entail a huge upset to the old mechanics. Speed was still defined as 'distance/time', but if speed (the speed of light, c) was now the new invariant, then distance and time intervals would have to relinquish this status.[33] But this upset would not only affect space and time; it would *necessarily* have *dynamic* implications—in other words, there would be consequences for our notions of mass and energy as well.[34] It turns out that not only must energy be conserved locally (as mentioned in the section on 'Electromagnetism', at the start of this chapter), but energy and momentum must be conserved *together*, and are melded into a continuum known as 'momenergy'. (So energy is not, in general, invariant between different frames of reference; it is, rather, the new quantity, $E^2 - p^2$, which is invariant (E and p are the total energy and momentum, using units in which c is taken as '1').) Thus the eighteenth-century fight for supremacy between Newton's momentum and Leibniz's 'energy', lasting almost 100 years, is finally resolved—both are indispensable.

We have saved the most famous until last. A requirement of satisfying Einstein's Principle of Relativity is that the speed of light, c, should not only be a constant, but it must never be surpassed. How this happens is that bodies become more massive as they get faster (i.e. as their kinetic energy increases), and at just such a rate that c can never be reached. In other words, there is a new feature, a *mass-dependence of kinetic energy*. The mass,[35] m, must increase in proportion to the energy, E. Everyone knows that the constant of proportionality is c^2:

$$E = mc^2 \qquad\qquad (17.4)$$

Einstein considered that this was the most profound result to come out of SR.[36] No longer was there Kant's divide between 'stuff' (mass or matter) and 'activity' (energy) (see Chapter 15). There had been earlier suspicions of this mass–energy equivalence (Thomson, Poincaré, and Hasenöhrl), but Einstein was the first to boldly postulate that it was universal, applying to *all* types of energy (not just kinetic but also heat, radiant, electrical, chemical, and so on) and *all* types of mass (gravitational, inertial, and any type of matter—feathers, cannon balls, etc.). This was not just aesthetically more beautiful—it *had* to be so: Einstein's Principle of Relativity applied to *all* physics (heat, optics, chemical reactions, etc.) and not just to mechanics; and the conservation of energy applied to energy in *all* its forms. Thus, from the twin requirements of the Principle of Relativity and the conservation of energy, Einstein had shown not merely that changes in energy and mass were equivalent, $\Delta E = \Delta mc^2$, but that *energy and mass were the same thing*, $E = mc^2$. (One neat consequence of this was that, as Einstein observed: 'the conservation of mass is now a special case of the conservation of energy'.[37])

If the kinetic energy is reduced to zero (i.e. the body is in its own 'rest-frame'), then the stationary body still has some mass 'left over'. This, the rest mass, m_0, is the usual mass[38] with which we are all familiar—the mass that may be weighed on kitchen scales, and so on. Einstein postulated that this leftover mass could be converted into energy even when the *kinetic* energy has gone to zero. This was not required by SR—the rest mass might have been indestructible. Later, Feynman showed that, in fact, the conversion of rest mass into energy *had* to be allowed, as it occurred in the annihilation of matter with antimatter.[39]

It has always seemed mysterious how the kinetic energy of a particle can change in value from one frame of reference to another. What we have just seen, however, is that there is a natural zero-point for kinetic energy—it is the rest energy, m_0c^2. What is still curious is that this zero-point energy is vastly larger than the commonplace kinetic energy.[40] (For example, the rest energy in a gram of matter is 90 trillion Joules, about the same as the energy released in the bomb at Nagasaki.) How is it that we are normally (thankfully) completely oblivious to this vast zero-point energy? It is simply due to the fact that this store of energy is, for the most part, remarkably stable[41] (see also Chapter 18).

Special Relativity describes the interactions of particles, light, and electromagnetism, while General Relativity brings in gravity—the effects due to a very large mass, such as the Sun or the Earth. New theories often arise in order to explain new data, but Einstein's theory of General Relativity (his theory of gravitation), beginning in 1915, was not motivated by the need to explain any experimental discrepancies. In fact, it is remarkable to think how good the agreement is between Newtonian and Einsteinian gravity (for falling apples, the orbit of the Moon, etc.) when we consider how utterly different their theories are. While

Newton has gravity as a force between two bodies, no matter how far apart they are, Einstein introduces a radical change: gravity is a field (cf. the beginning of this chapter), expressed as a 'curving' of spacetime, and then this curvature influences nearby masses.* Now, from SR, we have seen how all types of energy have mass—and so energy too should be affected by gravity. There soon came experimental confirmation of this; for example, rays of starlight were found to be bent by the Sun's gravity, showing that light is indeed 'massy'. Also, atomic clocks were found to run slower in a gravitational field, as if the clock 'spring' was heavier. To put this another way: anything weighty should be affected by gravity (that much was still true), but . . . everything *is* affected by gravity, and therefore everything—light, heat, and even inertial mass[42]—does have weight.

There is yet an even more radical prediction of GR. It is that not only does mass cause a curvature in spacetime, but a curvature in spacetime can lead to mass:† 'mass is an outgrowth of the field',[43] wrote Einstein. So, to recapitulate, in going from Newtonian mechanics to SR, we found that energy and 'stuff' are really the same sort of thing; but now, in going from SR to GR, we are finding that energy, 'stuff', *and curvature of spacetime* are really the same sort of thing. To repeat our quote of Einstein's (from the start of this chapter): 'Matter is where the concentration of energy is great, field [is] where the concentration of energy is small'.[44]

However, there is still a paradox as regards the exact location of this gravitational mass-energy, this energy-of-the-field. Electromagnetic energy explodes to infinity when confined to a point charge; likewise, there can be 'singularities' in the gravitational field (the Big Bang and black holes). On the other hand, gravitational field energy disappears altogether as soon as one looks too close—as, everywhere and 'everywhen', spacetime is *locally* flat (excluding the singularities).

Finally, on current theories (2015), there is another location problem—apparently, around 95% of the energy is missing. Everything we have ever seen or observed, on Earth, and in space, makes up a mere 5% of the total: the remaining 95% is dark energy and dark matter, and is still a mystery.

---

* To help visualize this, there is the often-used analogy of a trampoline surface distorted by a large mass, say, a cannonball, and then this surface affects the motion of small particles—say, ball-bearings.

† The trampoline analogy fails.

Chemical Energy

It is interesting to reflect that the whole of chemistry is due to just one type of energy—electrical energy.[45] For example, consider the chemical bond between a pair of atoms. The atoms are electrically neutral, and the bond can occur in three ways: (1) electrons can be donated from one atom to the other, and so the resulting 'ions' will be oppositely charged and will attract one another—this is known as 'ionic bonding'; (2) electrons can be shared between the atoms (as in 'covalent bonds'); and (3) there is something intermediate between the above examples (as in 'hydrogen bonds' in water). In all these cases, it is the *arrangement* of electric charge that is determinative—and so all the bond energies are a type of potential energy.

In a chemical reaction, total energy must be conserved, and if the energy required to make and break bonds doesn't balance, then some other forms of energy (heat, light, work, sound, etc.) make up the deficit. The enormous utility of chemistry to humankind is that not only can we make new substances, but we can make use of these various forms of energy: we can burn wood to generate heat; convert chemical energy into work (e.g. use an internal combustion engine, or fuel cell, etc., to make a massive object move from place to place); use the chemical energy stored in a battery to drive a current through a circuit; and so on. The 'reaction equation' takes care of the stoichiometry (the balance of particle types), but there is so much more to actual processes than just this equation. In reality, we are talking of trillions upon trillions of interacting atoms, and so we are inevitably dealing with questions of the statistical distribution of energy quanta, and therefore with entropy. Total energy must be conserved, but also entropy must increase. The beautiful way in which the two viewpoints are merged (the reaction equation for a 'handful' of atoms, and the reality of, say, 10^{30} atoms or molecules) is via the techniques of thermodynamics: the 'functions of state', such as the internal energy, U, and the entropy, S; the special condition known as 'equilibrium'; the universal temperature, T; and other thermodynamic parameters such as P, V, and so on. The simple thermodynamic relations that result belie the subtlety and difficulty of this subject (see Chapter 18).

A last word about the 'chemical potential functions' (so-called because they relate to the energy tied up in chemical bonds): as just stated, the entropy must always increase (in practice), and so we can mathematically set aside some energy* to ensure that this increase in entropy is accounted for. The energy leftover is then the energy that is free to do work—the 'free energy'. Instead of maximizing entropy, we can then talk equivalently of minimizing this 'free energy', and it may even be enlightening to think of 'free energy', and (the energy, $T\Delta S$, implicated in) entropy, as being in competition with each other.[46] The utility of reformulating the problem in this way is that there are widely recognized categories of experiment in which the experimental conditions conform to a certain type, and therefore these experiments can all be modelled in one way. For example, consider a 'hydrostatic' system with just two thermodynamic degrees of freedom—the pressure, P, and the volume, V. The Gibbs 'free energy', G, is defined as $G = U - TS + PV$, and is minimized (in a real process). It is a useful formulation in conditions of constant pressure and constant temperature, and many chemical reactions occur in just these conditions (e.g. when mixing reactants together in an open vessel, the pressure is effectively constant, and heat may be lost to the 'infinite' surroundings and so the temperature also remains constant). Moreover, if the system 'falls in free energy', then we know that the reaction will occur spontaneously. The Helmholtz 'free energy', A, is defined as $A = U - TS$, and during a change it has the differential form $dA = SdT - PdV$ (where $dU = TdS - PdV$). So, if heat is added in conditions of constant volume, then the change in A with T is a measure of the entropy: the Helmholtz energy is therefore particularly important in statistical mechanics. Finally, the enthalpy, H, defined as $H = U + PV$, is a useful energy function as it allows us to predict how much heat (sometimes 'latent heat') will be produced or consumed when a process occurs at constant pressure—a common condition in combustion reactions, vaporization, and melting.

Nuclear Energy

The nucleus is a tiny object ($\sim 10^{-15}$ m), about as big a reduction in the size of the atom ($\sim 10^{-10}$ m) as the atom is from a hair's breadth ($\sim 10^{-5}$ m). It is made up from nucleons: neutrons and protons. Now the protons are positively charged and so the nucleus would explode if it wasn't for the strong nuclear force which overcomes the repulsive electric force and binds the nucleons together. The more nucleons,

---

* The energy 'cost' of a change in entropy, ΔS, occurring at the temperature T, is $T\Delta S$.

the more strongly they are bound, until the most stable nucleus of all, iron (^{56}Fe), is reached (and there are some exceptionally strongly bound nuclei, such as ^{4}He, ^{12}C, and ^{16}O, along the way). After iron, the average binding energy per nucleon starts to gently decrease with increasing nuclear size. What is happening is that the repulsive electric force between protons is starting to win out over the much stronger nuclear force because the latter is short-ranged, and becomes comparatively feeble at the edges of large nuclei. Eventually, for nuclei larger than uranium, the strong force loses the battle, and such nuclei are unstable.*

There is still one more nuclear force, known as the weak force (it is orders of magnitude weaker than the strong or electromagnetic forces). Its effect is to modulate the number of neutrons and protons in the nucleus—it allows protons to change into neutrons, and neutrons to change into protons. Up until iron, the numbers of neutrons and protons are roughly equal, whereas after iron, the proportion of neutrons steadily increases. All nuclei are, in a sense, trying to reach the most energetically favourable states at the top of the 'hill' (Fig. 17.2). The heavier nuclei decay into smaller chunks by a process known as fission. This occurs naturally (in radioactivity) and artificially (in nuclear reactors and bombs), and much energy is released in the process (as heat, kinetic energy of the fragments, light, and sound, etc.). On the other hand, light nuclei (such as hydrogen, deuterium, tritium, helium, lithium, and carbon) can 'fuse' together to form heavier and heavier nuclei, a process that occurs naturally in stars, and also releases much energy (our Sun is powered by the fusion of hydrogen into helium). There is much hydrogen to be found in the oceans, and the by-product of fusion (helium) is wonderfully innocuous, and so artificially induced fusion would be a panacea for our energy needs. Much ingenuity and effort has been put into using fusion as a source of power (beginning in the late 1950s), and while the technical challenge of containing a tiny bit of the Sun on Earth has largely been met, still the experiments are a net *consumer* of energy.

The information in this section is summarized in Fig. 17.2.

A word about rest-mass energy: the rest-mass energy of a nucleus is less than the combined rest-mass energies of the constituent nucleons (the

* Strictly speaking, uranium itself is unstable, but it all depends on what you count as stable (the half-life of ^{238}U is around 4.5 billion years).

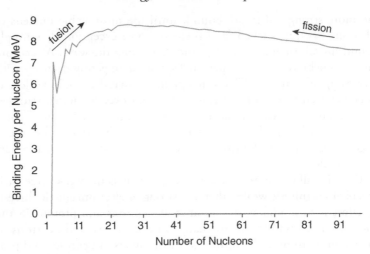

Fig. 17.2 The nuclear binding energy per nucleon against the number of nucleons in the nucleus.

difference is called the 'binding energy' of the nucleus). In an 'atomic' bomb, it is the nuclear binding energy that gets converted into destructive energy—there has not been a conversion of nucleon rest mass into energy as the nucleons remain intact.

A word about forces: the word 'force' is used to denote different qualitative domains (electromagnetic, strong nuclear, weak nuclear, and gravity); but all quantitative calculations and theories in modern physics involve *energy* rather than force.

Interactions between the Physicists: the Response to Big Ideas

As we said at the end of Chapter 16, by the middle of the nineteenth century, 'energy' had arrived in physics, yet it still took many years before all the implications, both cosmic and everyday, were appreciated. The following excerpts provide a fascinating glimpse into the thought processes of the physicists.

In 1880 a young American physicist, Henry Rowland, measured the mechanical equivalent of heat using Joule's paddle-wheel method (see

Chapter 14), and obtained a value slightly higher than Joule's. Joule wrote to Thomson:

I don't think my result can be wrong . . . I have received his [Rowland's] thermometer and shall as soon as possible compare it with mine . . . the fineness of the markings is much inferior to mine.[47]

Rowland had carried out Joule's paddle-wheel method over a range of temperatures up to 37 °C. The mechanical equivalent fell, but this was actually due to the fact that the specific heat capacity of water was not a constant over this temperature range. In other words, Joule's method could be used to determine the specific heat of water *assuming* 'energy', and its conversion and conservation. Physics had come full circle. There was still yet another hidden assumption, though—was the strength of gravity, for example between Manchester and Baltimore, a constant? Joule suggested repeating the experiment at Greenwich:

the latitude of Greenwich is impartial in just skirting the south of Ireland and the north of France and also the north of Austria; or at least it used to do so before the Devil with his wars displaced the countries—and I don't buy fresh maps every year.[48]

Despite the political correctness of Greenwich, the point is that no *absolute* determination can ever be found without taking something as a standard. The French mathematician and philosopher Poincaré (1854–1912) took this to its logical conclusion (see Chapter 18, 'Difficult Things'). He saw that the very generality of the principle of energy conservation meant that it had to be assumed true, or physics would fall apart.

The three Scots, Thomson, Tait and Maxwell, formed a cosy trio, communicating by postcard and referring to each other by the nicknames T, T', and dP/dt, respectively. Thomson and Tait were the spearheads of a crusade—to bring the new doctrine of energy and energy conservation into natural philosophy. They wrote a textbook, *Treatise on Natural Philosophy*,[49] published in 1867, and commonly referred to as T& T', which set the programme for classical physics in the nineteenth century. However, Tait was shamefully keen to appropriate the whole of physics to British, and above all Newtonian, origins. There is a risible episode in which he searched through the whole of Newton's *Principia*, saying that it (energy) had got to be in there somewhere.[50] ('It' wasn't: as we have shown in Chapters 3 and 7, Newton missed discovering energy.)

Tait totally misunderstood Clausius' concept of entropy, calling it the 'available energy', while Thomson was conspicuously silent. Maxwell relied on Tait to keep him up to date with the latest in science, and so managed to write a whole book (*Theory of Heat*[51]) misinformed about entropy, and not mentioning Clausius as its discoverer. Clausius protested, but it wasn't until Gibbs took up Clausius' cause that anything happened, and Maxwell 'was led to recant'[52] and to acknowledge Clausius' priority, writing: 'I mean to take such draughts of Clausiustical Ergon [work] as to place me in . . . [a] state of disgregation'.[53] Tyndall, translator of Clausius' works into English, was another physicist who explained to T, T', and dP/dt that Clausius was a 'genuine and good fellow'.[54] (Tyndall had also been active in promoting Mayer's cause; see Chapter 14.)

Maxwell and Boltzmann's separate work on the kinetic theory led to the Maxwell–Boltzmann distribution, but they never met or even corresponded. Maxwell respected Boltzmann's ideas, but found his work easier to follow by *not* reading Boltzmann's long papers, confiding to Tait '[I could] put the whole business in about six lines'.[55] Boltzmann, meanwhile, revered Maxwell, saying of Maxwell's (electromagnetic) equations 'Was it a God who wrote these signs?'[56] (a quote from Goethe's *Faust*).

Maxwell was 'absurdly and infuriatingly modest'.[57] In a survey talk in 1870, he praised Thomson's vortex theory of electricity at great length and then only briefly mentioned his own work, saying 'Another theory of electricity which I prefer . . . '[58] (Maxwell's theory of electromagnetism is now reckoned to stand alongside the work of Newton and Einstein as one of the most outstanding achievements in physics.)

In the nineteenth century, the mechanical world-view, introduced by Descartes in the seventeenth century (see Chapter 3), was struggling to survive. Maxwell's theory of electromagnetism seemed to need an ether in which the electromagnetic waves could travel, but the mechanical properties required of this ether were impossible. Another new feature was the arrival of scientific laws with a cosmic direction given to time. First, there was Fourier's law of heat conduction, then the Second Law of Thermodynamics, and finally—the biggest revolution outside the physical sciences—Darwin's theory of evolution in 1859. It is hard now to appreciate the novelty of this cosmic time-dependence. Laplace and Lagrange had carried out calculations confirming that the planetary orbits were stable; the uniformitarian geologist Lyell (1797–1875) proposed geological uniformity over very long—possibly infinite—timescales[59];

and even Clausius wrote, in 1868, 'One hears it often said that . . . the world may go on in the same way for ever'.[60]

Newton, as so often, was a rare exception, a solitary giant and less dogmatic than the Newtonian world-view that succeeded him. He *had* noticed that 'motion is always on the decay' (Query 31, *Opticks*), and was unable to resolve this except by suggesting that a Creator might patch things up when needed. Leibniz was scornful of Newton's need for a 'divine watchmaker' to wind up his clockwork universe from time to time.

Thomson was the first to fully embrace the possibility of an overall direction for time, presumably preferring it to either Laplacian determinism or blind chance—both were antithetical to religious sensibilities, not allowing room for a 'controlling intelligence'.[61] In 1851, quoting from the Bible, Thomson wrote: 'Everything in the material world is progressive . . . "The earth shall wax old &c"'.[62] The word 'progressive' is optimistic, but Thomson followed this up with his landmark synopsis of the Second Law in 1852: '[there is] a universal tendency to dissipation' and the earth will be, in some future era, 'unfit for the habitation of man' (see Chapter 12). Helmholtz was one of the very few who understood its import: 'we must admire the sagacity of Thomson, who . . . [using but a] little mathematical formula which speaks only of the heat, volume and pressure of bodies, was nevertheless able to discern consequences which threatened the universe . . . with eternal death'.[63]

There were some physicists who refused to countenance this bleakness. Rankine suggested that perhaps radiant energy could be re-concentrated at the boundary of the universe and then reflected back to us; and Loschmidt (see below) was another who objected to the 'terroristic nimbus'[64] of the Second Law. But there were no public outbreaks of despondency and despair—in fact, there was almost complete silence regarding the Second Law. The ideas (I believe) were simply too big to be understood in their own time. It wasn't until the popularizations of James Jeans and of Arthur Eddington in the twentieth century that the implications of the Second Law of Thermodynamics entered the public domain.

There were a few nineteenth-century philosophers who did pay attention to physics: one was Nietzsche (1844–1900), who formulated his theory of eternal recurrence, and another was Herbert Spencer (1820–1903), who incorporated the First and Second Laws of Thermodynamics into his ideas of evolution towards 'the greatest perfection and the most complete happiness' and 'alternate eras of Evolution and Dissolution'.[65]

Maxwell didn't comment on these particular conclusions, but thought that it was useful to have the views of non-scientists: 'Mathematicians, by guiding their thoughts always along the same tracks, have converted the field of thought into a kind of railway system, and are apt to neglect cross-country speculations'.[66]

Thomson had an input into the other two 'progressive' laws—the laws of cooling and of evolution. By using Fourier's law of heat conduction and law of radiative cooling, Thomson showed that the Earth had to be less than 200 million years old—not old enough to allow for Darwinian evolution. Thomson had an agenda—to discredit Lyell's and Darwin's theories. (Fourier, in 1820, had made a similar estimate, but thought it was absurd.[67]) Thomson's calculations were largely correct (given that he had to make a number of *ad hoc* starting assumptions) but, of course, he hadn't known about radioactivity. He lived just long enough to learn that radioactivity, discovered around 1900, acted as another source of terrestrial heat. This extended the age of the Earth sufficiently to allow for Darwinian evolution after all. Thomson (now Lord Kelvin) was not amused and never fully accepted that his 'most important'[68] work had been overturned. On the other hand, substituting a negative time, t, into Fourier's equations showed that the Earth must have cooled down from a special, high-temperature, initial state. Thomson thought that this amounted to a mathematical proof for a 'free will having power to interfere with the laws of dead matter'.[69]

The biggest paradox in thermodynamics was that concerning reversibility. The Second Law indicated an absolute direction in time, and yet it derived from the interactions of microscopic particles governed by laws that were *independent* of a direction in time. In other words, macroscopic irreversibility appeared to arise out of microscopic reversibility. The Scottish trio were the first to discuss these matters. In 1870, Maxwell wrote to T and T':

if you accurately reverse the motion of every particle . . . then all things will happen backwards . . . the raindrops will collect themselves from the ground and fly up to the clouds, &c &c and men will see all their friends passing from the grave to the cradle till we ourselves become the reverse of born, whatever that is.[70]

Maxwell fully resolved the paradox to his own satisfaction. He saw that while the microscopic laws of collision were completely reversible, his derivation of the distribution of velocities hadn't been obtained by these laws alone; rather, in dealing with the near-to-infinite numbers of molecules, he had had to make various assumptions about the starting velocities and positions. He realized that it was in these very assumptions that a macroscopic arrow of time had slipped in. As a 'thought experiment', Maxwell invented a fictitious microscopic creature, his 'demon', specifically to demonstrate that one could *not* cheat the system, and therefore the Second Law *had* to be statistical in nature. He ruefully watched the German school (Clausius and others) in their attempts to find a purely deterministic explanation, writing to Tait: 'It is rare sport to see . . . the German Icari flap their waxen wings in nephelococcygia [cloud-cuckoo-land]'.[71])

Sadly, Boltzmann's statistical interpretation of entropy was not appreciated in his lifetime (the first explanatory paper, by Paul and Tatyana Ehrenfest, appeared only in 1911, and the tombstone bearing Boltzmann's famous equation was erected in the 1930s). The barriers to understanding were manifold; as well as the reversibility conundrum, thinking statistically was still alien and the results not at all intuitive. There was, in addition, the problem that the German school of energeticists and positivists didn't even believe in atoms ('I don't believe that atoms exist!'[72] said Mach at a meeting in Vienna in 1897).

Boltzmann's work did get a friendly, if critical, reception in Britain, but in Germany he felt he was either misunderstood or ignored. To the great Dutch physicist Lorentz, he wrote (in 1887): 'I am very happy to have found in you a person who works to develop my ideas on the kinetic theory of gases. In Germany there is nobody who correctly understands these things'.[73] But Boltzmann did receive one very penetrating criticism from one almost-German, the physicist Loschmidt (from Bohemia in Czechoslovakia). (Loschmidt was most certainly not an anti-atomist, having, in fact, determined Avogadro's number; see Chapter 10.) Loschmidt puzzled over the same reversibility paradox as the Scots; namely, what happens when the motion of every atom is exactly reversed? In fact, it was in pondering Loschmidt's question that Boltzmann was led to try out his radically new combinatorial approach; and, in 1896, the German mathematician Zermelo (1871–1953) raised a similar paradox. Poincaré had shown with his 'recurrence theorem' that any mechanical system will eventually return to its initial state; and Zermelo saw that this could lead to a *reduction* in entropy, if you waited long enough. Boltzmann sarcastically countered—you should wait that long[74] (over $10^{10 \text{ billion}}$ years for a sample of air to spontaneously divide into its nitrogen and oxygen components—in other words, an eternity). Both Loschmidt and Zermelo could be answered in the same way (Boltzmann understood): reversed entropy-decreasing motions *can* occur, but they are extremely improbable. However, Planck couldn't accept this, as it meant that the Second Law wasn't *absolutely* true—it was just (very)$^{10\ \text{billion}}$ probably true.

Another German was Helmholtz, whom Boltzmann admired but found imperious and formal. When dining at the Helmholtz residence, Boltzmann apparently picked up the wrong piece of cutlery and Frau von Helmholtz said 'Herr Boltzmann, in Berlin you will not fit in'.[75]

Outside of the physical sciences, Darwin's theory of evolution was, as we have said, the greatest scientific advance of the nineteenth century. Boltzmann was a great admirer of this theory, perhaps seeing in it another example of a theory in which an innumerable succession of random events could lead to an irreversible outcome. He wisely observed: 'The overall struggle for existence of living beings is . . . a struggle for entropy [rather than for energy or raw materials]'.[76]

Willard Gibbs was an exceedingly quiet American—a professor at Yale, unmarried, and living his entire life (apart from foreign travels) in his childhood home, less than a block away from the university. He agreed with Maxwell and Boltzmann that 'the impossibility of an uncompensated decrease of entropy seems to be reduced to improbability'.[77] He was exceedingly modest as well as quiet, claiming that 'anyone with the same desires could have made the same researches'.[78] With this combination of modesty and 'extreme economy (one might almost say parsimony) in the use of words'[79] it is hardly surprising that Gibbs appealed to Maxwell. Maxwell made a clay and plaster cast of Gibbs' three-dimensional thermodynamic surface for water, and sent it to him in New Haven (1874). (It can still be seen, in a display case, at Yale University.)

As the positivists denied the possibility of atoms, it is hardly surprising that Boltzmann developed a strong antipathy towards philosophy. This is captured in his provisional title for a talk: 'Proof that Schopenhauer is a stupid ignorant philosopher, scribbling nonsense and dispensing verbiage that fundamentally and forever rots people's brains'.[80] (This was actually a paraphrase of Schopenhauer himself on Hegel.) Boltzmann also parried a criticism about his lack of mathematical elegance, saying that elegance was for tailors.

Maxwell was one who could perhaps have saved Boltzmann (his last researches were on Boltzmann's work) but he died young, aged 48 years, at exactly the same age and of the same condition (stomach cancer) as his mother. In 1906, Boltzmann committed suicide while on holiday with his family. While the rejection of his life's work must have played some part in this, it appears that Boltzmann suffered from some form of manic depression ('neurasthenia' to the Victorians). Only a few weeks after Boltzmann's death there was proof, in Einstein's and in Smoluchowski's analysis of Brownian motion, that atoms do exist.

Energy in the Public Domain

Apart from the physicist's conception of energy, energy has arrived in another way: we now have industrialized economies and power supply networks; and 'energy' has come into the public domain, and everyone is familiar with it. Did any of the nineteenth-century discoverers of energy anticipate this?

The prime minister, Gladstone, asked Faraday what was the use of electricity, and Faraday famously replied that one day the government would tax it. On the other hand, Joule's researches seemed to indicate that electric power would never compare to steam power as regards efficiency.

The nineteenth century saw the arrival of canals, trains, shipping, and heavy industry: their development was left to venture capitalists, but, unlike today, these eminent Victorians had *long-term* vision, and always built things to last, and with an eye on posterity.

Electric telegraphy began around mid-century, and William Thomson, hired by the telegraphy company, was instrumental in the cross-Channel and cross-Atlantic lines. However, these early lines mostly connected public buildings or central post offices. The idea of a communications network, requiring *electricity*, and serving almost everyone on the planet, was inconceivable (there is also the Catch-22 that a partial network isn't very useful, and so it's hard for a network to ever get started). In a similar vein, the Edwardian futurologist and science fiction writer, H.G. Wells, correctly predicted that an aircraft was technically feasible, but he also predicted that it would never be economically viable: the quantity of oil required was simply absurd, and the idea of a global politico-economic nexus arising specifically to extract this sticky brown stuff—fantastical.[81]

The economist Stanley Jevons wrote in 1865 that the rate of using coal had increased by 3.5% every year, and if the trend continued then Britain would run out of coal[82] by 1965. Osborne Reynolds, discoverer of the 'Reynolds number' in fluid dynamics, wrote that coal could always be imported (the problem of dwindling *global* supplies hadn't occurred to him). In general, Reynolds saw the energy problem as one of losses during transmission rather than of supplies at source. He compared transmission methods such as belting, shafting, compressed air, or hydraulic mains. When it came to the transmission of electricity, along wires, Reynolds was dismissive: 'electric transmission is far inferior to the flying rope'.[83] (The modern mind boggles at the thought of the countryside criss-crossed by miles of whizzing rope.)

Eventually, it was understood how losses during transmission could be overcome. Joule had shown that heating (the main loss) was proportional to I^2R; so if the current was reduced in transit, the losses would be lessened.[84] A lower current could be achieved by 'stepping up' the voltage, easily done by a transformer (invented in 1883), which in turn required alternating current (the a.c. alternator was invented by Tesla in 1888). All this came to pass—but what was the need for *electric* power as opposed to, say, piped coal-gas or deliveries of solid fuel?

The turning point was the discovery of the electric light bulb (independently by Swan and by Edison in 1879). The capital outlay per household was not enormous, and the light was easier to 'switch' on, safer, cleaner, and above all brighter, than the pre-existing lights (candles, oil lamps, and gas lamps). Once electricity was supplied to each home for *lighting*, then the market for home use of *all other kinds of electric appliances* was opened up, for the first time. The electricity was mostly generated in coal-fired power stations employing a very large steam engine. Even today, the steam engine, in the form of a steam turbine, is at the heart of most coal-fired, nuclear, and hydroelectric power stations. Thus, in evolved forms, the engines of both Watt and Joule live on.

H.G. Wells had been writing about aircraft, but he would have been equally astounded by the future of cars. The naturalist and broadcaster, David Attenborough, has shown that grass has led the evolution of elephants, in effect using them to limit the growth of forests so that grasslands can flourish. Has not a similar process been happening with cars and humans? In only 100 years, car numbers have increased exponentially, and they have insinuated themselves into human society, 'employing us' to set up 'food centres' (petrol stations), 'birthing centres' (car factories), and so on. They have become part of human courtship rituals, charged a toll on human lives, and caused some of us to become obese, which, in turn, has increased our reliance on them. Perhaps this is more than an analogy—perhaps they ('car-humans') really have 'speciated', like a virus, and then, like a virus, they will be hard to completely eradicate. Even improving the efficiency of cars may not solve the problem.

According to Jevons' Law (brought in by Stanley Jevons; see above), an improvement in the efficiency of a machine only leads to an overall *increase* in its use. Thus, while an improvement in efficiency is definitely a worthy goal, it has to be accompanied by controls (e.g. taxes on fuels and pollutants) in order to counteract the 'Jevons' paradox'.

The reader may get the impression that I am anti-car, but, actually, I am in awe of the technology (and likewise for airplanes, computers, and, indeed, all machines). The tale of the internal combustion engine (i.c.e.), in particular, is extraordinary, and not without irony. Until recently, the efficiency of a typical car was a mere 15%, a testimony to the absolute difficulty of getting work from heat (see Chapters 8 and 16). What happens to the remaining 85% of the fuel's energy? Answer: some goes into combatting air resistance, some is lost in the exhaust, and some is dissipated in the brakes, and in the tyre walls*-but the majority goes to heat up the heavy engine block (hence one must always wait for the engine to cool down before opening up the radiator cap). So cars are, in reality, not much more than 'boilers on wheels'. In Australia, where car air-conditioning is frequently used, then *cars are 'fridges on boilers on wheels'* (We will have more to say on the efficiency of engines at the end of Chapter 18.)

The history of the energy supply industry is interesting and important, but it is well documented, and would extend the length of the current book too much. In very brief outline: before the Industrial Revolution, the main fuels were wood and charcoal (and also wind, water, and muscle); post-Revolution, coke was required for the smelting of iron, and coal for the steam engine. Finally, in the modern era, the electricity supply industry uses coal, gas, nuclear, hydroelectric, and renewables (in that order); internal combustion engines mostly consume fossil fuels (oil, diesel, or gas); heat in homes is mostly generated by fossil fuels (especially gas in Europe), biofuels (especially in Africa and Asia), and geothermal in very localized areas; and cooling uses electricity (and also needs refrigerants or water). Also, some 10% of coal (including coke) is used in industry, mostly in blast furnaces. The true cost[85] of energy has never been charged: it seems that the sophistication of our institutions, whether political, governmental, financial, or social, has in no way been able to keep pace with our technological development. Even some of our

---

* But hardly any is lost to friction with the road surface, as there is no movement of the tyre relative to the road (!), except in a skid.

greatest thinkers have been unable to imagine the difficulties to come: Thomson, upon seeing the Niagara Falls, wished that every last drop of hydroelectricity could be exploited for human consumption;[86] and Feynman, in 1963, wrote: 'it is up to the physicist to figure out how to liberate us from [ever running out of] energy. It can be done'.[87]

18

Impossible Things, Difficult Things

Why, sometimes I've believed as many as six impossible things
before breakfast.

> White Queen, in *Alice's Adventures in Wonderland*,
> by Lewis Carroll

Impossible Things

From its history, we have found that the concept of energy has not been
easily won. But physics is not the same thing as history of physics: isn't
it sufficient to learn about energy from a physics textbook? Sufficient—
maybe; easy—no. A famous physicist—I think it was Gamow[1]—once
said that to try to understand thermodynamics he went through it three
times: the first time, he understood it but thought he'd better have another go, just to be sure; the second time, he realized he didn't understand it; the third time, he realized he never would understand it. In
mildly humorous vein, we look at two classic texts: *The Elements of Classical Thermodynamics*[2] by Pippard, and *Heat and Thermodynamics*[3] by
Zemansky.

Thermodynamics starts with a definition of what constitutes a system:
'a restricted region of space or a finite portion of matter',[4] which may be
separated from the 'surroundings' by walls that are diathermal or adiabatic (they do or don't let heat through, respectively). Now remembering
all the difficulties, first in the understanding of heat, and then in the discovering of energy, it is a little disconcerting to come across the reverse
viewpoint wherein energy is *assumed*, and heat is merely *defined* as the
difference between internal energy and work:

[heat is] equal to the difference between the internal-energy change and the
work done.[5]

Now, what is work? Work is anything that causes changes in the macroscopic distribution of matter, and such changes are ultimately equivalent to the raising or lowering of weights, the winding or unwinding of a spring, or, in general, the alteration of the configuration of some external mechanical device.[6] It is curious that something so anthropic and nonspecific as 'a device' comes into physics in this way.

Whenever the internal energy, U, of a system is changed by pure work, W, then we must have $\Delta U = \Delta W$, and this is true by whatever route the change has been accomplished.* In fact, this amounts to nothing less than the Principle of the Conservation of Energy, or the First Law of Thermodynamics. However, it is surprising to find it admitted that 'Unfortunately, it does not seem that [careful] experiments of this kind have ever been carried out',[7] and 'accurate measurements of adiabatic work along different paths between the same two states have never been made'.[8]

If it is found that $\Delta U \neq \Delta W$, then heat is defined as that which makes up for the discrepancy, and so it is evidently crucial to make a clear distinction between heat and work. However, this distinction can depend on the choice of system boundary (see Fig. 18.1): in (a) the system has had work done on it (by a falling weight), whereas in (b) the system has been heated (by the nearby immersion heater). (We have had to assume that the pulleys are frictionless, and that the generator and connecting wires have no resistance. It seems that adiabatic walls *do* allow work and electricity to pass through, and also a falling weight presupposes a gravitational field—which also crosses the walls.)

How can we be assured that no heat has been added to a system (and thus check up on the First Law)? Answer: by enclosing the system within adiabatic walls. Yes, but how do we know that such walls are truly adiabatic? Because they don't let any heat through . . . On the other hand, diathermal walls *do* let heat through, and this is how equilibrium is attained between adjoining systems. But if there are no changes in either system, is this because the walls were not diathermal or because the two systems were already in equilibrium?

What is equilibrium? Equilibrium is the state that has been achieved when there is no further change in the macroscopic properties of a given system or of any number of systems connected by diathermal walls. But how long do we wait? And if there *is* a change, then is it a departure from equilibrium or just a statistical fluctuation?

---

* 'Δx' means 'a small increment in x'.

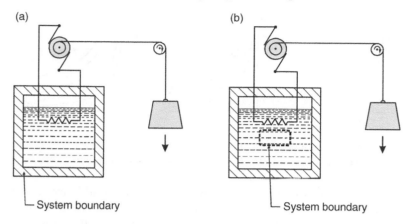

Fig. 18.1 Heat and work depend on the choice of system boundary (adapted from Zemansky, *Heat and Thermodynamics*, 5th edn).

From equilibrium, we come naturally to temperature. Pippard introduces temperature without any regard to heat. All we need to know is the definition of equilibrium, that there are such things as diathermal walls, and that the Zeroth Law of Thermodynamics applies:

Zeroth Law of Thermodynamics:
If, of three bodies, A, B and C, A and B are separately in equilibrium with C, then A and B are in equilibrium with one other.[9]

Consider as a 'thermometer'[10] the often-chosen scenario of a gas in a cylinder, characterized by its pressure, P, and volume, V. Bring this gas 'thermometer' into diathermal contact with the 'body' to be investigated—say, 'the lounge'—and adjust the P and V values until equilibrium has been achieved. Then, using this same gas 'thermometer', systematically vary its volume and, for each V, adjust the P until equilibrium has again been achieved. It will be found that all the resulting (P,V) pairs lie on a curve for which some function, $\varphi(P,V)$, has a constant value—see Fig. 18.2(a).

Now, using the gas 'thermometer' again, investigate another test body—say, 'bathwater'—and repeat the process. The new (P,V) pairs will lie on another 'curve of sameness' (Fig. 18.2(b)). Repeating the process again and again with other test bodies will eventually lead to a whole family of such 'curves of sameness'. Finally, label each isocurve with a number. *This number is the temperature.* The number can be 'chosen at

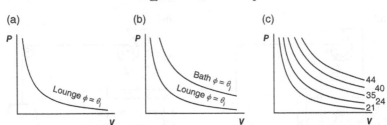

Fig. 18.2 'Curves of sameness': (a) one isocurve; (b) two isocurves; (c) a family of isocurves (after Pippard, *The Elements of Classical Thermodynamics*).

will . . . provided that there is some system, however arbitrary, in the labelling'.[11] However, once the labelling has been carried out, then it is fixed even if the gas 'thermometer' is replaced by a different sort of 'thermometer', such as a mercury-in-glass thermometer, a platinum resistor, and so on.

So far, we have not touched on any link between temperature and heat. In order to do this, Pippard brings in an extra premise: when two bodies are in 'diathermal contact', they come into equilibrium by the 'transfer of heat' from one to the other. As we don't yet know what 'heat' actually is, we can't be sure which way it is being transferred. It is up to us to choose, but once we decide that heat is transferred *from* hot to cold, then subsequent experiments are all consistent with this choice (so if we choose that heat is transferred from hot to cold, then it always will be). It then only remains to set up a scale of hotness such that bodies that are hotter are also at a higher temperature. We won't elaborate but, in order to prove that setting up such a scale is possible, Pippard uses a *reductio ad absurdum* argument along with the converse of the Zeroth Law. One has the feeling that some arcane game of logic is being played out.

So much for the Zeroth and First Laws. In order to establish the Second Law of Thermodynamics, an impossible scenario must be imagined—the Carnot Cycle, or ideal heat-engine (see Chapter 12). The gas must be ideal; the piston must fit perfectly, and move without friction and infinitely slowly (so as not to cause turbulence or discontinuities in the gas density); the pressure outside the cylinder must at all times be only infinitesimally different from the pressure within; the cylinder walls and piston must be perfectly insulating, except when thermal contact with a heat reservoir is required; and the reservoirs themselves must be infinite (i.e. unaffected by the heat added to or taken away from them) and only infinitesimally higher or lower in temperature than the gas-engine. When all these conditions are satisfied, the changes are said to be 'quasi-static'.[12]

When we come to the kinetic theory (say, of gases), we have some further idealizations to make. The volume element under consideration must be small when compared to the total volume, but large enough to include a very large number of microscopic components (molecules, atoms, ions, or whatever). At usual temperatures and pressures, a millionth of a cubic centimetre of air contains around 10^{13} molecules—and this is deemed a satisfactory size for a volume element of gas.[13] There are also assumptions to do with the ideal nature of the gas: the molecules are taken to be small hard spheres that are in perpetual random motion and exert no forces on each other except at the instant of collision. Between collisions, they therefore move with uniform rectilinear motion. The molecular diameter is assumed small compared to the average separation of neighbours (typically, molecules are a few ångstroms across ($1 \text{ Å} = 10^{-8}$ cm) and their average separation is 50 times this), collisions are assumed to be perfectly elastic, and the container walls perfectly smooth. Actual gases most closely approach this ideal in the limit, when the pressure is reduced to zero. (In other words, when there is no gas whatsoever, then it is ideal.)

In the early kinetic theories (such as that of Daniel Bernoulli), the molecules were considered to be point particles that never collided: equilibrium was slow in coming—it could be arrived at only by collisions with the walls. An even worse problem arises in the case of radiation in a cavity consisting of perfectly internally reflecting walls (the so-called black-body radiation). Not only doesn't the radiation interact with itself, but doesn't interact with the walls either. The only way in which equilibrium can be achieved is to (hypothetically) introduce 'an extremely small piece of matter such as a grain of coal'.[14]

It is one thing whether a thought experiment is difficult to carry out, but can it be allowed that it is impossible, even in principle?[15]

Difficult Things

We move from the impossible to the merely difficult, and consider a miscellany of perennially difficult topics. This is not a textbook and the aim is to demystify and explain in woolly words and analogies, rather than with mathematics: to quote from Feynman, 'A physical understanding is a completely unmathematical, imprecise, and inexact thing, but absolutely necessary [for a physicist]'.[16]

The laws of thermodynamics—are they empirical?

Let's begin by explaining that the First Law of Thermodynamics and the Law of the Conservation of Energy are really the same thing, except that the former applies specifically to heat and work, and the latter to all forms of energy.

Both the First and the Second Laws of Thermodynamics started from the common experience that perpetual motion of the first and second kinds is impossible (perpetual motion of the first kind refers to an engine, with no fuel, perpetually generating work; perpetual motion of the second kind refers to heat flowing from cold to hot, unaided). It is because the laws started from such common knowledge that they are said to be empirically based, and it was, in fact, the extraordinary broadness of this empirical base that led to the extraordinary broadness of the reach of these laws.

However, the early investigators, in particular Mayer and Helmholtz, considered that the conservation of energy was not just confirmed by experiment but *had* to be true because of the deeper law of 'cause equals effect'. Max Planck (1858–1947) made a similar point:

the impossibility of perpetual motion of the first kind is certainly the most direct of the general proofs of the principle of energy. Nevertheless, hardly anyone would now think of making the validity of that principle depend on the degree of accuracy of the experimental proof.[17]

Henri Poincaré (1854–1912), a philosopher-physicist, generalized this to all the laws of thermodynamics, and to the Principle of Least Action as well:

[these laws] represent the quintessence of innumerable observations. However, from their very generality results a consequence . . . namely, that they are no longer capable of verification.[18]

In other words, physics as we know it would fall apart without the conservation of energy, and we must now simply take energy and its conservation as givens. In fact, we would sooner invent new forms of energy than sacrifice the law of the conservation of energy. This is exactly what did happen in 1930. Wolfgang Pauli (1900–58) was troubled by problems with the β-decay spectrum, and chose not to follow his predecessor Debye's advice ('Oh, it's better not to think of it at all, like new taxes'.[19]) Instead, Pauli invented a new particle, the neutrino, as 'a desperate remedy to save . . . the law of conservation of energy'.[20] (Pauli describes his motivations in a strange letter that begins 'Dear Radioactive Ladies and Gentlemen'.)

Kinetic and potential energy—which is more fundamental?

Vis viva, or kinetic energy, as it's now called, was first postulated by Leibniz, in 1686, and then the idea of potential energy evolved gradually over the succeeding 100 years—an indication of the fact that potential energy is the subtler concept. It's unfortunate that the adjective 'potential' has stuck (this dates from Daniel Bernoulli, in 1738), as it sounds as if the energy is not really there but is in abeyance. And, indeed, it is mysterious how a boulder can lie on a mountain top for a millennium, with its energy seemingly locked away. But it is only the potential *kinetic* energy that is locked away (like snakes, we prioritize motion). The actual potential energy, the 'energy of configuration', is right there in a continuous 1,000-year-long agony of keeping the boulder away from the centre of the Earth. It's easy not to notice something that is always there; for example, the pressure on the soles of our feet, or on our bottoms, almost every hour of our lives. (In fact, it took a Newton to notice, and to put this into his law of 'Action and Reaction'.)

But why are there just these two categories of energy? This is due to the fact that any given particle can be viewed, simultaneously, in two distinct ways: on the one hand, the particle is an 'individual', and on the other hand it is 'part of a system'. In the latter view, the particle has an energy of configuration or, more generally, an energy of interaction with other parts of the system. This is the potential energy. But, whether interacting or not, an individual particle always has the 'energy proper to itself'—its rest-mass energy and, if it is moving, its kinetic energy. The distinction between these forms of energy becomes blurred in modern field theories and, in any case, they are not so antithetical as is sometimes made out. Even in classical physics, the distribution of energy between the two forms, kinetic and potential, depends on how the system has been modelled, and on the choice of frame of reference.

Imagine, for example, that we anchor our frame of reference to a wagon on a roller-coaster and, furthermore, that we keep our eyes tightly shut. As our eyes are closed, we are not aware of our motion— in effect, our kinetic energy has vanished. But, we do notice that we have to hold on to the wagon more tightly in certain sections, our stomachs are wrenched, and our ears may pop: in other words, the interaction strength changes, and in exactly those places where an outside observer would see that our speed or direction is changing rapidly. (More remote from the everyday, but essentially the same scenario as the roller-coaster, consider planetary orbits: the potential energy is proportional to mM/r as expected, but if we ignore the angle (in other words, the view is unchanging), then there is an *extra* potential energy term proportional to $1/r^2$.)

Another example occurs in electrodynamics. Consider the following argument, adapted from Feynman's *Lectures on Physics*.[21] In a stationary frame of reference, S, an electron is moving to the right, parallel to a stationary wire in which a current, I, is flowing. The speed of the exterior electron is the same as the average drift speed, v, of the conduction electrons. In the moving frame of reference, S', the electron is stationary but the wire is moving to the left with speed, v, and the current in the wire is now I' (see Fig. 18.3).

Now in frame S, the wire is surrounded by a magnetic field, and the moving exterior electron is attracted to the wire, and gently curves down towards it. In frame S', the exterior electron is stationary, and so will not be affected by magnetism—but will the same thing still happen? Will the electron still curve down gently towards the wire? Yes, it will; S and S' differ only by the fact that one moves uniformly with respect to the other, and therefore the *same* physical happenings *must* occur in both frames (by the Principle of the Relativity of Motion). However, in S', the electron moves towards the wire for a new reason—electrostatic attraction. In S, the wire is electrically neutral (the density of negative charge from the conduction electrons, ρ_-, is balanced by the density of positive charges, ρ_+) and so there is no electrostatic attraction. In S', the moving wire has shrunk, ever so slightly, due to a special relativistic effect known as 'length contraction', and so the density of positive charge has increased from what it was before: $\rho'_+ > \rho_+$. Also, the

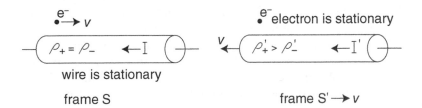

Fig. 18.3 An electron and a wire viewed from two frames of reference (adapted from Feynman's *Lectures on Physics*, Volume II, Section 13–6).

conduction electrons in S' are now stationary (ignoring the miniscule thermal motions), and so we also have $\rho'_- < \rho_-$. On both counts, the resultant density of positive charge in the wire in frame S' increases. So, finally, we see that the negatively charged electron will be electrostatically attracted to this positively charged moving wire.[22]

We have gone from S (no electrostatic attraction) to S' (no magnetic attraction); but we can also view this as a shift from S (the exterior electron has kinetic energy, the wire has no kinetic energy, there is zero electrostatic potential energy, and there is magnetic field energy) to S' (the exterior electron has zero kinetic energy, the wire has finite kinetic energy, there is an electrostatic potential energy, and there is no magnetic field energy). Although, strictly speaking, we should carry out a field analysis, there is no mistake: there has been a change in the attribution of kinetic and potential energies between two, totally equivalent, frames of reference. (Note that, as mentioned in the similar example given in Chapter 17, the speed, v, is tiny—it's the average drift speed of the conduction electrons, around 0.0001 m s^{-1}—and yet relativistic effects are manifest.)

There's still the question of which is more fundamental: kinetic or potential energy. Maxwell answers that it is the kinetic energy: it has only one form, $\frac{1}{2}mv^2$ (and a simple extension for rotations), whereas potential energy comes in a variety of forms, even while they all depend just on the relative configuration (the interactions) of the constituent parts. As Maxwell writes:

we have acquired the notion of matter in motion, and know what is meant by the energy of that motion, [and] we are unable to conceive that any possible addition to our knowledge could explain the energy of motion or give us a more perfect knowledge of it than we have already.[23]

On the other hand, with regard to potential energy, he writes:

the progress of science is continually opening up new views of the forms and relations of different kinds of potential energy.[24]

In other words, there is only one elemental form for kinetic energy, whereas potential energy, pertaining as it does to system aspects, must be reformulated for each new system. In this sense, kinetic energy *is* more fundamental.

Kinetic energy—why is it $\frac{1}{2}mv^2$?

(This section is rather technical, and may be skipped without any loss of continuity in the development of the concept of energy.) Having just stated, in the previous section, that kinetic energy has the form $\frac{1}{2}mv^2$ (for one particle of mass m and velocity v), let us now ask *why* it must have this form. We may simply be satisfied that this (Leibniz's) formulation arises from Huygens' analysis of Galileo's free-fall experiments (see Chapter 3); or from the understanding that the work done by a force, $\int F.ds$ (see Chapter 7), is equivalent to $\int mvdv$, and this integral has the value $\frac{1}{2}mv^2$. However, we would like to push our knowledge to the fundamental limits, starting with as few prior conceptions (about free fall or 'work done') as possible.

The Russian physicists Landau and Lifshitz[25] argue that something as elemental as kinetic energy cannot depend on direction, and therefore it must depend on v^2 (when squaring a vector, it becomes a directionless quantity). But this doesn't clinch the matter—after all, we could still satisfy the requirement of 'no absolute direction' by having kinetic energy depend on $\sqrt{v^2}$ instead of on v^2. (Also, what about the dependence on mass?)[26]

The independent physics researcher Maimon[27] has a better answer. Suppose there exists a universal kinetic energy function, T, which depends on m and v in some, as yet, unknown way, $T = T(m,v)$. Now consider a ball of putty of mass m and speed v. If the ball is brought to a standstill by crashing into a wall, then, by energy conservation, all[28] of its kinetic energy must be converted into heat: $T(m,v) = \Delta heat$. Also, we have the empirical knowledge that the amount of heating depends on the mass of the ball (a ball of n times the mass will generate n times the amount of heat). Therefore, we must have the following:[29]

$$T(m,v) = \Delta heat = mT(v) \qquad (18.1)$$

Then, Maimon looks at an inelastic collision between two balls of putty, each of mass m, and each approaching the other in a straight line with a constant speed, v. This collision is viewed from two frames of reference: a stationary frame, S, and another frame, S', that moves with the same direction and speed as one of the balls. By symmetry in S, the balls must stick together after the collision, and, by the Principle of Relativity, this must be true in S' as well; also, the heat from the collision must be equal to twice the heating of one ball (in effect, each ball acts as a 'wall' for the other). Now, by the Principle of Relativity, the heating must be the same in both frames of reference:[30] $\Delta heat = \Delta heat'$. Also by the Principle of Relativity, energy conservation must be satisfied in both frames of reference. All this information is summarized in Table 18.1.

1

Table 18.1 Before and after an inelastic collision, viewed from two frames of reference.

Frame of reference	Before collision	After collision
S	$E_{before} = mT(v) + mT(v)$ $= 2mT(v)$	$E_{after} = \Delta heat$
$\xrightarrow{S'}$ v	$E'_{before} = mT(2v)$	$E'_{after} = 2mT(v) + \Delta heat'$

Conservation of energy in the moving frame of reference requires the following:

before collision   after collision

$$mT(2v) \quad = \quad 2mT(v) + \Delta heat' \qquad (18.2)$$

while conservation of energy in S requires:

$$2mT(v) = \Delta heat \qquad (18.3)$$

But as $\Delta heat = \Delta heat'$, then, combining (18.2) and (18.3), we have

$$mT(2v) = 2mT(v) + 2mT(v) \qquad (18.4)$$

or

$$T(2v) = 4T(v) \qquad (18.5)$$

The only way this can be true is if T is a function of the *square* of speed. In fact, we lose nothing by saying that T actually *is* v^2. Finally, bringing back the dependence on m, we have $T = mv^2$.

We have almost reached our starting requirement, $T = \frac{1}{2}mv^2$. This last requirement, the factor '½', is brought in by the work of Ehlers, Rindler, and Penrose.[31] These authors show that the factor of '½' is necessary in order to make the special relativistic and classical (non-relativistic) definitions agree with each other (i.e. agree in the limit of low speeds).[32]

Both proofs concern colliding bodies (or particles), but what about non-interacting bodies? Consideration of this question leads to an enigma: it turns out to be impossible to prove (to derive from more elemental principles) the formula for the kinetic energy of a single isolated free particle. This is because the speed and direction of the isolated particle are relative to 'absolute' space whereas, for colliding particles, their velocities are, in effect, relative to each other.[33] The curious end result is that the kinetic energy of an isolated free particle must simply be *defined* to be $\frac{1}{2}mv^2$. On the other hand, a system consisting entirely of nothing but one free particle is freakishly simple; and its very freakishness[34] means that it doesn't really matter how its energy has been defined.

Kinetic energy—where does the energy go?

Considering frames of reference again, it seems strange that the kinetic energy of a body can disappear merely by switching from one frame to another. Take the example of billiard balls on a billiard table. Now pluck an individual ball out of this system, and send it hurtling through space at high speed—say, 60 km/hour. From a special frame of reference travelling with the ball, the ball's speed is zero. But we feel sure that it still has energy—after all, it could land on someone's head and bring out a large bump.

It seems that the kinetic energy of this billiard ball is crying out to be tied to us (to someone's head). Even our use of the phrase 'at rest' is loaded with anthropic connotations. And yet this elemental example is freakish in its simplicity—how often do we see a truly isolated billiard ball hurtling through space? We may paraphrase the legal maxim 'hard cases make bad law' into a physics maxim, 'odd scenarios make bad intuition'. The temptation to bring ourselves into the system (of the *isolated* ball) is irresistible: we cannot easily grant that the billiard ball, all on its own, already comprises the entire system, and instead we demand to know where its kinetic energy has gone *relative to us*.

But this is only a partial answer. The complete answer to the question is the recognition of the fact that energy is *not* an invariant[35] quantity. Yes, it is conserved in an isolated system and yes, as potential energy, it usually does remain unchanged between frames of reference (because of the fact that potential energy most commonly refers to quantities defined *within* the system, howsoever that system is viewed)—but, most generally, energy does vary between frames of reference, and it is, rather, the *total action* between definite end-states that is the true invariant quantity.[36]

Rest energy—an absurdly large zero-point energy?

Concerning frames of reference yet again: it always seemed puzzling to me that in changing frames of reference in the Galilean way (i.e. before Einstein), no account was taken of the mass. A frame was given a 'linear boost' (a spurt in speed) but it didn't make any difference whether that frame had masses in it or not, or whether the frame itself was an actual physical thing with axes made of balsa wood or steel. In Special Relativity (SR), this is rectified—the mass *does* make a difference. What comes out of SR is that the zero-point of kinetic energy is no longer arbitrary[37]—it is equal to the 'rest energy', $E_0 = m_0 c^2$ (where m_0 is the mass that an object has when stationary, i.e. in its own 'rest frame'—see Chapter 17). Instead, the puzzle is now: why is the zero-point in energy so incredibly large? (It's an amazing 10^{17} times the typical everyday energy.) The answer is that the rest energy is somewhat like a store of energy. We are normally blissfully unaware of this vast store because, for one thing, we're usually dealing with speeds nowhere near c; and, for another, the rest energy is tied up in a rather stable form—as the rest mass of fundamental particles. (Note that nuclei are not fundamental: their rest mass is a kind of potential energy, which depends on the interaction of the nucleons within; and some of the nuclear rest energy trickles away—that is, gets converted into heat—when radioactive nuclei decay.)

Now, magnetic energy, as we have seen (in Chapter 17), is usually a very small fraction of the electrical energy, and may be considered as a relativistic correction to it. Likewise, kinetic energy is a very small fraction of the rest mass energy, so may it also be considered as a relativistic 'correction' . . .?

The rest energy of light?

We have been talking of how the energy of a material body is split between its rest energy and the energy of its motion (above and in Chapter 17). In one special case, the case of pure radiation, the energy arises solely from motion, as the rest mass, and therefore the rest energy, is zero. The 'energy books'[38] only balance if the speed of the motion has the highest possible value—the speed of light, of course.

Heat

What is heat? Heat is a new 'block' of energy and brings a completely new approach into physics: we can't answer the question of what it is just by reference to the macroscopic relations of classical thermodynamics, even those comprehensive statements such as the First and Second Laws, and this is because heat is statistical energy. For a complete answer, we are compelled to delve into the (once speculative) microscopic arena. Then we find, as suspected all along by Bacon, Boyle, Hooke, Daniel Bernoulli, and others, that heat *is* the motion of the 'small parts'— the net effect of the random individual motions of an extremely large number of extremely small particles.

There are a number of qualifications to make. 'Random' doesn't mean lawless (the particles still obey Newtonian or quantum laws, as appropriate) but means that the motions are not coordinated in a macroscopic sense. (So the *bulk* motion as occurs, say, in wind, or in a sound wave, doesn't count as heat.) 'Particles' are almost always microscopic (for example, atoms, molecules, ions, electrons, 'Brownian particles', and so forth—as relevant to the investigation) but could exceptionally be as large as stars.[39] The important thing is that the 'particles' interact, somehow (else equilibrium could never be achieved). Finally, an 'extremely large number' means that there are so many of them that a statistical description is absolutely essential. A collection of, say, five molecules cannot manifest heat. The statistical outlook was a radically novel departure from what had gone before.

What of radiant heat? This can be modelled in two ways—as a collection of particles (photons) or as electromagnetic radiation. Either of these models leads to the same result: at equilibrium, the radiation has a characteristic distribution of wavelengths, and the peak wavelength (the colour) is inversely proportional to the temperature. More typically, we could have a mixture of radiation and matter—say, photons and electrons. When the radiation and the matter are in equilibrium with each other, then the hot electrons jostle around, have many collisions, and so continually change direction (accelerate) and thereby generate electromagnetic radiation. The temperature of this generated radiation (as evidenced by its colour) will be the *same* as the temperature of the hot electrons. The radiation, in its turn, will 'keep the electrons warm'; that is, keep them at this same equilibrium temperature. In other words, the radiant and matter components of the system are maintaining the one consistent overall temperature. This is rather a simplified picture (a full treatment needs quantum electrodynamics) but the gist is correct, and we see that the concepts of equilibrium and temperature are creeping in. Before these are discussed, an old chestnut will be examined.

Does heat only exist in transit?

Classical thermodynamics is premised on the fact that a given state of a system can be completely specified by just a few macroscopic thermodynamic parameters. Whatever changes may be undergone, to whatever degree, and in whatever order, then, once the initial values are resumed, the system will have exactly the same colour, smell, sensation of warmth, viscosity, conductivity, taste, phase, or any other property. This goes for heat too—the system has exactly the same amount of heat in it (the 'internal heat') when the same state is resumed. On the other hand, the transferred heat, ΔQ, is variable and dependent on the route taken.

Now, here's the funny thing: thermodynamics makes a beeline for this variable, route-dependent quantity, ΔQ, while snubbing the fixed, internal, heat-in-a-body. The trouble is that the heat-in-a-body is not a very useful quantity for the very reason that it's too body-specific. Although we know that it is constant for a given state, we don't know what that constant value actually is—that would require microscopic theories, and we are eschewing them in classical thermodynamics. ΔQ, on the other hand, while different for all the different routes, can at least be experimentally tracked for all these routes. (For example, the system could be surrounded by a water jacket, and the total mass of water, and its temperature change, measured. Taking the specific heat of water as a given, and assuming no heat leaks unaccounted for, then ΔQ can be determined.)

While the 'internal heat' is thus the less useful quantity, I wouldn't go so far as to say that it doesn't exist: I don't know what's keeping my soup in the vacuum flask warm if it isn't heat. Ironically, in the one scenario so often exploited in thermodynamic demonstrations—the ideal gas—the internal heat *is* known; in fact, it's identical to the total internal energy, U. (Furthermore, U is related to the specific heat at constant volume, C_V, and this is, at last, a quantity that experiment can determine.)

There is yet another curious fact about the transfer of heat. Infinitesimal additions or subtractions of heat, dQ, can occur under either quasistatic or realistic conditions but, whatever these conditions, the amount of heat is not described by a mathematical function; in other words, dQ is not the 'd of Q'.[40] But how, then, shall we be able to combine ('integrate over') these infinitesimal chunks? An analogy helps to explain how this can be achieved. Imagine three guinea pigs, Tom, Dick, and Harry (perhaps, rescued from Laplace's ice 'bomb'), nibbling their way through a biscuit. The amount of biscuit consumed in time, dt, cannot be given as a function (the 'd of Tom', etc.), but if we could characterize each guinea pig by its own unique bite-size, then we would be able to add up all the separate bites, and determine how much biscuit had been eaten in a given finite time interval. Back to heat: the temperature (in fact, $1/T$) determines the size of dQ (the 'bite-size'), and so enables us to integrate over all the separate inexact differentials and obtain an exact result.[41] We say that temperature has acted as an 'integrating factor' for heat—but what actually is temperature?

What is temperature?

Temperature is at once both mysterious and yet intuitive. For example, when measuring the length of a table, we get the reading off the tape measure straightaway, with no hanging about; whereas when measuring the temperature of the bath we know that we must stir the water, bide our time, and wait for the thermometer reading to settle down. Also, when told of the Zeroth Law of Thermodynamics (that if A is in equilibrium with B and with C, then B and C are in equilibrium with each other), we don't show amazement that no one has told us what A, B, and C actually are, or complain if A is a thin column of mercury in a sealed glass tube, B is a bath, and C is my armpit (well, I might complain). Looking up the definition of temperature in Feynman's *Lectures on Physics*,[42] he says: 'What we mean by equal temperatures is just . . . the final condition when things have been sitting around interacting with each other long enough'. But how long is 'long enough'? And, in fact, we know that if we could look extremely closely, we would *not* see an unchanging picture (there would be a constant buzz of activity at the microscopic level). There is also the fact that there's a bewildering variety of definitions and occurrences of the temperature parameter, and yet they all refer back to one and the same universal temperature, T. This is *the* 'mystery' of thermodynamics.[43] Let us start by listing the multifarious definitions and occurrences of temperature:

- The sensation of hotness.
- The calorimetric equations, $\Delta Q = mc\Delta T$ at constant volume and $\Delta Q = m\lambda\Delta V$ at constant temperature.*

---

* ΔQ is the amount of heat transferred, m is the mass, c is the specific heat, and λ is the expansion coefficient for the given substance.

- Pippard's 'curves of sameness'.
- Celsius, Fahrenheit, and other empirical scales.
- The absolute temperature scale, T (in Kelvin).
- The Ideal Gas Law, $PV = nkT$.
- T in radiation laws, such as $\lambda \propto 1/T$ (Wien's Law), and energy flux density $\propto T^4$ (the Stefan–Boltzmann Law).
- From the kinetic theory, $T \propto \langle \frac{1}{2}mv^2 \rangle$.
- Boltzmann's factor, $e^{-(\beta \text{energy})}$, where $\beta = 1/(kT)$.
- $\langle$thermal energy$\rangle$/(degree of freedom) $= \frac{1}{2}kT$.
- T in the Carnot cycle (for quasi-static additions of heat): $dQ = TdS$. T in thermodynamic relations: $dU = TdS - PdV$, and others.
- T as an 'integrating factor' for dQ, as in $\int dQ/T = \Delta S$.

How can we show that T is the same in all these cases (or, in other words, why do we have no need of separate parameters: T_{Carnot}, T_{IdealGas}, T_{kinetic}, $T_{\text{Boltzmann}}$, T_{Hg}, and so on)? The story of how these scales occur and are cross-calibrated is given in many textbooks and, in highly condensed form, in the notes;[44] but there is still the question of *why* they are all equivalent. The fundamental reason for this equivalence is that all processes are ultimately microscopic and, moreover, microscopic *in the same way*.[45] All relate to billions of tiny energy exchanges in 'bumper-car' collisions between 'springs' (even while the bumper cars and springs are themselves analogies). At equilibrium, the mean microscopic energy transfer per unit of time depends on the temperature; more emphatically, it depends *only* on the temperature, and not on this or that gas, this or that material, or which working substance is used in an engine, and so on. This is the reason *why* the unit of thermal heat is $\frac{1}{2}kT$, Boltzmann's factor is $1/(kT)$, the 'integrating factor' is $1/T$, and why the specific details of the bodies A, B, and C, used in the Zeroth Law, don't need to be given—and so on. Even where macroscopic results are obtained, the analysis and the experiments assume the quasi-static, *microscopic* conditions: waiting long enough for equilibrium to be achieved; temperature differences across diathermal boundaries that must be infinitesimal; the fact that all introduced changes, and the rate of such changes, must be infinitesimal; and so on (see above). We are finding that T is indeed a subtle idea, but it is one that presupposes the very conditions that will guarantee its workability.

To summarize, the reasons why there is just one universal temperature are as follows:

• T, by its definition, only makes sense at equilibrium, and presupposes the very experimental conditions it requires.

• Equilibrium, or 'thermalization', is achieved in *one* fundamental way—by random, microscopic, energy exchanges of size $\frac{1}{2}kT$.

• The molecules, or other thermalized entities, go around singly and not in gangs of say, millions or billions.

• There is only one equilibrium parameter, T—we are not trying to equilibrate some other attribute, such as 'fabulousness', at the same time.

Which is more fundamental, temperature or pressure?

Could we have constructed thermodynamics using isobars instead of isotherms,[46] and using P instead of T as the fundamental parameter? We have seen earlier that $P \propto T$ (from a microscopic analysis—the kinetic theory, from the Ideal Gas Laws, and from experiment), and so it would seem that P and T are of equal status—but this is wrong. Feynman explains all.[47] Consider two compartments separated by a moveable wall (see Fig. 18.4),[48] and suppose that in one compartment we have n_A molecules of mass m_A, and in the other compartment we have n_B molecules of mass m_B.

The moveable wall gets battered on both sides by collisions from gas molecules, until eventually it settles down at some position at which the gas pressure on both sides is equal. There may be many massive sluggish molecules on the left, balancing the pressure of fewer, lighter, but faster molecules (on average) on the right:

$$\text{Pressure equalization, } n_A < (\tfrac{1}{2})m_A v_A^2 > \; = \; n_B < (\tfrac{1}{2})m_B v_B^2 > \qquad (18.6)$$

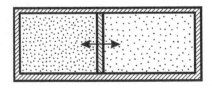

Fig. 18.4 Gas compartments separated by a moveable wall.

But is this the true, final equilibrium? Pressure is a *bulk* phenomenon, and so the equalization of the two pressures happens relatively quickly. (Because of the huge surface area of the wall, pressure equalization occurs via *n*-body collisions, where *n* is of the order of billions of molecules.) However, concurrently, there are individual two-body collisions—between the moveable wall and *one* gas molecule. Each one of these collisions has only a very miniscule effect on the massive wall but, more gradually than for pressure, the cumulative result is that the motion of the whole moveable wall becomes 'thermalized': the wall is then never completely stationary, but has an incessant, tiny, thermal motion, as often to the right as to the left, and in which we find:[49]

Temperature equalization,

$$< (\tfrac{1}{2})m_1 v_1^2 > \; = \; < (\tfrac{1}{2})m_2 v_2^2 > \; = \; < (\tfrac{1}{2})m_{wall} v_{wall}^2 > \qquad (18.7)$$

In effect, we have been able to regard the whole moveable wall as if it was one giant gas molecule of a rather odd shape.*

Experiment confirms that our prediction is correct (first the pressure equilibrates and then the temperature equilibrates as well: also, Avogadro's Law is confirmed). Temperature is therefore unquestionably more fundamental than pressure, as Carnot intuited (also, everything has a temperature, but not everything has a pressure) (see also the section on 'The Second Law of Thermodynamics and global warming').

Temperature, thermodynamics, and statistical physics all involve very difficult[50] ideas. The reason why these ideas are so difficult, and often lead to counterintuitive results, is that they are so far from common experience. In the same way as we never see one truly isolated billiard ball hurtling through space (see the section 'Kinetic energy—where does the energy go?'), we also never truly experience billions of collisions—we see the net outcome, yes, but we never *directly* experience them one by one by one . . . right up to a billion.

Entropy and the distribution of energy

What is entropy? (Entropy has been explained so well by Atkins, in *The 2nd Law: Energy, Chaos, and Form*,[51] that we shall not go into great

---

* Furthermore, combining Equations 18.6 and 18.7 implies that $n_A = n_B$ and is the true microscopic explanation of Avogadro's Law.

detail.) We have seen that a thermalized unit of heat is universal, depending only on the temperature; and temperature has to do with a universally defined condition known as equilibrium; but, finally, the entropy, S, is something that does depend on the actual details of the actual system under investigation.[52] It is up to the physicist or chemist to look at the given system and discover what the 'architecture' of the energy levels is, whether it be the energy levels for conduction electrons in a metal, the molecules of a gas, a vapour–liquid mixture, and so on. Both the number of energy levels and their occupancy (e.g. how many electrons are allowed) must first be determined before changes in the entropy can be determined.

Boltzmann defines entropy as the tally, $S = k \log W$, of all the ways in which the microscopic energy units can randomly distribute themselves between the prescribed energy levels for the given system (see Chapter 17). If there are many different arrangements that give rise to *one* macroscopic energy profile, then that profile has a high chance of occurring, and a correspondingly high entropy. By analogy, if the 10^{24} molecules in a one-litre container of gas had individual names (Tom, Dick, Harry, Abdul, Ahmed, etc.[53]), then rearranging the molecules by name would make no difference whatsoever to the macroscopic energy profile. But, as there are 10^{24} factorial possible arrangements—a very very . . . very big number—then this profile has a relatively very very . . . very high entropy.

The entropy is high, but relative to what? Consider now that there is a thimble in one corner of the container. If, at one time, it so happens that all the molecules find themselves within this thimble, then this would be a *manifestly* different energy profile—for example, the container might implode. Our experience and our intuition tells us that this 'thimble-state' is very unlikely to occur—by chance, but we can use the laws of chance, as did Boltzmann, to determine how much less likely. It turns out that the first energy profile (molecules filling the whole container) is $10^{10^{25}}$ times more likely than the 'thimble-state'. This is a stupendously big number, so, realistically, the 'thimble-state' has zero chance of occurring spontaneously.

Sometimes entropy is referred to as 'disorder', and entropy maximization is then equivalent to the maximization of disorder. However, this renaming can lead to counterintuitive results. For example, a gas, as we have seen, is more 'disordered' when it completely fills the container, whereas matter on a cosmic scale is more 'disordered' when it is clumped (into planets and stars as opposed to a uniform spread of dust or gas).[54]

The confusion is resolved when we note that it is not *spatial* distribution alone that is being thermalized; there is also the thermalizing of the *momentum* of the particles. As particles fall together under gravity, then their average speed increases, and so the temperature and the entropy both increase. This always more than compensates for the 'orderliness' of the clumping. (The fact that spatial *and* momentum parameters are implicated is confirmation of Hamilton's modelling of systems using both q and p parameters.)

But what, exactly, is the driver for change? We know that many physical processes—for example, melting, boiling, and the free expansion of a gas—continue even when the temperature, T, remains constant. Gibbs started his great work of 1878 with the words:

. . . when the entropy of the system has reached a maximum, the system will be in a state of equilibrium. Although this principle has by no means escaped the attention of physicists, its importance does not appear to have been duly appreciated.[55]

Gibbs was drawing attention to the fact that at equilibrium, the distribution of entropy against energy is 'stationary' (flat) with respect to small changes in energy: and this, finally, is the driving force behind all statistical processes in physics—the energy is trying to get as evenly distributed as possible. We can at last explain why heat *must* flow from hot to cold. The basic unit of thermal energy is proportional to kT, and so this unit is smaller as the temperature is lower: the energy can therefore be distributed more finely at lower temperatures.

Thus, even when the temperature has reached its equilibrium value, and even when no heat is added or work is done, there is still this one agency for change. Take the mysterious example of the 'free expansion' of an ideal gas (see Chapter 10, Part II). The temperature and internal energy of the gas remain constant, and no work is done against intermolecular forces or external pressures. However, the volume does increase, and so the energy levels of the gas become more numerous, and closer together. (Quantum mechanics is needed to explain this, but it's a bit like standing waves on a rope: as we consider longer and longer ropes, then there are more and more possible standing wavelengths, and so the energies are more and more crowded together.) Therefore, as the gas expands freely, its energy levels become more and more finely graded and numerous, and its entropy increases. To sum up, temperature equilibration smooths out the energy as best it can, but when a common temperature has been reached then entropy maximization carries on with the job and leads to an energy distribution that is smooth to *greater and greater precision*.

The Second Law of Thermodynamics, reversibility, and structure

In 1909, The Greek mathematician Carathéodory (1873–1950) finally succeeded in ridding the Second Law of anthropic constructs such as heat-engines, and putting it into its most economical form:

In the neighbourhood of any equilibrium state of a system there are states which are inaccessible by an adiabatic process.[56]

'Neighbourhood' is reminding us that entropy increases locally and incrementally—we can't save up the increase for later or for somewhere else. 'Adiabatic' means 'without transfer of heat' (see Chapter 10), so there always exist close states that are *only* accessible *with* heating. In other words, work alone isn't enough to undo or set up the sort of messing (randomizing) that only heat can do. For example, when doing the house*work*, I can never clean away every last bit of dust, but also, I could never set up the initial arrangement of dust—one mote here, another mote there, and so on.*

Now, when considering just two or three microscopic particles, the interactions are reversible (we can replace '*t*' by '*–t*'), but considering processes on everyday macroscopic scales, with billions of particles, then time is not reversible. Maxwell was one of the first physicists to appreciate that this macroscopic irreversibility was statistical in origin. (We have already quoted Maxwell's poetic example of raindrops flying upwards and people becoming unborn; see Chapter 17.) The fact that we don't see macroscopic time running backwards is therefore not because this would be impossible in an absolute sense, but because the starting conditions in these time-reversed cases are very specific and incredibly unlikely to occur at random.

However, it's not the fault of entropy that, out of a multitude of starting states and final states, only a handful are special—to us. For example, think of the number of ways in which colours can be spread between pixels in a photograph. For a random distribution, then billions of different photos will come out uniformly brown, and the photo with my daughter and her cat will be lost amongst this crowd. I can't help attaching special significance to this photo, but the 'entropy nose' is only tuned to various extensive aspects—in this example, to the distribution and conservation of the total amount of ink.

---

* Of course, I wouldn't want to even try.

If the direction in macroscopic cosmic* time is not due to any asymmetry in the microscopic *laws* of physics,[57] it must be due to an asymmetry in the microscopic *conditions* (that is, the starting conditions). The question then becomes: why, actually, is the starting condition so special? There are (at least) two possibilities: either the current state of the universe represents a large statistical fluctuation or the universe started in a special way. Physicists have adopted the second option ever since there has been convincing experimental evidence[58] of a 'Big Bang' origin of the universe.

But how can macroscopic structures ever arise (via this process of continual thermalization, continual randomization); in other words, why do we have stars, planets, and black holes, and so forth instead of, say, uniform radiation or uniform dust? There are three steps to 'structure'. First, some tiny deviations from featurelessness occur by statistical fluctuations. Second, 'structure breeds more structure' by statistical default—like a crystal that attracts nearby atoms and so grows in size; the grading of gravel into different layers as a barrel is bounced around on a truck; or the drawing that emerges as sand dribbles out of a hole in a swinging bucket. Third, an additional mechanism can come into play that serves to 'protect' the structure. In physics, a particular arrangement of matter (e.g. planets, stars, and galaxies) may be 'protected' due to each body being in a 'well' of potential energy; in chemistry, arrangements of atoms in large molecules may be protected by large activation-energy barriers; and in biology, we have extremely complicated biological entities that manage to survive if they are 'the fittest'. None of this contradicts the Second Law of Thermodynamics.[59] In fact, both the First and Second Laws are *overall* directives; in order to explain what happens *in detail*, we need . . . the whole of the rest of science.

The Second Law of Thermodynamics and global warming

What is the link between global warming and the Second Law of Thermodynamics?

The 'temperature' of the planet (that is, the surface temperature) is a tricky statistic, as misleading as other artificially constructed statistics, such as 'gross domestic product'. This is because the Earth never reaches equilibrium: just as it's beginning to warm up on the day side, the Sun is 'switched off'. Also, how shall we combine the various temperatures of

* 'Cosmic' means pertaining to the whole universe.

the oceans, the land masses, and the ice caps, and take account of winds, currents, different reflectivities, internal sources of heat, and so on. (It's much more complicated than simply putting a clinical thermometer in your mouth and 'taking your temperature'.) Nevertheless, global warming is really happening, and, rather than temperature, we need to take into account the *energy* uptake of the oceans, the atmosphere, the forests, and so on.

Yet, it's interesting that Carnot's insight is not wrong, and 'temperature' is, after all, a telling statistic. Now Carnot, as we remember (see Chapter 12), put the Second Law in terms of heat-engines: every engine must throw away heat at a lower temperature; and the efficiency of any engine is greater the higher the source temperature, and the lower the sink temperature. We must appreciate that the Second Law is a universal law, and that 'heat-engines' are to be understood as 'any physical process that does work'. Looked at in this way, global warming has enormous repercussions: cars and airplanes are engines, but forests and brains are 'engines' too. As the heat sink (the planet) becomes warmer—has a higher average temperature—so the efficiency of every 'engine', every macroscopic physical process, is reduced[60]. . .

19

Conclusions: What Is Energy?

What is energy? The physicist at the start of this book can now make up a very long list of the various forms of energy: the motion energy of a cricket ball, the energy stored in food, a battery, the binding energy of nuclei, the bond energy of chemicals, heat, the energy of a flowing river or a trade wind, the energy transported by electromagnetic waves, the stress energy in a loaded beam or in the gravitational field, the energy stored in a dam or a millpond, the energy in steam, the energy stored in a capacitor, the energy required to move a magnet in a magnetic field, or electrons round a circuit, the energy in a flywheel, or raised weights, or a squashed/stretched spring, the rest mass of fundamental particles, the energy used by a light bulb in an hour, the energy implied by a certain curvature of spacetime, and so on—and modern examples, such as the energy of the Higgs boson or of a rotating black hole. Upon careful examination, all these various forms fall into just two main types of energy: kinetic, the energy of motion; and potential, the energy of interaction of parts of a system. In certain situations, there is a smooth interchange between these two types, a 'dance to the music of time'.[1] The kinetic component is perhaps the more fundamental—it is of one basic form, '$\frac{1}{2}mv^2$', and also explains the mechanism by which an enormous collection of tiny chunks of energy is distributed randomly. The form of the potential energy is various, and depends on the details of the given system. But there can be some ambiguity, some overlap, between the two basic types, kinetic and potential. Finally, there are three overarching laws that govern energy: it is conserved within an isolated system (the whole universe is the biggest example); its cosmic distribution is becoming as randomized as possible, as finely as possible; and, through time, it takes the path that uses up the least action.

But what actually is energy? Non-physicists may wonder why the enquiry isn't broadened to cover questions such as 'What does "energy" mean for humans and for society?', 'What is energy philosophically speaking?', 'How does "energy" occur in literature', and so on. These questions have their own interest, but 'What is energy?' is a *physics* question, and needs a physics answer. Moreover, when the physics is right, then the philosophy is right, no matter how long it takes the philosophers to come around.[2] Now, if the starting question was 'What is lasagne?' then we could answer it thus: lasagne is an Italian dish, made from alternating layers of pasta, meat sauce, and cheese sauce, and the meat sauce is made from minced beef, onions, tomatoes, garlic, oil, and seasonings. In other words, we would give a list of ingredients. But the question 'What is energy?' cannot be answered in this way, by a list of ingredients. However, Feynman's *Lectures on Physics*—in my estimation, the best book of physics explanation since Galileo's *Two New Sciences*—says it all: energy is that-which-is-conserved. More than that, energy is that which is *defined* by its property of being conserved. There are other conserved quantities in physics—linear momentum, angular momentum, electric charge—but these things are not defined by the property of conservation.[3] (For example, the amount of electric charge on an electron can be determined by measuring its deflection as it moves through a given magnetic field.)

However, the history shows us that it has not been easy to discover energy through this property of conservation. Consider: when you have a medical problem, you can never assume that there is only *one* thing wrong with you—there could be two, three, or more interrelated causes for your symptoms. Likewise, for energy, the unknown factors must be disentangled in order to arrive at the *individual* 'blocks', the actual formulae: $\frac{1}{2}mv^2$, $F.dr$, qV, mc^2, $\frac{1}{2}CV^2$, $\frac{1}{2}kT$, $dQ = TdS$, $\Delta Q = mc\Delta T$. . . . Finally, it was crucial to employ a *closed* system—but, when you don't yet know what energy is, how can you be sure that the system is indeed closed?

But the one-line definition of energy, as 'that-which-is-conserved', is not always at our disposal: sometimes energy is *not* conserved (the system being investigated is not isolated), and also, while it brings in the actual formulae of the blocks, it doesn't explain *why* energy is transformed between these blocks (for that, we need other laws, such as the Principle of Least Action). Now, the energy in, say, a flammable gas, exists in many forms: we have the chemical bond energy binding the atoms into molecules; the thermal energy of these molecules; bulk motion energy (for example, the energy in a directed jet of gas); the weak bond energy between molecules; and there may be external force fields, such as electromagnetic or gravitational.[4] Instead of listing all these various sources of energy, we can ask a different question: what *work* can the system do for us? We could burn the gas, and thereby bring water to the boil, capture the steam, and use it to push a piston that is geared up to raise some weights. This example leads us to another, more practical, definition of energy: the energy of a system is the capacity of the system to do work. This is probably the best one-line definition of energy, applicable in a large number of cases, especially in classical physics. However, even this one-liner has limitations: it is less useful in modern physics, as 'force', and 'work done by a force', are no longer the most appropriate ways of modelling the system; and sometimes the energy is locked away and so no work *can* be done. Consider, for example, a universe that has reached its 'heat death': there is still energy in the system, but it can do no work. (The amount of energy left in the system is more than a mere end-state convention; for example, we might wish to compare the energies of different universes that have 'died' at different temperatures.) We must just accept that there is no universal one-line definition for energy. As Einstein said—an explanation must be as simple as possible, *but no simpler.*

We often forget our victories as soon as we have won them, but the history can remind us to be appropriately awestruck. Take, for example, our intuitive understanding of gravity as something that makes apples fall to Earth, and planets orbit around the Sun. This 'intuition' is actually a sophisticated intellectual understanding. As Hamilton said: 'Do you think that we [actually] *see* the attraction of the planets? We scarcely see their orbits'.[5] (In other words, even to know that a planet has an orbit, elliptical in shape, requires advanced observations and theories.) Let us pause for a brief review of the history of energy.

'Energy' was not simply waiting to be discovered, like a palaeontologist might find the first ichthyosaur or a prospector stumble across the Koh-i-noor diamond.[6] The concept of energy had to be forged, and out of ingredients both physical and metaphysical.

First, we saw how the futile quest for perpetually acting machines hinted that *something* was being conserved. Then Leibniz identified 'mv^2' as that something—except that it was only conserved in those collisions where it was conserved. There followed a 50-year controversy over which was the important measure of 'motion'—Newton's momentum or Leibniz's 'mv^2'.

Amontons built a small-scale prototype 'fire mill', and was able to quantify and compare the work capabilities of men, horses, and his hypothetical heat-engine.

Then, the concept of potential energy began its gradual evolution and also its merging with the engineer's 'work'. The Bernoullis, father and son, were instrumental in understanding that kinetic and potential energy were conserved *jointly*—as mechanical energy. Daniel Bernoulli, in particular, was the first to understand energy in a modern way—as a fuel that could make engines do work.

Meanwhile, the most extraordinary technological invention of all time—the steam engine—was discovered in Britain, by Newcomen, and Watt.

Then, in the last quarter of the eighteenth century, the mechanics of Lagrange, the acme of analytical perfection, brought in a new outlook: not just conservation but also the most economical path in time.

And then, a most surprising thing happened (in the first decades of the nineteenth century). On the one hand, Lagrange's mechanics was successfully applied to everything (everything *mechanical*) and especially celestial motions; on the other hand, Watt's improbable steam engine was pulling locomotives and powering the industrial revolution; and on the other other hand, Lavoisier and Laplace's subtle heat theory explained much chemistry, phase changes, gases, and the speed of sound. But (and this is the real surprise) these three domains—the conservation of energy in 'clockwork' mechanisms, the power revolution, and the theory of heat—all remained completely separate from one another. So, while the countryside began to be criss-crossed by steam trains on railway lines, *still* 'energy' had not been discovered.

The subtle-fluid theory of heat was remarkably successful, right up until the great unravelling in the mid-nineteenth century. This subtle-fluid theory (the caloric theory) had its roots in chemistry: chemical substances were always conserved, and so the heat-fluid should likewise always be conserved. There were really only a few niggling details that the theory couldn't explain convincingly: the 'inexhaustible' amount of heat that could be generated by friction, the transmission of smells across a room, and the sublimation of ice into a vacuum. It must not be thought, however, that the main impetus for the discovery of energy came from any such discrepancies. Like the Copernican theory, and Einstein's theory of General Relativity, the theory of energy was motivated mostly by its superior explanatory power and its beauty. It was in this spirit that Mayer and Helmholtz appealed to the age-old wisdom that 'cause equals effect', and Joule, likewise, asserted that the 'Creator' would not sanction an absolute creation or loss of 'force'. But this was still a hard step to take as it required not only a break with the ruling caloric theory but overriding a 'category error'—heat and work were as different as p.s.i (pounds per square inch) and PSA (Pleasant Sunday Afternoon).

Fifty years earlier, the remarkable Rumford, and also the early kinetic theorists, had started to cross this bridge. One crucial advance in understanding came with the progression from a static to a dynamic outlook: from a static balance of forces to the perpetual 'dance to the music of time'; from the static heat-fluid finding its level to the dynamic equilibrium of Prévost; from the static to the dynamic descriptions of a gas. As Thomson wrote in 1872: 'I learned from Joule the dynamical theory of heat and was forced to abandon at once many, & gradually from year to year all other, statical preconceptions'.[7]

Hamilton discovered a theory that blended light and particle mechanics, and which had an energy function, 'the Hamiltonian', at its core: it was not appreciated in his lifetime but was to pay rich dividends in another century.

Clausius and Thomson developed the first two Laws of Thermodynamics. These Laws were unique in a number of ways: they were given in words, for public consumption; they were cosmic (they applied to the whole universe); and, in the case of the Second Law, a direction (an inequality) was brought into physics for the first time.[8]

The history also shows that niggling details don't go away. The remedy was both to develop theories, and to make many and various *quantitative* measurements. And there was no telling, in advance, what effects had to be taken notice of: the rate of transmission of a smell across a room; the air temperature during a thaw; the rate of melting of snow; shadows being unaffected by breezes; the precession of planetary orbits; a warming effect beyond the red end of the spectrum; the size of hailstones; drops of colourless liquid in a flask; the rate at which a magnet is moved towards a coil; the dip and orientation of a compass needle; sublimation in a vacuum; the brightness and direction of comets' tails; bubbles at electrodes and sparks at commutators; where cannonballs land relative to masts and towers; the heating of carriage wheels and axles; the impossibility of reflecting cold; the temperature of the sea's surface after a storm; the viscosity of gases; the warmth of a gun barrel fired with and without a bullet; the clicking sound as billiard balls collide; the redness of blood in the Tropics . . .

Also, it was essential that mathematics advanced hand in hand with experiment. Nature had to be hypothetically sliced, chopped, stretched, smoothed and generally beaten into mathematical shape. Energy could never have emerged without the calculus, or probability theory, but, even earlier, it was Galileo who first understood the importance of mathematics, and how to winkle out an ideal law from real observations. Which is better, Newton's force view or the modern energy view? It is the whole-system energy view that has been found to be more profound. The problems of absolutes are avoided (energy, and also space, time, and mass, are now all 'relative', that is to say, they are all referred to the given system). Also, Einstein extends 'uniform motion' to include accelerations;[9] and, instead of hypothesizing a pre-existing empty space and universal time, he takes the presence of a large centre of gravitation as a given. (This is more 'correct'—after all, all our observations really *have* been made in the presence of a large centre of gravitation.) But what, if anything, is determined absolutely? Consider the three most important principles[10] in the whole of physics: the principle of Energy Conservation, the Principle of Least Action, and (Einstein's) Principle of Relativity. The last principle (the Principle of Relativity) stands in contrast to the others—it is strangely metaphysical—how can such a metaphysical principle have anything physical to say? Answer: its physicality lies in the fact that it insists that the true universal aspects must be distinguished from the artifice of a particular point of view. So, as the kinetic and potential energy contributions change in going from one observer to another, what finally is that kernel of physical reality that can be plucked out of these shifting sands? It is not the total energy, it is that more complicated thing—the 'change in action between given end conditions'. Why, then, do we have more to do with 'energy' than with 'action'—in physics, and in everyday life? (Everyone has heard of energy; who but the physicist has heard of action?) This can't be answered, in our present state of understanding, but it might have something to do with the fact that 'action' is too high-level a measure for most practical purposes—like saying 'the answer is 42'. We are nevertheless bound to use this high-level parameter when we don't have complete information, or when the system is too complicated for a complete solution—this occurs especially in quantum mechanics and statistical mechanics.

Energy has extensive (entropy) and intensive (temperature) attributes. While action may be divided up as (p,q) or (E,t), it is in the latter pairing where the extensive and intensive aspects are most polarized. But there appears to be some overlap in the duty of the parameters E (energy) and t (time). Consider a body travelling all alone in 'outer space' at constant speed and in a straight line. What if, all of a sudden, it spontaneously combusts, like a haystack, or it decays, like a radioactive nucleus? We shall then have to admit that either the original system wasn't truly isolated, or the body had internal structure (internal potential energy) of which we were previously unaware. We shall have to keep on and on and on watching the body—through *time*—to make sure that it really is isolated. Only when time is uniform and featureless, in other words, the same everywhen, and not bunched up or stretched out, can we be sure that energy really is conserved. But if time is completely homogeneous then how can we measure its passing?[11] Another analogy for the inter-dependence of 'energy' and 'time' is set in the quantum mechanics regime. Imagine that we are watching a wave go by our 'quantum-mechanical window'. If we close the curtain to a mere crack then we shall know at exactly what time a wave-crest goes by; but if we want to know the energy of the wave, then we shall have to sacrifice this time-certitude, and watch for a long window of time in order to determine the wave's frequency (the wave energy is proportional to the frequency).[12]

It is interesting that the discovery of energy was so Eurocentric. While the hearth, 'fire lathe', lever, wheel, windmill, wheelbarrow, water-wheel, rowing boats, and sailing boats, and so on cropped up in most places all over the world, the steam engine, and the electric motor occurred in only one place, at only one time. Doubtless there are geographical, socio-political, and economic factors, but the one-off nature of these discoveries is ultimately a testimony to the rule of the Second Law of Thermodynamics—it really is very difficult to get work from heat. This is evidenced by the extraordinary intricacy of these engines—so much more complicated than the prototypes of Hero's steam kettle, Pixii's dynamo, or dal Negro's motor.

One thing is sure and that is that the concept of energy is here to stay. It is not sufficient for a concept just to be mathematically defined, measurable, and leading to consistent results—it must also get used. There is no doubt that 'energy' meets these requirements.

This is not the place to address our current woes concerning diminishing resources and global warming. I will, nevertheless, offer two thoughts. One is that there is no safe form of energy—it is energetic, after all. The other is that, as the sink for all our activities becomes warmer, so all our 'engines'—biological as well as man-made—will work less and less efficiently (we remember Carnot's directive: for maximizing efficiency, the sink of heat must have as low a temperature as possible). Cars already run less efficiently in the summer,[13] plants will have to consume even more water for cooling, the human 'engine' will work less well, and so on.

While the challenge of global energy use, and global warming, has more to do with holding on to our humanity than with anything technological, it is nonetheless seductive to try and think of some 'hi-tech' solutions. One idea of mine was to seed the Earth's surface with rocket-thrusters, and then to synchronize the firing of these thrusters (those on the sunny side coming on, just as those at dusk were being turned off) in order to move the Earth into a new orbit, very slightly further away from the Sun. However, a back-of-the-envelope calculation soon showed that the amount of energy required was astronomical (in fact, planetary). We must heed Pauli's advice that in order to have one good idea, one must have many ideas. I hope this book will stimulate many readers to have many ideas.

In one sentence, energy is: the ceaseless jiggling motion, the endless straining at the leash, even in apparently empty space, the rest mass and the radiation, the curvature of spacetime, the foreground activity, the background hum, the *sine qua non*.

APPENDIX I

Timeline

300 BC	*Mechanica*, Peripatetics, law of the lever
250 BC	Archimedes, law of the lever
200 BC	first water-wheels, China
100 BC	Illyria (Albania), water-powered wheels for grinding corn
1	Ko Yu, invents wheelbarrow, China
60	Hero of Alexandria, steam-powered toy
600	Persia (Iran), first windmills
624	Brahmagupta's perpetual wheel, India
1086	5,624 water-wheels listed in the Domesday Book, England
1150	Bhaskara II's perpetual wheel, India
1225	depiction of wheelbarrow, Chartres Cathedral, France
1235	de Honecourt, France, perpetual overbalancing wheel
1250	Jordanus de Nemore, law of the lever
1400	windmills used in Holland for land drainage
1586	Stevin's 'wreath of spheres', Antwerp, Flanders (Belgium)
[1618–48	The 'Thirty Years' War']
1620	Bacon, heat is a motion of the small parts in *Novum organum*
1620s–1630s	Galileo, 'v^2 proportional to h'; thermoscope; Galilean relativity of motion
1629–33	Descartes, 'The world'
1632	Galileo, 'Dialogue concerning the two chief systems of the world—Ptolemaic and Copernican'
[1634	Thomas Hobbes visits Galileo]
1637	Descartes, *Discours de la méthode,* including the *Dioptrique* Descartes (in letter to Huygens), 'force' is 'weight × height'
1638	Galileo, *Discourses Concerning the Two New Sciences* [John Milton visits Galileo]
[c. 1640	Poussin's painting, *A Dance to the Music of Time*]
1644	Torricelli's barometer and the 'ocean of atmosphere'
1646	Pascal's 'barometric' experiments with water and wine at Rouen
[1648	Thomas Hobbes meets Descartes]
1656	Huygens, *De motu corporum ex percussione*

1657	von Guericke's demonstration with evacuated copper hemispheres
1660	Boyle, *New Experiments Physico-mechanical, Touching the Spring of the Air, and its Effects*
[1660s–1670s	Benedict Spinoza]
1662	Boyle's Law and Fermat's Principle of least time (for the path of light)
1663	Marquis of Worcester, 'water-commanding' engine
1665	Boyle distinguishes motion of heat and bulk motion in *New Experiments and Observations Touching Cold*
[1665–66	The Great Plague of London, the fire of London]
1673	Huygens, *Horologium oscillatorium sive de motu pendularium*, dedicated to Louis XIV, and *De vi centrifuga*
1679	Mariotte, 'Essay du chaud et du froid'; Papin's 'digester'
1686	Leibniz, *Brevis demonstration erroris memorabilis Cartesii*—he discovers mv^2 or *vis viva*
1687	Newton, *Principia*
[1690	John Locke, *An Essay Concerning Human Understanding*]
1692	Leibniz, *Essay de dynamique*
1695	Leibniz, *Specimen dynamicum*
1697	Johann Bernoulli's solution of the brachistochrone problem, using the path of light through a medium of variable refractive index
1698	Savery's 'fire-engine' or 'miner's friend'
1699	Amontons's 'fire-mill' and definitions of work and friction
[*c.* 1700	Becher and Stahl, phlogiston theory]
1701	Newton's law of cooling (published anonymously)
1703	Jakob Bernoulli on the compound pendulum as a hypothetical lever
1704	Antoine Parent, 'Theory of the greatest possible perfection of machines'
1712	Newcomen's 'atmospheric' steam engine
1715	Brook Taylor, 'Methodus incrementorum directa et inversa' Johann Bernoulli's 'Principle of virtual velocities/work' in a letter to Varignon; first use of word 'energy' in physics
1716	Jakob Hermann, first kinetic theory in *Phoromania*; death of Leibniz
1720	's Gravesande, *Mathematical Elements of Natural Philosophy, Confirm'd by Experiments*
1720s	Experiments on balls falling into clay Fahrenheit carries out experiments on thermometers and on heat

1723	Brook Taylor uses the 'method of mixtures' to develop a scale of temperature
1724	*Vis viva* controversy continues with competition of the Académie Royale des Sciences on the 'communication of motion'
1727	Hales, *Vegetable Staticks*; Newton dies
1727–28	Johann Bernoulli's work on vibrating strings; Voltaire in exile in London
1733	Daniel Bernoulli's work on vibrating strings and trigonometric series
1735	Boerhaave, *Elementa chemiae* and his subtle-fluid theory of heat
1736	Euler, 'Mechanica, sive motus scientia analytice exposita', in which Newtonian mechanics is put in analytical form
1737	Algarotti, *Newtonianism for the Ladies* Emilie du Châtelet, 'Dissertation sur la nature et la propagation du feu' Voltaire, 'Elements of the philosophy of Newton'
1738	Daniel Bernoulli, *Hydrodynamica,* including the kinetic theory of gases, the idea of *vis potentiali*, and the conservation of 'live force' Daniel Bernoulli's paper on the Sun–Earth–Moon system, demonstrating the route-independence of *vis viva* between fixed end-points Martine finds that 'quicksilver is [exceptionally] ticklish'
1740	Du Châtelet, *Lessons in Physics*
1741–46	Maupertuis develops his 'Principle of Least Action'
1743	D'Alembert, *Traité de dynamique* Clairaut, *Théorie de la figure de la terre*, the start of potential function theory
[1748	David Hume, *An Enquiry Concerning Human Understanding*]
1749	du Châtelet translates Newton's *Principia* into French
1751–72	Diderot, *Encyclopédie* (some with d'Alembert)
1752	Déparcieux uses reversibility argument to quantify the efficiency of water-wheels [Voltaire, *The Diatribe of Dr Akakia, Citizen of St Malo*]
1756	Cullen, 'Of the cold produced by evaporating fluids and of some other means of producing cold'
1759	Smeaton, overshot water-wheels are more efficient than undershot wheels
1760s	Black develops his theories of latent and specific heat
1763	Boscovich, 'A theory of natural philosophy'
1765	Watt's idea of separate condenser

1769	Wilcke, latent and specific heat
1770s	Irvine's theory of heat
1773–80	Lagrange, intimations of potential function theory
1776	First commercial Boulton and Watt steam engine [Adam Smith, *The Wealth of Nations*] [American War of Independence]
1777	Scheele, 'A chemical treatise on air and fire', discovers radiant heat
1779	Crawford, first to consider the specific heat of gases in *Experiments and Observations on Animal Heat*
[*c.* 1780s on	The Industrial Revolution]
[1781	Immanuel Kant, *Critique of Pure Reason*]
	Hornblower's two-cylinder compound engine
1782	Watt's rotative engine
1783–84	Legendre functions (used by Laplace)
	Lavoisier and Laplace, *Memoir on Heat,* use of ice calorimeter
	Watt's first double-acting engine
	Lazare Carnot, 'Essai sur les machines en general'
	[First flight in a hot-air balloon, made by the Montgolfier brothers]
1785–89	Ingen-Housz, 'Nouvelles éxperiences et observations sur divers objets de physique', speed of heat-conductivity experiments
[1786	First ascent of Mont Blanc, by Balmat and Paccard]
[1786–89	Mozart composes 'The Marriage of Figaro', 'Don Giovanni', and 'Cosi Fan Tutte']
1787?	Cavendish's kinetic theory in 'Heat', only discovered in 1969
[1788	Hutton, *Theory of the Earth*]
	Lagrange, *Analytique mécanique*
1789	Laplace's theory of spheroidal attraction, applied to the rings of Saturn, leads to 'Laplace's equation'
	Lavoisier, *Traité élémentaire de chimie,* in which *'calorique'* is one of the elements [French Revolution]
	Rumford's generation of heat by boring cannons
1790	Pictet, 'Essai sur le feu'
1791	Prévost, 'De l'equilibre du feu'
1792	Davies Gilbert shows that $\int P \, dV$ is the work done by a steam engine
1794	Lavoisier loses his head at the guillotine
1796	Southern's 'indicator'
[1798	Coleridge, 'The Rime of the Ancient Mariner']
1799	Trevithick's first high-pressure steam engine Davy's ice-rubbing experiment
1799–1825	Laplace, *Mécanique céleste*
1800	William Herschel discovers infrared radiation Volta invents the voltaic pile [Watt's patent expires]

1800s	Dalton, thermal expansion of gases, atomic theory
1801	Young puts forward the wave theory of light and also of heat
1802	Ritter discovers ultraviolet radiation; Gay-Lussac, expansivity of gases with heat
1804	Fire-piston demonstration at Lyon
1806	Berthollet and Laplace found the Societé d'Arcueil
1807	Young, in his 'A course of lectures in natural philosophy and the mechanical arts', proposes the term 'energy' instead of 'living force'
	Gay-Lussac's law of simple proportions and the twin flasks experiment
[1810	Goethe, *Theory of Colours*]
[1810	Column-of-water engine, von Reichenbach]
1811	Avogadro's Hypothesis
1812	Clément and Desormes' measurement of γ
[1815	Battle of Waterloo]
1816	Laplace's 'adiabatic' correction to the speed of sound
	Herapath's kinetic theory of gases
	[Coleridge's poem 'Kubla Khan']
1819	Dulong and Petit's Law on the constancy of specific heat per atom
1820	Oersted, compass needle moves when near 'galvanic current'
1821–60	Faraday's experiments on the 'unity of force'
1821	Seebeck, link between electricity and heat
1821 (1822?)	Poisson's equation
1822	Fourier, *The Analytical Theory of Heat*
	[Charles Babbage starts making his first calculator, or 'difference engine']
1824	Sadi Carnot, *Réflexions sur la puissance motrice du feu*
[1824	First performance of Beethoven's Ninth Symphony]
[1827	'Congreves' friction matches]
1828	Green defines the potential function (in 'An essay on the mathematical analysis of electricity and magnetism')
[1829	Stephenson's 'Rocket', Newcastle, England]
	Coriolis defines work as the integral of force over distance
1830	Hippolyte Pixii's dynamo
1831	dal Negro's electric motor
1833	Hamilton's mechanics
1834	Peltier, link between electricity and heat
	Lenz's Law; Clapeyron, 'Memoir on the motive power of heat'
1835	Ampère's wave theory of heat
[1837	Start of Queen Victoria's reign]
1841–48	Julius Robert Mayer's conservation of 'force' and mechanical equivalent of heat

[1842   Quetelet, *Treatise on Man*]
1843   James Joule, the mechanical equivalent of heat (from magneto-electricity) and the 'I^2R' law
1843–45   Waterston, kinetic theory of gases
1844–45   Joule's twin cylinder and paddle-wheel experiments
[1845   Neptune predicted by Le Verrier (Paris) and Adams (Cambridge)]
1846   Groves, 'On the correlation of physical forces'
1847   Joule's ideas noticed by William Thomson (Lord Kelvin) at the British Association meeting in Oxford; Helmholtz, 'Über der Erhaltung der Kraft'
[1848   Year of revolutions in Europe]
1850   Clausius, the First and Second Laws of Thermodynamics, in 'On the motive power of heat, and on the laws which can be deduced from it for the theory of heat'
1851   Thomson accepts dynamical theory of heat and conservation of energy; alternative statement of Second Law; introduces term 'energy'
1852   Thomson, 'On a universal tendency in nature to the dissipation of mechanical energy'
1854   Helmholtz, 'heat death' of the universe
1857   Clausius, kinetic theory of gases
[1859   Darwin's theory of evolution]
1865   Clausius coins term 'entropy' and states two principles: the energy in the universe is constant; the entropy of the universe tends to a maximum
1865–73   Maxwell's theory of electromagnetism
1867   Maxwell's kinetic theory of a gas
1872   Boltzmann factor, Maxwell–Boltzmann distribution
1875–78   Gibbs, 'On the equilibrium of heterogeneous substances'
[1876   Charles Dodgson (Lewis Carroll), 'The Hunting of the Snark']
1876   Boltzmann, equipartition theorem
1877   Boltzmann, microscopic formulation of entropy
1884   Poynting, defines flux of electromagnetic radiation
1886   Hertz discovers radio waves
1900   Planck, quantum of radiation energy, $E = hv$
1905   Einstein, Special Theory of Relativity, $E = mc^2$, light is particulate
1915   Einstein, General Theory of Relativity, leading later to the stress-energy tensor

APPENDIX II

Powers of Ten for Energy[1]

Example	Energy in joules
Energy of a typical microwave oven photon	1.6×10^{-24}
Thermal energy of molecule at 25 °C (approx.)	4×10^{-21}
Energy range of photons of visible light	$(3-5) \times 10^{-19}$
Energy range of X-ray photons	$(2-2000) \times 10^{-17}$
Rest mass energy of an electron	8.2×10^{-14}
Average total energy released in the nuclear fission of one ^{235}U atom	3.4×10^{-11}
Rest mass energy of a proton	1.5030×10^{-10}
Rest mass energy of a Higgs boson	2×10^{-8} (125.3 GeV)
Kinetic energy of a mosquito in flight	1.6×10^{-7}
Kinetic energy of a small coin falling 1 m	1×10^{-1}
Energy to heat 1 g of cool dry air by 1 °C	1
Work done by a force of 1 ft-lb through 1 ft	1.4
Most energetic cosmic ray ever detected (in 1991)	5×10^1
Flash energy of a typical pocket camera	1×10^2
Energy to melt 1 g of ice	3.3×10^2
One horsepower applied for one second	7.5×10^2
Kinetic energy of a rifle bullet	1.8×10^3
Energy to vaporize 1 g of water into steam	2.3×10^3
Energy in an alkaline AA battery	9×10^3
Energy released by the metabolism of 1 g of fat	3.8×10^4
Kinetic energy of a 1 g meteor hitting the Earth	5×10^5
Recommended daily food intake for a man	1.1×10^7
Energy from burning 1 m^3 of natural gas	4×10^7
Energy in an average lightning bolt	1.1×10^9
Energy in a full petrol tank of a large car	2×10^9
Kinetic energy of an Airbus A380 at 289 m s^{-1}	2.3×10^{10}
Total energy released in the nuclear fission of one gram of ^{235}U	8.8×10^{10}
Energy released by the atomic bomb used in the Second World War at Nagasaki	8.8×10^{13}
Energy of an average hurricane, per second	6×10^{14}
Energy from the Sun striking the Earth per second	1.7×10^{17}
World energy consumption in the year 2010	5.0×10^{20}

APPENDIX III

Extras

1) Scientists listed in Chapter 1, 'Introduction':
    Persecuted geniuses: Galileo, Voltaire, Mayer, and Lavoisier
    'Royal' patronage: Daniel Bernoulli (Peter the Great), Euler and Mauper-
    tuis (King Frederick), Lagrange (Louis XVI), Leibniz (House of Hanover),
    Lazare Carnot (Napoleon Buonaparte), Francis Bacon (King James I),
    Count Rumford (Elector of Bavaria), and William Thomson, a.k.a. Lord
    Kelvin (Queen Victoria)
    Ivory-towered professors: Black, Newton, Maxwell, Helmholtz, Thomson,
    Tait, Lagrange, Gibbs, and so on
    Lowly laboratory assistants: Hooke, Watt, and Faraday
    Social climbers: Rumford—and also Leibniz, and Voltaire
    Other worldly dreamers: Einstein, Newton, Mayer, Descartes, Hamilton,
    and Cavendish
    Richest man in all Ireland: Boyle
    Richest man in all England: Cavendish
    One who made a loaf of bread last a week: Faraday
    Feuding families: the Bernoullis
    Prodigal sons: Mayer—to the East Indies and back; also Fahrenheit (and,
    to a lesser extent, Watt and Davy)
    Thomson and Thompson 'twins': William Thomson (later Lord Kelvin)
    and Benjamin Thompson (later Count Rumford)
    Foundling: d'Alembert
    Titled: Bacon, Boyle, Newton, Rumford, Cavendish, and Kelvin
    Revolutionary aristocrats: Lavoisier and Lazare Carnot
    Entrepreneurs: Rumford, Watt, Boulton, Kelvin, and, to a lesser extent,
    Galileo, and Helmholtz
    Industrialists: Boulton, Watt, and Roebuck
    Clerics: Hales, Clausius, Prévost, and Boscovich
    Lawyers: Bacon, and Lavoisier
    Academics: the Bernoullis, Lagrange, Thomson, Clausius, Tait, Black,
    Helmholtz, Boltzmann, Gibbs, Maxwell, and many others
    Engineers: Parent, Newcomen, Smeaton, Watt, Lazare, and Sadi Carnot
    Savants: d'Alembert, Diderot, and Voltaire
    Doctors: Boerhaave, Black, Dulong, Mayer, and Helmholtz
    Pharmacists: Scheele, and Mayer

Diplomats: Huygens and Rumford
Soldiers: Descartes and Rumford
Teacher: Dalton
Spy: Rumford
Taxman: Lavoisier
Brewer: Joule
Miller: Green
Persecuted: under house arrest (Galileo), imprisoned (Voltaire), guillotined (Lavoisier), and ignored (Mayer, Joule, and Waterston)
Ladies' man: Rumford
Pathologically shy: Cavendish
Gregarious American: Rumford
Very quiet American: Gibbs
Two wives: Mme Lavoisier (later, Mme Rumford) and Mrs Maxwell
One mistress: Emilie (la Marquise) du Châtelet
English eccentrics and gentlemen scientists: Cavendish and Joule; also Erasmus Darwin and other 'lunatics'
Scottish Humeans: Black and Hutton
Scottish Presbyterians: Watt and, in terms of their ancestry, Thomson, Maxwell, and Tait
French *philosophes*: d'Alembert, Voltaire, and Diderot
French professional scientists: Laplace, Gay-Lussac, Lagrange, Fourier, Biot, and many others
German Romantics: Oersted, Goethe, Johannes Muller (physiologist), and Ritter (also Humphry Davy)
German materialists (e.g. Helmholtz) and positivists (e.g. Mach)

2) For quasi-static processes, $\Delta S = 0$ (in 'the system'); for non quasi-static 'real' processes $\Delta S \neq 0$. Moreover, in these real processes, the finite ΔS goes to 'the surroundings' (the Universe). Finally, there is the astonishing empirical result that, for real processes, ΔS is not only finite but >0.

3) A schematic representation of Pascal's 'vacuum-in-a-vacuum':

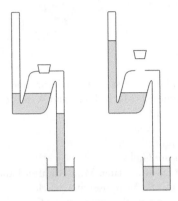

Fig. AIII.1 Pascal's 'vacuum-in-a-vacuum' (adapted from *The Construction of Modern Science: Mechanisms and Mechanics*, 1977, Richard Westfall).

4) Clausius' three-temperature cycle:

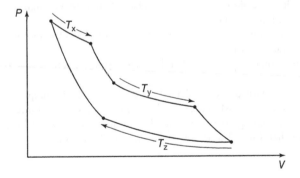

Fig. AIII.2 Clausius' three-temperature Carnot cycle (adapted from *The Mechanical Theory of Heat—with Its Applications to the Steam Engine and to Physical Properties of Bodies*, 1867, Fourth Memoir, Fig. 7).

APPENDIX IV

Miniature Portraits

Simon Stevin	Francis Bacon	Galileo Galilei	René Descartes	Robert Boyle
1548–1620	1561–1626	1564–1642	1596–1650	1627–91

Christiaan Huygens 1629–95	Gottfried Leibniz 1646–1716	Johann Bernoulli 1667–1748	Robert Hooke 1635–1703

Thomas Savery 1650–1715	Willem 's Gravesande 1688–1742	Hermann Boerhaave 1688–1738	Daniel Bernoulli 1700–82	Pierre Louis Maupertuis 1698–1759

Jean le Rond
d'Alembert
1717–83

Leonard Euler
1707–83

Joseph–Louis
Lagrange
1736–1813

James Watt
1736–1819

Joseph Black
1728–99

Antoine Lavoisier
1743–94

Wilhelm Scheele
1742–86

Henry Cavendish
1731–1810

Count Rumford/
Benjamin
Thompson
1753–1814

Lazare Carnot
1753–1823

Alessandro Volta
1745–1827

Humphry Davy
1778–1829

Thomas Young
1773–1829

Joseph Fourier
1768–1830

Sadi Carnot
1796–1832

William Rowan
Hamilton
1805–65

James Prescott
Joule
1818–89

Julius
Robert Mayer
1814–78

Herman von
Helmholtz
1821–94

William
Thomson/
Lord Kelvin
1824–1907

Rudolf Clausius
1822–88

Ludwig Boltzmann
1844–1906

James Clerk
Maxwell
1831–79

J. Willard
Gibbs
1839–1903

Max Planck
1858–1947

Albert Einstein   Erwin Schrödinger Richard Feynman
1879–1955   1887–1961   1918–88

Images for Amontons, Green, and Waterston do not exist. Thank you to Wikimedia Commons for all images prior to Planck: a donation has been made. For the remaining photographs, permission and thanks are as follows:

Planck, Archiv der Max-Planck-Gesellschaft, Berlin-Dahlem, photo by Tita Binz, 1936

Einstein, ETH-Bibliothek Zurich, Image Archiv

Schrödinger, Special Collections & Archives Research Center, Oregon State University, photographer unknown

Feynman, courtesy of the Estate of Richard P Feynman, via Basic Books.

Notes and References

Two sources will be abbreviated as follows:

DSB = *Dictionary of Scientific Biography*, editor-in-chief Charles C. Gillispie, Charles Scribner's Sons, New York (1970–80) and since December 2007 available as an e-book.

Cardwell, W to C = Cardwell, Donald S.L., *From Watt to Clausius: the Rise of Thermodynamics in the Early Industrial Age*, Heinemann, London, 1971.

CHAPTER 1, PAGES 2–4

1. The people referred to are identified in Appendix III.
2. Lanczos, Cornelius, *The Variational Principles of Mechanics*, University of Toronto Press, Toronto, 4th edn, 1974, Preface, p. x.
3. Feynman, Richard, *Lectures on Physics*, with R. Leighton and M. Sands, vol. I, Addison-Wesley, Reading, MA, 1966.
4. Feynman, *Lectures on Physics*.

CHAPTER 2, PAGES 5–14

1. There were also mathematical advances arising from astronomy, warfare, and trade—but those arising from machines had an especial relevance to energy.
2. Coverage based on Hiebert, Erwin N., *Historical Roots of the Principle of the Conservation of Energy*, Arno Press, New York, 1981; and Ord-Hume, Arthur W.J.G., *Perpetual Motion: the History of an Obsession*, St. Martin's Press, New York, 1980.
3. Aristotle's *Mechanica*, 1–3, as found in Hiebert, *Historical Roots*, this and following quotes.
4. The contraries were 'motion and rest' (the line of the circle moves while the point at the centre is at rest); 'the concave and the convex' (the line of the circle defines both a concave surface and a convex surface); and one circle turning in one direction causes an adjacent circle to rotate in the opposite direction.
5. Needham, Joseph, *Science in Traditional China: a Comparative Perspective*, Harvard University Press, Cambridge, MA, 1981.
6. Hero's *Mechanica*, as referenced and found in Hiebert, *Historical Roots*.
7. Sarma, S.R., 'Astronomical instruments in Brahmagupta's Brahmasphuta-sidhanta', *Indian Historical Review*, XII, 1986–7, p. 69, as found at www.hp-gramatke.de.net/perpetuum/english/page0220.htm

8. Ord-Hume, *Perpetual Motion*, p. 63.
9. Somerset, Edward, Marquis of Worcester, 'A century of inventions, written in 1655', printed by J. Grismond, 1663; electronic version at www.history.rochester.edu/steam//dircks/ (with thanks to Fran Versace and Sean Singh).
10. Tallmadge, G.K., 'Perpetual motion machines of Mark Anthony Zimara', *Isis*, 33(8), 1941, pp. 8–14. The artist Burton Lee Potterveld was commissioned by Professor Tallmadge to depict Zimara's self-blowing windmill.
11. Wilkins, John, *Mathematical Magick*, London, 1648 (as given in Ord-Hume, *Perpetual Motion*).
12. Stevin, Simon, *De Beghinselen der Weeghconst* (*The Laws of Statics*), 1586 (the microfiche copy in the National Library of Australia, ANL).

CHAPTER 3, PAGES 15–45

1. Not in chronological order.
2. One thing that Galileo didn't appreciate was that sliding friction can be made negligible—idealized away—but a rolling ball is different (it uses up energy to roll).
3. I owe this perceptive insight to Charles Gillispie in Gillispie, Charles C., *The Edge of Objectivity*, Princeton University Press, Princeton, NJ, 1973.
4. Gillispie, *The Edge of Objectivity*.
5. Galilei, Galileo, *Two New Sciences, Including Centres of Gravity and Forces of Percussion* (1638), translated with introduction and notes by Stillman Drake, University of Wisconsin Press, Madison, WI, 1974, p. 166.
6. Galilei, Galileo, *Dialogue Concerning the Two Chief World Systems— Ptolemaic and Copernican*, 2nd rev. edn, translated with notes by Stillman Drake, foreword by Albert Einstein, University of California Press, Berkeley, CA, 1967, p. 239.
7. Drake, Stillman, 'Galileo's discovery of the law of free fall', *Scientific American*, 228(5), 1973, p. 84.
8. Galileo, *Two New Sciences*, p. 68.
9. Drake, Stillman, *Galileo*, Oxford University Press, Oxford, 1980; and Koestler, Arthur, *The Sleepwalkers*, Penguin, London, 1986. (Koestler, in his otherwise excellent book, has Kepler as the favourite and Galileo as the un-favourite.)
10. In Koestler's assessment; Koestler, *The Sleepwalkers*.
11. Drake, Stillman, *Discoveries and Opinions of Galileo*, Doubleday, New York, 1957.
12. Galileo, *Two Chief World Systems*, p. 28.
13. Galileo, *Two Chief World Systems*, p. 147.
14. Galileo, *Two Chief World Systems*, p. 116.
15. Galileo, *Two Chief World Systems*, pp. 186–7.
16. Galileo, *Two Chief World Systems*, the postil on p. 273.

17. In scaling up from the small animal to the giant, Galileo had not scaled the Earth, or the constituent atoms.
18. Drake, Stillman, *Galileo at Work*, University of Chicago Press, Chicago, 1978, p. 298.
19. Galilei, Galileo, *Dialogues Concerning Two New Sciences* (1638), translated by H. Crew and A. de Salvio, edited by A. Favaro, Dover Publications, New York, 1954; reprint of 1914 edition; 'Fourth Day', p. 293.
20. Galileo, *Dialogues*, p. 271. Note the word 'energia'—however, this is used adjectively, not in a physics sense.
21. Sorrel, Tom, *Descartes*, Past Masters, Oxford University Press, Oxford, 1987.
22. A distinction that would be important in the 'variational mechanics' (Chapter 7, Part IV).
23. Descartes, René, 'The World', in *Discourse on Method and Related Writings* (1637); Penguin Classics, London, 1999, translated and introduced by Desmond Clarke.
24. Descartes, René, *The Principles of Philosophy* (1644), Part II, in Haldane, E.S. and Ross, G.R.T., *The Philosophical Works of Descartes*, Dover Publications, New York, 1955.
25. Descartes, 'The World'.
26. Descartes, *The Principles of Philosophy*.
27. Westfall, Richard, *Force in Newton's Physics*, American Elsevier and Macdonald, London, 1971, ch. II, this and the following quotes in Descartes' letters to Mersenne and Huygens.
28. Descartes, René, letter to Mersenne, July 1638, in *Oeuvres de Descartes*, vol. II, ('ma physique n'est autre chose que géometrie').
29. Huygens, Christiaan, *Horologium oscillatorium*, 1673; original Latin on Gallica website of the French National Library; English translations by Richard Blackwell (1986) and Ian Bruce (2007).
30. Huygens, Christiaan, *De motu de corporum ex percussione* (1656), in *Oeuvres complètes*, Martinus Nijhoff, La Haye, 1888–1950, vol. 16, p. 181 (my translations).
31. Bell, A.E., *Christiaan Huygens and the Development of Science in the Seventeenth Century*, Longmans, Green & Co., New York, 1947.
32. Huygens, *De motu* (I have translated the hypotheses and propositions from old French, itself a translation from the Latin).
33. Gabbey, Alan, 'Huygens and mechanics', in Bos, H.J.M., Rudwick, M.J.S., Snelders, H.A.M. and Visser, R.P.W. (eds), *Studies on Christiaan Huygens Symposium on 'The Life and Works of Christiaan Huygens', Amsterdam, 1979*, Swets and Zeitlinger, Lisse, 1980.
34. Huygens, *De motu*.
35. Westfall, *Force in Newton's Physics*, ch IV.
36. Huygens also found this formula cropping up, and remaining constant, in his experiments and theoretical analysis of compound pendulums.

37. Gabbey, 'Huygens and mechanics'. The gravitational force also increases proportionately with the mass, while the acceleration remains the same regardless of the mass.

38. Crew, H., in Bell, *Christiaan Huygens*.

39. Newton, Isaac, *The Mathematical Principles of Natural Philosophy* (1687), translated 1729 by Andrew Motte, published by Daniel Adee, New York, 1848; and in Hawking, Stephen (ed.), *On the Shoulders of Giants*, Running Press, Philadelphia, PA, 2002.

40. Westfall, Richard, *Never at Rest: a Biography of Isaac Newton*, Cambridge University Press, Cambridge, UK, 1983.

41. Newton, *The Mathematical Principles*.

42. Newton, *The Mathematical Principles*.

43. This new measure, 'acceleration', could only be defined because of a revolution in mathematics—Newton's calculus (although, in fact, Newton didn't use his calculus in the *Principia*).

44. As it is called today.

45. Newton, *The Mathematical Principles*.

46. Newton, *The Mathematical Principles*, the paragraph following Law III.

47. Newton, *The Mathematical Principles*, Scholium after corollary VI.

48. Maxwell, James Clerk, *Matter and Motion*, Dover Publications, New York, 1952.

49. Newton, *The Mathematical Principles*, Scholium after corollary VI.

50. Thomson, W. and Tait, P.G., *Treatise on Natural Philosophy*, Macmillan, London, for the University of Oxford, 1867.

51. Newton, *The Mathematical Principles*, Scholium.

52. Home, Roderick, 'The Third Law in Newton's mechanics', *British Journal for the History of Science*, 4(13), 1968, p. 39.

53. Newton, Isaac, 'Queries' at the end of *Opticks: or, a Treatise of the Reflections, Refractions, Inflections and Colours of Light*, 2nd edn, 1717, translated by Andrew Motte, revised by Florian Cajori, as in Great Books of the Western World, no. 32, Encyclopaedia Britannica, Inc., University of Chicago Press, 2nd edn, 1990.

54. Newton, 'Queries', Query 31, p. 541.

55. Newton, 'Queries', Query 31, pp. 540–1.

56. At least, I couldn't find anything further in Westfall, *Never at Rest*.

57. Newton, *The Mathematical Principles*; see all of Book II.

58. Newton, 'Queries', Query 28, p. 528.

59. Newton, 'Queries', Query 31, p. 540.

60. Newton, 'Queries', Query 31, p. 542.

61. But I am in no way suggesting that Newton actually anticipated the modern physics.

62. Macdonald, Ross, *Leibniz*, Past Masters, Oxford University Press, Oxford, 1984.

63. Leibniz, Gottfried Wilhelm, *Philosophical Writings*, edited by G.H.R. Parkinson, Dent, London, 1973.
64. And therefore it *was* always possible to turn 'big money into small change'.
65. Leibniz, *Philosophical Writings*.
66. Leibniz, *Philosophical Writings*.
67. Leibniz, Gottfried Wilhelm, 'Brief demonstration . . .' (1686), in *Philosophical Papers and Letters*, translated and edited by Leroy Loemker, Chicago University Press, Chicago, 1956.
68. Leibniz, Gottfried Wilhelm 'Essay on dynamics', in Costabel, Pierre, *Leibniz and Dynamics: the Texts of 1692*, Hermann, Paris/Cornell University Press, Ithaca, NY, 1973.
69. Westfall, *Force in Newton's Physics*, p. 297.
70. Westfall, *Force in Newton's Physics*, p. 297.
71. Westfall, *Force in Newton's Physics*, discussion on pp. 297–303.
72. For readers with knowledge of integration, Leibniz defined *vis mortua* as mv so that the integral of it with respect to v would come out as mv^2; in fact, as $\frac{1}{2}mv^2$.
73. DSB on Bernoulli, Johann.
74. Westfall, *Force in Newton's Physics*, p. 288.
75. Westfall, *Force in Newton's Physics*.
76. Leibniz, *Philosophical Writings*, p. 158.
77. Westfall, *Force in Newton's Physics*, p. 295.

CHAPTER 4, PAGES 46–62

1. Bacon, Francis, *The New Organon*, 1620; translated in 1863 by James Spedding, Robert L. Ellis and Douglas D. Heath; 'The Aphorisms', Book Two, XX.
2. Gassendi, Pierre, *Epicuri philosophiae syntagma*, 1649.
3. Boyle, Robert; this and the following quotes of Boyle excerpted from Boas Hall, Marie, *Robert Boyle on Natural Philosophy*, Indiana University Press, Bloomington, IN, 1965; and Boas, Marie, *Robert Boyle and Seventeenth Century Chemistry*, Cambridge University Press, Cambridge, 1958.
4. Maxwell, James Clerk, *Theory of Heat* (1871); reissued by Dover Publications, New York, 2001, with a new introduction and notes by Peter Pesic; see p. 303.
5. Robert, Boyle, *New Experiments and Considerations Touching Cold*, printed for J. Crook, London, 1665.
6. Hooke, Robert, *Micrographia*, 1665, p. 46 of original; see also Dover Phoenix Editions, New York, 2003.
7. Cardwell, W to C, ch. 1.

8. Knowles Middleton, W.E., *A History of the Thermometer and its Uses in Meteorology*, Johns Hopkins University Press, Baltimore, MD, 1966, p. 8.

9. Knowles Middleton, *History of the Thermometer*, p. 27.

10. Boyle, Robert, in Boas Hall, *Robert Boyle on Natural Philosophy*.

11. Water is therefore not useful as a thermometric substance in this region: the relation of volume to temperature may not be linear, but at least it should be single-valued.

12. Why Italy? Glass-blowing was an advanced art in Italy at this time, but this can't be the only answer; rather, this was both a cause and a consequence of a rising scientific spirit in northern Italy.

13. Boyle, Robert, in Boas Hall, *Robert Boyle on Natural Philosophy*.

14. Note that this is the mechanical philosophy as it evolved *after* Descartes: Descartes had a continuum of matter (no atoms, no vacuum), whereas Boyle and his successors did allow atoms and did allow a vacuum.

15. Galileo, *Two New Sciences, Including Centres of Gravity and Forces of Percussion* (1638), translated with introduction and notes by Stillman Drake, University of Wisconsin Press, Madison, WI, 1974; 'First Day'.

16. Knowles Middleton, W.E., *The History of the Barometer*, Johns Hopkins University Press, Baltimore, MD, 1964, p. 5.

17. Torricelli, Evangelista, 'we live submerged at the bottom of an ocean of air' (1644), in a letter.

18. You may like to try to imagine what the experimental arrangement could be—and then see the schematic diagram in Appendix III.

19. Boyle, Robert, *New Experiments, Physico-Mechanical, Touching the Spring of the Air, and its Effects*, 1662; 2nd edn, 'Whereunto is added a defence of the author's explication of the experiments, against the objections of Fransiscus Linus and Thomas Hobbes', printed by H. Hall for Tho. Robson, Oxford.

20. Boyle, *New Experiments*.

21. See, via a computer search, the painting by Joseph Wright, *An Experiment on a Bird in the Air Pump* (1768), which captures the spirit of this enquiry (although it is set in the eighteenth century).

22. Mariotte, Edmé, 'Essai du chaud et du froid' (1679), in *Oeuvres de Mariotte*, vol. I, P. van der Aa, Leiden, 1717, p. 183.

23. Knowles Middleton, W.E., *The Experimenters, a Study of the Accademia del Cimiento*, Johns Hopkins University Press, Baltimore, MD, 1971.

24. Cardwell, W to C, ch. 1.

25. Papin, Denis, *A Continuation of the New Digester of Bones*, London, 1687.

26. Amontons, Guillaume, 'Moyen de substituer commodement l'action du feu . . .', *Mémoires de l'Academie des Sciences*, 1699, p. 112.

27. Amontons, Guillaume, 'De la résistance causée dans les machines . . .', *Mémoires de l'Academie des Sciences*, 1699, p. 206.

28. Amontons, Guillaume, 'De la résistance causée dans les machines'.

CHAPTER 5, PAGES 66–79

1. Boyle, Robert, *New Experiments, Physico-Mechanical, Touching the Spring of the Air, and its Effects*, 1662; printed by H. Hall for Tho. Robson, Oxford.
2. Newton, Isaac, *The Mathematical Principles of Natural Philosophy*, translated by Andrew Motte, revised by Florian Cajori, as in Great Books of the Western World, no. 32, Encyclopaedia Britannica, Inc., University of Chicago Press, 2nd edn, 1990, Book II, Proposition 23.
3. He spelled it 'aether'.
4. Newton, Isaac, 'Queries' at the end of *Opticks*, 2nd edn, 1717, translated by Andrew Motte, revised by Florian Cajori, as in Great Books of the Western World, no. 32.
5. Newton, Isaac, 'A scale of the degrees of heat', *Philosophical Transactions of the Royal Society*, 1701, p. 824.
6. Westfall, R., *Never at Rest: a Biography of Isaac Newton*, Cambridge University Press, Cambridge, 1983.
7. Hales, Stephen, *Vegetable Staticks* (1727), Experiment LXVII, ch. VI, p. 101; reissued by the Scientific Book Guild, Oldbourne Book Co. Ltd, 1961, with a foreword by M.A. Hoskin.
8. Leicester, Henry M., *The Historical Background to Chemistry*, Dover Publications, New York, 1956.
9. Pope, Alexander, *The Works of Alexander Pope*, edited by J. Wharton, 1797.
10. Hales, *Vegetable Staticks*.
11. Boerhaave, Hermann, *The Elements of Chemistry*, translated from the original Latin by Timothy Dallowe, London, printed for J. & J. Pemberton, Fleet Street; J. Clarke, Royal Exchange; A. Millar in the Strand; and J. Gray in the Poultry, London, 1735.
12. Boerhaave, *The Elements of Chemistry*, this and subsequent quotes until endnote 13.
13. DSB on Brook Taylor.
14. Boerhaave, *The Elements of Chemistry*; and 'Fahrenheit, Daniel Gabriel', in Van der Star, P. (ed.), *Fahrenheit's Letters to Leibniz and Boerhaave*, Amsterdam, Rodopi, 1983.
15. Martine, George, *Essays and Observations on the Constitution and Graduation of Thermometers and on the Heating and Cooling of Bodies*, 3rd edn, Edinburgh, 1780.
16. Hermann, Jakob, *Phoromania*, Amsterdam, 1716.
17. Knowles Middleton, W.E., 'Jacob Hermann and the kinetic theory', *British Journal for the History of Science*, 2, 1965, pp. 247–50.
18. Bernoulli, Daniel, *Hydrodynamica* (1738); reissued by Dover Publications, New York, 1968; all subsequent quotations in this section relate to ch. X.
19. Bernoulli, *Hydrodynamica*.
20. Boerhaave, *The Elements of Chemistry*.

21. In this section, and later on Watt (Chapter 8), I am indebted to Cardwell for his rare wisdom, and his ability to imbue the reader with due awe for the achievements of the past.
22. Cardwell, Donald S.L., *Technology, Science and History*, Heinemann Educational, London, 1972.
23. Cardwell, *Technology, Science and History*.

CHAPTER 6, PAGES 81–91

1. Uglow, Jenny, *The Lunar Men: the Friends Who Made the Future, 1730–1810*, Faber and Faber, London, 2002.
2. Ross, I.S., *The Life of Adam Smith*, Oxford Scholarship Online, Oxford University Press, New York, 1995.
3. Guerlac, Henry, *Essays and Papers in the History of Modern Science*, Johns Hopkins University Press, Baltimore, 1977, ch. 18.
4. Guerlac, *Essays and Papers*; and also in Black, Joseph, *Lectures on the Elements of Chemistry* (lecture notes taken by John Robison, 1803), Edinburgh; and as extracts in Magie, W.F., *A Source Book in Physics*, McGraw-Hill, New York, 1935. This and subsequent quotes until endnote 5.
5. Heilbron, J.L., *Electricity in the 17th and 18th Centuries: A Study of Early Modern Physics*, University of California Press, Berkeley, CA, 1979, p. 85.
6. Guerlac, *Essays and Papers*; Magie, *A Source Book in Physics*; this and subsequent quotes until endnote 7.
7. Martine, George, *Essays and Observations on the Constitution and Graduation of Thermometers and on the Heating and Cooling of Bodies*, 3rd edn, Edinburgh, 1780.
8. Guerlac, *Essays and Papers*.
9. Heilbron, *Electricity in the 17th and 18th Centuries*; and McKie, D. and Heathcote, N.H., *The Discovery of Specific and Latent Heat*, Edward Arnold, London, 1935.
10. Cardwell, W to C, p. 64.
11. Crawford, Adair, *Experiments and Observations on Animal Heat and the Inflammation of Combustible Bodies*, John Murray, London, 1779; this and subsequent quotes taken from the 2nd edn, 1788.
12. Scheele, Carl Wilhelm, 'A chemical treatise on fire and air' (1777), in *Collected Papers of Carl Wilhelm Scheele*, translated by L. Dobbin, G. Bell and Sons, London, 1931; this and the following quotes.
13. Lambert, J.H., *Pyrometrie oder uom maasse des Feuers und der Waàrme*, 1779; and Cardwell, W to C, p. 90.
14. Heilbron, *Electricity in the 17th and 18th Centuries*.
15. Cardwell, W to C, ch. 4.
16. Pictet, Marc-Auguste, *Essai sur le feu*, Geneva, 1790, and English translation by W. Belcombe, London, 1791.

17. Pictet, *Essai sur le feu*.

18. Evans, James, and Popp, Brian, 'Pictet's experiment', *American Journal of Physics*, 53(8), 1985, pp. 737–53.

19. Prévost, Pierre, *Du calorique rayonnant*, Paris and Geneva, 1809; and 'Mémoire sur l'equilibre du feu', *Journal de Physique* (Paris) 28, 1791, pp. 314–22.

20. A body hotter than its surroundings will cool down rather than become even hotter; a body cooler than its surroundings will heat up rather than become even cooler. This is because the (average) microscopic speeds are higher for hotter bodies.

CHAPTER 7, PAGES 93–139

1. Newton, Isaac, *The Mathematical Principles of Natural Philosophy* (1687), translated by Andrew Motte, Daniel Adee, New York, 1848; and in Hawking, Stephen (ed.), *On the Shoulders of Giants*, Running Press, Philadelphia, PA, 2002.

2. Hankins, Thomas, 'Eighteenth century attempts to resolve the *vis viva* controversy', *Isis*, 56(3) 1965, pp. 281–97.

3. Hankins, 'Eighteenth century attempts to resolve the *vis viva* controversy'.

4. Laudan, L.L., 'The *vis viva* controversy, a post mortem', *Isis*, 59, 1968, p. 139.

5. Westfall, Richard, *Never at Rest: a Biography of Isaac Newton*, Cambridge University Press, Cambridge, 1983.

6. Hankins, 'Eighteenth century attempts to resolve the *vis viva* controversy'.

7. Newton, Isaac, 'Queries' at the end of *Opticks*, 2nd edn, 1717, translated by Andrew Motte, revised by Florian Cajori, as in Great Books of the Western World, no. 32, Encyclopaedia Britannica, Inc., University of Chicago, 2nd edn, 1990, Query 31.

8. Note that there is nothing illogical in Newton's suggestion; it's just that the world isn't like this.

9. DSB on Bernoulli, Johann.

10. Newton, 'Queries', Query 31.

11. Johann also considered the thought experiment of attaching perfectly elastic springs to perfectly hard bodies.

12. Using the integral calculus.

13. In vector notation, we write it as $m\mathbf{v}$.

14. Kinetic energy has v *squared* in it, and this becomes vanishingly small compared to v.

15. Feynman's *Lectures on Physics*, with R. Leighton and M. Sands, vol. 1, Addison-Wesley, Reading, MA, 1966, middle of p. 39–9 (Feynman's analysis applies to translational kinetic energy, and to gas molecules colliding with a moveable wall).

16. Feynman's *Lectures on Physics*, vol. I, Fig. 39–3 and bottom of p. 39–8.

17. Newton, *The Mathematical Principles of Natural Philosophy*, Proposition XXXIX of Book I. Richard Westfall (in *Force in Newton's Physics*, American Elsevier and Macdonald, London, 1971) says: '[the quantities denoting energy] had no significance whatever in Newton's eyes' (p. 489) and 'he [Newton] was unable to grasp the importance of the work–energy equation [that] he had implicitly derived. Above all, the quantity mv^2 held no meaning for him' (p. 489).

18. Bernal, J.D., *Science in History*, Penguin, London, 1969.

19. Parent, Antoine, 'Theory of the greatest possible perfection of machines', 1704.

20. Westfall, *Force in Newton's Physics*; and Parkinson, G.H.R. (ed.), *Philosophical Writings*, Dent, London, 1974.

21. Westfall, *Force in Newton's Physics*; and Parkinson, *Philosophical Writings*.

22. Westfall, *Force in Newton's Physics*; and Parkinson, *Philosophical Writings*.

23. Kline, Morris, *Mathematical Thought from Ancient to Modern Times*, Oxford University Press, New York, 1972.

24. Clairaut, Alexis-Claude, *Théorie de la figure de la terre tirée des principes de l'hydrostatique*, Paris, 1743.

25. Bernoulli, Daniel, *Hydrodynamica* (1738), translated by T. Carmody and H. Kobus, preface by Hunter Rouse-Ball, Dover Publications, New York, 1968.

26. Bernoulli, *Hydrodynamica*, Part II.

27. Bernoulli, *Hydrodynamica*.

28. Certainly this is the view of the historian Clifford Truesdell—but he was overly partisan and guilty of anachronistic criticisms.

29. Euler, Leonhard, *Mechanica, sive motus scientia analytice exposita*, two vols, St Petersburg, 1736; in *Opera Omnia Leonhardi Euleri*, 2nd series, I and II (1911 onwards), Berlin.

30. For example, the rotational kinetic energy is $\frac{1}{2}I\omega^2$ for a spinning body, where I is the 'moment of inertia' and ω is the spin rate. Also, rotational kinetic energy could be added to ordinary kinetic energy, and in any order. (Angular momentum is conserved *separately* from linear momentum and, curiously, the order of performing turning operations *does* matter, except where the turning operations are infinitesimal.)

31. Euler, Leonhard, *Letters to a German Princess*, written between 1760 and 1762, translated by Henry Hunter, 1795 edition, notes by David Brewster, vol. 1; reprinted by Thoemmes Press, Bristol, 1997.

32. Heilbron, J., *Electricity in the 17th and 18th Centuries: a Study of Early Modern Physics*, University of California Press, Berkeley, CA, 1979, p. 45.

33. Boscovich, R.G., *A Theory of Natural Philosophy* (1763), translated by J.M. Child, The MIT Press, Cambridge, MA, 1966.

34. Pearce-Williams, L., *Michael Faraday, a Biography*, Da Capo Press, New York, 1987.

35. Darwin, Erasmus, 'The temple of nature' and 'The loves of the plants', published together as *The Botanic Garden* (1803); reissued by Scholar Press, Menston, 1973.
36. But see the qualifications in Chapter 17, 'Electromagnetism and Other Kinds of Energy-in-the-Field'.
37. Bernoulli, *Hydrodynamica*, for this and all subsequent quotes until note 38.
38. D'Alembert, Clairaut, Euler, and Lagrange.
39. For so-called 'ideal' gases, see our Chapter 10. (The main difference between Bernoulli and today is that he makes no distinction between speed and average speed.)
40. Truesdell, Clifford, 'The rational mechanics of flexible or elastic bodies, 1638–1788', introduction to *Leonhardi Euleri, Opera Omnia*, 2nd series, vols X and XI, Fussli, pp. 173–4. This and subsequent quotes.
41. Euler, Leonhard, 'Additamentum I de curvis elasticis', added to *Methodus inveniendi lineas curvas maximi minimive proprietate gaudentes*, Lausanne and Geneva, 1744; in *Leonhardi Euleri, Opera Omnia*, 1st series, vol. XXIV, pp. 231–97; the English translation is in Oldfather, W.A., Ellis, C.A. and Brown, D.M., 'Leonhard Euler's elastic curves', *Isis*, 20, 1933, pp. 72–160.
42. Euler, Leonhard, letter to Daniel Bernoulli from St Petersburg, 13 September 1738, in *Die Werke von Daniel Bernoulli*, Band 3, *Mechanik*, general editor David Speiser, Birkhauser Verlag, Basel, 1987, p. 72; translated by Sabine Wilkens.
43. This case could be taken as 'central', with a force centre removed infinitely far away.
44. Bernoulli, *Hydrodynamica*, preface to the English translation, p. xi.
45. When maximized, this is always a 'point of inflection' rather than a true maximum—see note 76.
46. Goldstine, H., *A History of the Calculus of Variations, from the 17th through the 19th Century*, Springer Verlag, New York, 1980.
47. Lanczos, C., *The Variational Principles of Mechanics*, 4th edn, University of Toronto Press, Toronto, 1974, Chapter X, 'Historical survey'.
48. DSB on Maupertuis.
49. In the 'variational mechanics' to follow (Lagrangian mechanics), the displacements ('*variations*'), while they are *virtual* (imagined) and in *arbitrary directions*, must nevertheless adhere to some strict conditions: they must all occur at the *same time*, be *instantaneous, infinitesimal, reversible*, and occur in ways that are consistent ('*in harmony*') with the constraints. For example, we don't go rushing in and break the lever-arm or knock it off its perch. Rather, we tweak it, mathematically speaking, and let it rotate infinitesimally away from balance.
50. Bernoulli, Johann, 26 February 1717, published posthumously in Varignon's *Nouvelle mécanique* of 1725.
51. Or, if moving (uniformly), then the rigid bodies must be electrically neutral.

52. We shall sometimes talk of particles or bodies instead of masses.

53. As well as defining the name, the font has also been specially chosen: bold font means a vector quantity (something with magnitude and direction), whereas italic font means just the magnitude.

54. Lanczos, *The Variational Principles of Mechanics*, Chapter IV, 'D'Alembert's Principle'.

55. Dugas, René, *A History of Mechanics*, Dover Publications, New York, 1988, foreword by Louis de Broglie, translated by J.R. Maddox, p. 247; quotations of d'Alembert from his *Traité de dynamique* (1743), edition of 1758.

56. Dugas, *A History of Mechanics*, p. 246; and in d'Alembert's *Traité de dynamique*.

57. Truesdell, 'Rational mechanics', p. 186.

58. DSB on d'Alembert.

59. This reminds one of Einstein's celebrated Principle of Equivalence, but d'Alembert's Principle came first.

60. Lagrange, J.L., *Analytique mécanique*, 1788, translated by Auguste Boissonnade and Victor Vagliente, Kluwer Academic, Boston, 1997.

61. DSB on Hamilton.

62. Lagrange's *Analytique mécanique*, Lagrange's own introduction.

63. Euler, Leonard, p. 19 of 'Life of Euler', in *Letters to a German Princess*.

64. DSB on Lagrange, p. 569.

65. DSB on Lagrange, p. 561 (Euler's 'beautiful theorem').

66. The details can be found in my forthcoming book *The Lazy Universe: the Principle of Least Action Explained* (Oxford University Press, Oxford, 2015). The book is not a textbook; however, undergraduate-level physics has been assumed.

67. This is not a requirement in Hamilton's later version, known as 'Hamilton's Principle'; see Chapter 13.

68. Naming conventions: if it is just a question of minimizing an integral, then these are now known as the Euler–Lagrange equations; if the problem relates specifically to mechanics, then these are known as the Lagrange Equations of Motion.

69. For generalized coordinates such as voltages, the coefficients of sine waves in a Fourier series, and so on, then we are looking beyond Lagrange's own era.

70. Except where, in especially simple cases ('rectangular coordinates'), the Lagrangian and Newtonian descriptions coincided.

71. The following examples come from *Feynman's Lectures*, vol. II, ch. 19. Note that, strictly speaking, we are considering Hamilton's Principle, in which $\int (T - V)\mathrm{d}t$ is minimized between fixed end-times, as well as fixed end-positions.

72. This also follows from the mathematical result that the 'square of the mean' is less than the 'mean of the squares'.

73. The application of what came to be known as Hamilton's Principle (the minimization of $\int(T - V)dt$ between definite end-states) is what came to be known as Lagrangian mechanics.

74. There are some cases in physics where change is discontinuous or even chaotic—but these cases cannot be treated by the methods of Lagrangian mechanics.

75. The Lagrange equations are a set of equations, one for each generalized coordinate ('degree of freedom'). They are invariant not when taken one degree of freedom at a time, but when considered as the *whole set*.

76. This analogy answers another puzzling question: why do we refer to a Principle of *Least* Action? (Our answer is adapted from the paper by Gray, C.G. and Taylor, Edwin F., 'When action is not least', *American Journal of Physics*, 75(5), May 2007.) In Lagrangian mechanics, we have a *path* that we are optimizing, rather than, say, optimizing the height of a mountain (this is because we are dealing with a line or *path* integral). Now a path between two given end-points always has a shortest length— the length of the straight line between those points—whereas it has no definitive maximum length (we can always add more wiggles). The same applies in Lagrangian mechanics—the minimum path-length is unique, whereas the maximum is never a 'true maximum'. It is for this reason that we talk of *least* action, even while maxima or 'saddle points' can sometimes occur.

77. In Lagrangian mechanics, this virtual landscape is known as 'configuration space'. It's a bit like going to the imaginary Land of Narnia and coming back with *real* treasures! This is explained further in *The Lazy Universe*.

78. First, let's correct ourselves straightaway and realize that $T - V$ isn't very significant—it's the whole mathematical object, 'the minimization of $\int(T - V)dt$ between definite end-states', that is the new quantity of importance.

79. See note 59.

80. The slogan arises in Misner, C.W., Thorne, K.S. and Wheeler, J.A., *Gravitation*, W.H. Freeman and Co., San Fransisco, 1973. (Note that it is only a slogan, and, strictly speaking, we must substitute 'mass-energy' for 'matter', and 'spacetime' for 'space'.)

81. Nowadays, we would say the *energy* of the explosion.

82. Using the energy method, we answer by noting that the closed truck, the weighbridge, and the Earth's gravity represent a *closed* system, and therefore the total energy is conserved, and so what happens inside the truck is irrelevant. (Actually, the stored muscle energy of the pigeons *could* make a detectable difference if the pigeons all launched themselves off their perches at the same time.)

83. This smoothness arises out of the requirement, in the variational mechanics, that T and V are in *functional* form and, moreover, functions that must satisfy certain criteria of smoothness, continuity, and finiteness.

84. This is done by a suitable choice of Lagrange's generalized coordinates (for example, the lever-arm can be modelled as a 'rigid body' instead of as a collection of mass-points).
85. Specifically, Lagrange's 'variational' mechanics. Note that the 'variational' mechanics insists that the variations are infinitesimal and *local*.
86. That is to say, extensive enough to fully encompass the system.
87. This is too big a topic to discuss here, as it would require an excursion into the invariants of geometry, but some flavour of it is given in the analogy of extremals in physical geography, at the end of the section on Lagrangian mechanics; see also the comment at the end of note 75 (see Lanczos' *The Variational Principles of Mechanics*).
88. However, we shall find that when we arrive at the laws of thermodynamics (Chapter 16), dissipative effects *can* be treated in the systems view, and this heralds the arrival of another 'block' of energy—heat. This approach is profound and ends up leading to a new cosmic law of nature (the Second Law of Thermodynamics), and is an endorsement of the systems/energy view.

CHAPTER 8, PAGES 141–159

1. Guerlac, Henry, 'Some aspects of science during the French Revolution', in *Essays and Papers in the History of Modern Science*, Johns Hopkins University Press, Baltimore, 1977.
2. Uglow, Jenny, *The Lunar Society*, Faber and Faber, London, 2002.
3. Marsden, Ben, *Watt's Perfect Engine*, Icon Books, Cambridge, 2002, p. 90.
4. Bernal. J.D., *Science in History*, Penguin, London, 1969.
5. Uglow, *The Lunar Society*.
6. Robison, John, letter 213, Robison to Watt, in E. Robinson and D. McKie (eds), *Partners in Science, Letters of James Watt and Joseph Black*, Harvard University Press, Cambridge, MA, 1969, p. 321.
7. The following account has been adapted from Cardwell, W to C, and *The Fontana History of Technology*, Fontana, London, 1994.
8. Marsden, *Watt's Perfect Engine*, p. 58.
9. Cardwell demonstrates that this is nonsense (Cardwell, W to C, pp. 42–5).
10. Marsden, *Watt's Perfect Engine*, p. 132.
11. Cardwell, W to C, p. 80.
12. Marsden, *Watt's Perfect Engine*, p. 119.
13. DSB on Watt.
14. DSB on Watt.
15. Cardwell, W to C, p. 79.
16. Uglow, *The Lunar Society*.
17. Uglow, *The Lunar Society*; and Robinson and McKie, *Partners in Science*.
18. Bronowski, J., *The Ascent of Man*, BBC Publications, London, 1973.
19. Bernal, *Science in Society*.

20. Doherty, Howard, of Doherty's Garage, St Andrews Street, Bendigo, personal communication.
21. Carnot, Lazare, *Essai sur les machines en général*, 1783.
22. We use the first name to distinguish Carnot from his son, Sadi.
23. Gillispie, C.C., *Lazare Carnot, Savant*, Princeton University Press, Princeton, NJ, 1971.
24. DSB on Lazare Carnot, p. 78.
25. Gillispie, C.C., *The Edge of Objectivity: an Essay in the History of Scientific Ideas*, Oxford University Press, 1960, p. 205.
26. Lavoisier, Antoine-Laurent, *Memoir sur la chaleur*, presented to the Royal Academy of Sciences in Paris in 1783 and published in 1784; translated by Henry Guerlac.
27. Mendoza, E., Introduction to Sadi Carnot's *Reflections on the Motive Power of Fire*, Dover Publications, New York, 1960, p. xv.
28. Biographical notes to Robert Kerr's translation of Lavoisier's *Elements of Chemistry*, edn, vol. 45, Britannica Great Books, Chicago, 1952.
29. Biographical notes in Lavoisier, *Elements of Chemistry*.
30. Guerlac, Henry, *Antoine-Laurent Lavoisier, Chemist and Revolutionary*, Charles Scribner's Sons, New York, 1973.
31. Grattan-Guiness, Ivor, in Gillispie, C.C., *Pierre-Simon Laplace, 1749–1829*, Princeton University Press, Princeton, NJ, 2000.
32. Gillispie, Charles C., Laplace's biographer, personal communication, 2007.
33. Lavoisier, *Memoir sur la chaleur*, this and subsequent quotes.
34. Of course, it wouldn't be the same guinea pig.
35. The *Encyclopaedie* was edited jointly by Diderot and d'Alembert (see our Chapter 7, Part IV). Gough J.B., 'The origins of Lavoisier's theory of the gaseous state', in *The Analytic Spirit, Essays in the History of Science, in Honour of Henry Guerlac*, edited by Harry Woolf, Cornell University Press, Ithaca, NY, 1981.
36. Lavoisier, *The Elements of Chemistry*, 29th edn, vol. 45, Britannica Great Books, Chicago, 1987, Second Part, this and subsequent quotes.
37. Guerlac, *Antoine-Laurent Lavoisier*, op. cit.
38. Jungnickel, Christa and McCormmach, Russell, *Cavendish, The Experimental Life*, Bucknell, Lewisburg, PA, 1999, p. 491.
39. Jungnickel and Russell, *Cavendish, The Experimental Life*, pp. 304–5, also p. 329 *re* house alterations.
40. Jungnickel and Russell, *Cavendish, The Experimental Life*.
41. MacCormmach, Russell, 'Henry Cavendish on the theory of heat', *Isis*, 79(1), 1988, pp. 37–67.
42. MacCormmach, 'Henry Cavendish on the theory of heat', note 4 on p. 39; Cavendish, Henry, 'Obs on Mr Hutchins's expts for determining the degree of cold at which quicksilver freezes', *Philosophical Transactions of the Royal Society*, 73, 1783, pp. 303–28.

43. MacCormmach, Russell, *The Speculative Truth: Henry Cavendish, Natural Philosophy and the Rise of Modern Theoretical Science*, Oxford University Press, Oxford, 2004.
44. Cf. Chapter 7, Part III; Daniel Bernoulli's analysis of *n* attracting bodies, in 1750.
45. MacCormmach, *The Speculative Truth*.
46. Jungnickel and McCormmach, *Cavendish, The Experimental Life*, p. 409.
47. MacCormmach, Russell, *The Speculative Truth*.

CHAPTER 9, PAGES 163–173

1. Brown, Sanborn, *Benjamin Thompson, Count Rumford*, The MIT Press, Cambridge, MA, 1979.
2. Brown, *Benjamin Thompson, Count Rumford*.
3. Brown, G.I., *Scientist, Soldier, Statesman and Spy, Count Rumford*, Alan Sutton, Stroud, 2001, p. 120.
4. Cardwell, W to C, p. 98.
5. Vrest, Orton, *The Forgotten Art of Building a Good Fireplace*, Yankee Books, Emmaus, PA, 1974.
6. Brown, Sanborn, 'An experimental inquiry concerning the source of the heat which is excited by friction', in *Benjamin Thompson—Count Rumford: Count Rumford on the Nature of Heat*, Pergamon Press, Oxford, 1967, ch. 4, Essay IX, p. 55, this and subsequent quotes.
7. Brown, 'An inquiry concerning the weight ascribed to heat', in *Benjamin Thompson—Count Rumford*, ch. 6, p. 98.
8. Brown, 'An inquiry concerning the weight ascribed to heat', in *Benjamin Thompson—Count Rumford*, ch. 6, p. 90.
9. Figure 9.2 depicts a lecture at the Royal Institution. The picture is by the satirical cartoonist James Gillray, and shows Young administering vapours to a subject, with Davy, holding a bellows (?), standing next to Young, and Rumford, standing alone on the right, near to the open door.
10. Davy, Humphry, 'Essay on heat and light', in Beddoes, Thomas, *Contributions to Physical and Medical Knowledge, Principally from the West of England*, printed by Biggs & Cottle for Longman and Rees, Paternoster Row, London, 1799.
11. Andrade, E.N. da C., 'Two historical notes', *Nature*, cxxxv, 1935, p. 359.
12. Azimov, Isaac, quoted in Robinson, Andrew, *The Last Man Who Knew Everything*, Pi Press, Pearson Education, New York, 2006, p. 2.
13. Feynman, Richard, *Feynman's Lectures on Physics*, Addison-Wesley, Reading, MA, 1963, vol. I, end of Section 37–31.
14. DSB on Young, p. 567.
15. DSB on Young, p. 568, reference 64 therein.
16. Whittaker, Sir Edmund, *A History of the Theories of Aether and Electricity*, vol. 1: *The Classic Theories*, Harper & Bros, New York, 1960.

17. Robinson, *The Last Man Who Knew Everything*, p. 7.
18. Young, Thomas, *A Course of Lectures on Natural Philosophy and the Mechanical Arts*, four vols, Thoemmes, Bristol, 2003; a facsimile reprint of the original 1807 edition.
19. Davy, Humphry, First Bakerian Lecture, 'On some chemical agencies of electricity', 1806.
20. Cardwell, W to C, p. 112.
21. Davy, Humphry, *Elements of Agricultural Chemistry*, Longman, London, 1813.
22. Crosland, Maurice, *Gay-Lussac, Scientist and Bourgeois*, Cambridge University Press, Cambridge, 2004, p. 86.
23. Cardwell, W to C, pp. 111–114.
24. I owe this quote and this insight to Cardwell, W to C, p. 105 and reference therein: *Francis Trevithick, The Life of Richard Trevithick* (London, 1874), p. 113.

CHAPTER 10, PAGES 174–195

1. Dalton, John, *A New System of Chemical Philosophy*, Part II, Manchester, 1810, 'Simple atmospheres and compound atmosphere'; in Greenaway, Frank, *John Dalton and the Atom*, Heinemann, London, 1966.
2. Each gas, with its own characteristic particle shape and particle size, should have its own characteristic buoyancy.
3. At this stage, 'atom' = 'one molecule of the gas'.
4. In 1818, the Swedish chemist, Berzelius, measured and compiled such data.
5. Avogadro's constant was actually discovered by Loschmidt, in 1865. Strictly speaking, it's the number of molecules in one mole of gas. For example, in 16 g of oxygen, at room temperature and atmospheric pressure, there are 6.024×10^{23} molecules.
6. Also, how could 'Avogadro's number' ever be determined?
7. At a given temperature.
8. Holmyard, E.J., *Makers of Chemistry*, The Clarendon Press, Oxford, 1953, p. 222. Sir Henry Roscoe (1833–1915) was a chemist, founder of a college of London University, and uncle of Beatrix Potter. However, Gillispie is dubious about the accuracy of Roscoe's observations (personal communication with Charles Gillispie).
9. Dalton, John, Fourth 'Experimental essay', entitled 'On the thermal expansion of gases', presented in 1800 and in the *Memoirs of the Manchester Literary and Philosophical Society*, 1802.
10. Crosland, Maurice, *Gay-Lussac, Scientist and Bourgeois*, Cambridge University Press, Cambridge, 2004.
11. Cardwell, W to C, p. 131.

12. Crosland, *Gay-Lussac, Scientist and Bourgeois*.
13. We shall have to revise this prejudice against 'extensive' quantities later (see the end of Chapter 18) when entropy enters the picture—not to mention energy itself.
14. Fox, Robert, 'The fire piston and its origins in Europe', *Technology and Culture*, 10(3), 1969, pp. 355–70.
15. Not counting changes of state.
16. The gas *inside* the expanding chamber.
17. The weight of a molecule was itself linked to the density of the gas (where the density in this example is the total weight divided by the total volume, and not the density of distribution of molecules in space).
18. Supposedly, the reduced volume would have less capacity for holding heat, and the excess heat would then be expelled and detectable as a temperature rise.
19. Cardwell, W to C, p. 136.
20. Mendoza, E., *Introduction to Sadi Carnot's Reflections on the Motive Power of Fire*, Dover Publications, New York, 1960; and Cardwell, W to C, p. 137.
21. Fox, Robert, 'The background to the discovery of Dulong and Petit's Law', *British Journal for the History of Science*, 4(1), 1968, pp. 1–22.
22. Fox, 'The background to the discovery of Dulong and Petit's Law'.
23. It involves more than just entropy maximization.
24. Fox, 'The background to the discovery of Dulong and Petit's Law'.

CHAPTER 11, PAGES 197–202

1. Herivel, J.W., 'Aspects of French theoretical physics in the nineteenth century', *British Journal for the History of Science*, 3(10), 1966, p. 112.
2. Biographical notes before Fourier, Joseph, *The Analytical Theory of Heat* (1822), vol. 45, Britannica Great Books, Chicago, 1952.
3. Herivel, 'Aspects of French theoretical physics', p. 113.
4. Fourier, Joseph, *The Analytical Theory of Heat* (1822), vol. 45, 29th edn, Britannica Great Books, Chicago, 1987, preliminary discourse, p. 169.
5. Fourier, *The Analytical Theory of Heat* (1987), section I, first chapter, p. 177.
6. Fourier, *The Analytical Theory of Heat* (1987), preliminary discourse, p. 171.
7. Cardwell, W to C, ref. 64, p. 117.
8. Cardwell, W to C, p. 117.
9. Fourier, *The Analytical Theory of Heat* (1987), preliminary discourse, p. 174.
10. In fact, Euler had done something similar, many years before, when modelling the elasticity of a solid: it is not known whether Fourier knew of Euler's work.
11. $\partial^2 T/\partial x^2 = (c/\kappa)\partial T/\partial t$, where T is the temperature, c is the specific heat capacity, κ is the coefficient of conductivity, x is the distance along the bar, and t is the time.

12. We consider two more ways in which Fourier's heat-conduction equation is curious. First, the temperature varies only to *first* order in time. This is because it is describing the flow of a *massless* fluid ('*calorique*'). But how can the equation be correct, and yet be derived from the flow properties of a fictitious weightless fluid? The answer is that, as found with the caloric theory (see Chapter 10), the right physics can sometimes come through even while the metaphor is wrong. The second strange feature is that the solutions show that some heat is transmitted at *infinite* speed. This is absurd, and also, it is at odds with the fact that heat conduction is a rather slow process. However, the physicality is saved as the *proportion* of heat so transmitted is vanishingly small. (In fact—curiouser and curiouser—it tails off exponentially, in similar fashion to the Maxwell–Boltzmann distribution curve, to be explained in Chapter 17).

13. Gillispie, C.C., *Pierre-Simon Laplace, 1749–1827*, Princeton University Press, Princeton, NJ, 1997, p. 249.

14. An exponent is a superscript indicating what the 'power' is; for example, in the expression for live force, mv^2, the exponent of v is 2.

15. DSB on Fourier, p. 97.

16. Biographical notes preceding Fourier, *The Analytical Theory of Heat* (1952), p. 164.

17. Herapath, John, 'Mr Herapath on the causes, laws and phenomena of heat, gases etc.', *Annals of Philosophy*, 9, 1821, pp. 273–93; see p. 279.

18. Herapath, 'Mr Herapath on the causes . . .', p. 282.

19. 'X', 'Remarks on Mr Herapath's theory', *Annals of Philosophy*, ii, 1821, p. 390.

20. Priority is usually wrongly given to Joule: Cardwell, W to C.

CHAPTER 12, PAGES 204–225

1. In Feynman's opinion; Feynman, Richard, *Lectures on Physics*, with R. Leighton and M. Sands, vol. 1, Addison-Wesley, Reading, MA, 1966.

2. Carnot, Sadi, *Réflexions sur la puissance motrice du feu* (*Reflections on the Motive Power of Fire*), translated by R.H. Thurston, edited and with an introduction by E. Mendoza, Dover Publications, New York, 1960; this and subsequent quotes.

3. DSB on Lazare Carnot.

4. Carnot, *Réflexions*. The cycle instructions are from this source.

5. Gillispie, Charles, quoted in Robert Fox's edition and translation of Sadi Carnot's *Réflexions sur la puissance motrice du feu*, Manchester University Press, Manchester, 1986.

6. This latter is because none of the expansions are 'free'; that is, the expanding air pushes back on the surrounding external air and does work all the while (see Chapter 10).

7. The work done is the area under a curve, but the sign depends on the direction (running the engine forward, i.e. clockwise round the cycle, expansions are *positive* but compressions are *negative*). So, we simply subtract the area under the lower curves, from the area under the higher curves (remembering that the area goes right down to the V-axis). The net result is—the *enclosed* area.

8. Carnot, *Réflexions* (Mendoza).

9. Of course, *non*-ideal water-engines work at less than 100% efficiency.

10. I am not sure if the monitoring of heat all the way round a Carnot cycle has *ever* been done (see also quotes in Chapter 18, 'Impossible Things').

11. Carnot, *Réflexions* (Mendoza).

12. Carnot, *Réflexions* (Mendoza), 'Appendix: selections from the posthumous notes of Sadi Carnot'.

13. Carnot, *Réflexions* (Fox), p. 17.

14. Thompson, Sylvanus, *The Life of William Thomson, Baron Kelvin of Largs*, Macmillan, London, 1910, vol. 1, p. 133.

15. Thompson, *The Life of William Thomson*.

16. Apologies to those experts in French history.

CHAPTER 13, PAGES 226–238

1. Green, George, *An Essay on the Mathematical Analysis of Electricity and Magnetism*, Nottingham, 1828; also in Green, George, *Mathematical Papers*, edited by N.M. Ferrers, Macmillan, London, 1871 (reprinted by Chelsea, New York, 1970).

2. St Andrew's University 'McTutor' biographies of mathematicians, at http://www-history.mcs.st-andrews.ac.uk/Mathematicians/Green.html

3. Cannel, D.M., *George Green, Mathematician and Physicist 1793–1841: the Background to His Life and Work*, Athlone Press, London, 1993.

4. Confusingly, this is also denoted *V*.

5. Cannel, *George Green*, p. 170.

6. In spherical polar coordinates.

7. By determining the partial rate of change of *V* with *x,y,z* at the position of the test particle.

8. Cannel, *George Green*, p. 70.

9. Cannel, *George Green*, p. 146; also in Thompson, Sylvanus, *The Life of William Thomson, Baron Kelvin of Largs*, Macmillan, London, 1910, pp. 113–19.

10. Hankins, Thomas L., *Sir William Rowan Hamilton*, Johns Hopkins University Press, Baltimore, MD, 1980.

11. Hankins, *Sir William Rowan Hamilton*.

12. Hamilton, William R., 'On a general method in dynamics, by which the study of all free systems of attracting or repelling points is reduced to the

search and differentiation of one central relation or characteristic function', *Philosophical Transactions of the Royal Society*, 1834, pp. 247–308.

13. Hamilton's Principle: a mechanical system will maintain a 'stationary' value for $\int (T - V)dt$, when this integral is 'varied' between fixed end-conditions (both end-positions *and* both end-times must be fixed). (Hamilton attached no teleological significance to these minimum principles: the 'pretensions [of the law of Least Action] to a cosmological necessity, on the ground of economy in the universe, are now generally rejected. And the rejection appears just.' Hankins, *Sir William Rowan Hamilton*, p. 427, note 13.)

14. In fact, this geometric property, also known as the 'ray property', was a consequence of Fermat's Principle of Least Time (see Chapter 7), which stated that the time interval between a pair of surfaces would be the least possible.

15. Hankins, *Sir William Rowan Hamilton*, p. 92, notes 15 and 16.

16. In Hamilton's optical theory, it could be shown that the light propagated itself by infinitesimal wavelets starting at each wavefront, in the manner postulated by Huygens some 150 years earlier.

17. In full, the British Association for the Advancement of Science. Charles Dickens satirized it as the association for the advancement of 'umbugology and ditchwateristics', *The Times*, 1837; Hankins, *Sir William Rowan Hamilton*, p. 142.

18. Hankins, *Sir William Rowan Hamilton*, p. 164.

19. Hamilton, William, R., 'On a general method of expressing the paths of light, and of the planets, by the coefficients of a characteristic function', *Dublin University Review and Quarterly Magazine*, I, 1833, pp. 795–826; also Hamilton, 'On a general method in dynamics'.

20. 'Algebraic' meant that the function could have terms that were squared, cubed, and so on, but there were no differentials with respect to time.

21. The words 'positions' and (next paragraph) 'momenta' are in quotes because they denote *generalized* positions and momenta (cf. Chapter 7, 'Lagrangian Mechanics'). Note that the principal function could also depend on the time, t, but it nevertheless determined only the paths and not the speed of the particles along these paths.

22. The p are defined as $\partial L/\partial \dot{q}$, where L is 'the Lagrangian', $T - V$, given as a function of q, $\dot{q}$, and t (note that $\dot{q}$ is another way of writing dq/dt).

23. The principal function and H are connected together in each of two (first-order) partial differential equations; see Lanczos, Cornelius, *The Variational Principles of Mechanics*, 4th edn, University of Toronto Press, Toronto, 1970.

24. In the simplest case—there were other more complicated possibilities (notation: a dot over q means dq/dt).

25. The other part of the transformation is that T_{new} was given by $\Sigma p\dot{q}$, where the summation is over all n particles.

26. This phase space, while imaginary, was different to Lagrange's imaginary configuration space.
27. Hankins, *Sir William Rowan Hamilton*, p. 164.
28. Kilmister, C.W., *Hamiltonian Dynamics*, Longman, London, 1964.
29. Hankins, *Sir William Rowan Hamilton*, p. 204.
30. Hankins, *Sir William Rowan Hamilton*, p. 208, note 32.
31. Hankins, *Sir William Rowan Hamilton*, p. 64, note 7.
32. In Hamilton's Principle, a path between definite end states is 'varied', and the path with the 'stationary' (usually a minimum—see Chapter 7, note 76) action is chosen.
33. The nuisance of having physical quantities that change value with the frame of reference is avoided by employing a purely mathematical (virtual) test-space; explained further in my forthcoming book *The Lazy Universe: the Principle of Least Action Explained* (Oxford University Press, Oxford, 2015).

CHAPTER 14, PAGES 240–257

1. When browsing through an old *Chambers Dictionary*, I found 'PSA' listed as an acronym for Pleasant Sunday Afternoon.
2. Lindsay, Robert Bruce, *Julius Robert Mayer, Prophet of Energy*, Selected Readings in Physics, General Editor D. ter Haar, Pergamon Press, Oxford, 1973, pp. 7–8.
3. Mayer, Julius, R., 'Comments on the mechanical equivalent of heat' (1851), in Lindsay, *Julius Robert Mayer*, p. 203.
4. Mayer, Julius, R., 'The motions of organisms and their relation to metabolism: an essay in natural science' (1845), in Lindsay, *Julius Robert Mayer*, p. 99.
5. Mayer, 'The motions of organisms', p. 74.
6. Mayer, 'The motions of organisms', p. 72.
7. Mayer, 'The motions of organisms', p. 86.
8. Cardwell, D.S.L., *James Joule, a Biography*, Manchester University Press, Manchester, 1989, p. 32.
9. Not just a loop of wire, but a circuit with a battery or electrolytic cell in it.
10. Joule, James, 'On the calorific effects of magneto-electricity and on the mechanical value of heat', *Philosophical Magazine*, 23, 1843, pp. 263–76.
11. Joule, 'On the calorific effects'.
12. Joule, 'On the calorific effects'.
13. Joule, 'On the calorific effects'.
14. I don't know how this was done.
15. Joule, 'On the calorific effects', postscript; and Cardwell, *James Joule*, p. 58.
16. Later, this was understood to be strictly true only for ideal gases.
17. Cardwell, *James Joule*, p. 76.

18. Cardwell, *James Joule*, p. 61.
19. This was not an especially intuitive example for Joule to choose: the 'living force' is provided by the muscles in the fingers, and the 'attraction through space' is increased as the spring is more compressed.
20. Cardwell, *James Joule*, p. 99.
21. Cardwell, *James Joule*.
22. Cardwell, *James Joule*.
23. Also, he shouldn't have gauged the 'work of the motor' by spin-rate alone, but also included the number of weights that were spinning.
24. Cardwell, *James Joule*, p. 96.
25. DSB on Waterston.
26. DSB on Waterston.
27. DSB on Waterston.
28. Cardwell, *James Joule*, this and subsequent quotes.
29. Cardwell, *James Joule*, p. 76.
30. Cardwell, *James Joule*, p. 43.
31. Cardwell, *James Joule*, p. 208.

CHAPTER 15, PAGES 259–275

1. Pearce Williams, L., *Michael Faraday, a Biography*, Da Capo Press, New York, 1987, ch. 1, p. 28, ref. 47.
2. Pearce Williams, *Michael Faraday*, p. 29.
3. DSB on Faraday.
4. Quoted on the title page of Grove, W.R., ed., *The Correlation and Conservation of Forces*, containing essays by Grove, Helmholtz, Mayer, Faraday, Liebig, and Carpenter, published by Appleton and Co, New York, 1867.
5. Pearce Williams, *Michael Faraday*; this and all subsequent quotes until note 6.
6. By contrast, electric charge can be present (equal amounts of opposite charges) and the net charge can be zero.
7. Grove, *The Correlation and Conservation of Forces*, in the article by Faraday 'Some thoughts on the conservation of force'.
8. Grove, *The Correlation and Conservation of Forces*.
9. M'Kendrick, J.G., *Hermann Ludwig Ferdinand von Helmholtz*, Longmans, Green & Co., New York, 1899, school report, p. 6.
10. Königsberger, Leo, *Hermann von Helmholtz*, translated by F.A. Welby, preface by Lord Kelvin, Dover Publications, New York, 1906, p. 17.
11. Königsberger, *Hermann von Helmholtz*, p. 14.
12. Kahl, Russell, ed., 'An autobiographical sketch', *Selected Writings of Hermann von Helmholtz*, Wesleyan University Press, Middletown, CT, 1971, ch. 17, p. 470.
13. Königsberger, *Hermann von Helmholtz*.

14. Kahl, 'An autobiographical sketch'.
15. Kahl, 'An autobiographical sketch'.
16. Bevilacqua, F., in Cahan, David (ed.), *Hermann von Helmholtz and the Foundations of Nineteenth Century Science*, University of California Press, Berkeley, CA, 1993, p. 300.
17. M'Kendrick, *Hermann Ludwig Ferdinand von Helmholtz*, p. 35.
18. Königsberger, *Hermann von Helmholtz*, ch. V.
19. Helmholtz, Hermann von, 'The conservation of force: a physical memoir' (1847), in Kahl, *Selected Writings of Hermann von Helmholtz*.
20. Helmholtz, 'The conservation of force'; this and all subsequent quotes until note 21.
21. Jungnickel, C. and McCormmach, R., *The Intellectual Mastery of Nature*, University of Chicago Press, Chicago, 1986, vol. 1, p. 159.
22. Helmholtz, 'The conservation of force'.
23. Helmholtz, 'The conservation of force'.
24. Darrigol, Olivier, *Electrodynamics from Ampère to Einstein*, Oxford University Press, Oxford, 2000, p. 218.
25. Helmholtz, 'The conservation of force'; this and subsequent quotes until note 26.
26. Kahl, 'An autobiographical sketch', p. 471.
27. Darrigol, *Electrodynamics from Ampère to Einstein*, p. 216.
28. Smith, Crosbie, *The Science of Energy*, The Athlone Press, London, 1998, p. 126.
29. Smith, *The Science of Energy*.
30. Maxwell, J.C., 'Hermann Ludwig Ferdinand Helmholtz', *Nature*, 15, 1877, pp. 389–91.
31. Königsberger, *Hermann von Helmholtz*; this and subsequent quotes until note 32.
32. Maxwell, 'Hermann Ludwig Ferdinand Helmholtz'.
33. Maxwell, 'Hermann Ludwig Ferdinand Helmholtz'.
34. Smith, *The Science of Energy*.
35. Heidelberger, M., 'Force, law and experiment', in Cahan, *Hermann von Helmholtz*, p. 180.
36. Helmholtz, 'The conservation of force', Appendix 4, added to the memoir in 1881, p. 52.

CHAPTER 16, PAGES 278–296

1. Clapeyron, Émile, 'On the motive power of heat' (1834), in Carnot, Sadi, *Réflexions sur la puissance motrice du feu* (*Reflections on the Motive Power of Fire*), translated by R.H. Thurston, edited with an introduction by E. Mendoza, Dover Publications, New York, 1960.

2. Thompson, Sylvanus, *The Life of William Thomson, Baron Kelvin of Largs*, Macmillan, London, 1910.
3. Absolute degrees are called Kelvin and symbolized 'K'. There is a conversion between all non-absolute scales and the absolute scale; for example, for the Celsius scale, 0 °C = 273.15 K, and the interval 1 °C is equal to the interval 1 K.
4. Thomson, William, 'An account of Carnot's theory of the motive power of heat, with numerical results deduced from Regnault's experiments on steam', *Transactions of the Royal Society of Edinburgh*, 16, 1849, pp. 541–74.
5. Clausius, Rudolf, 'On the moving force of heat and the laws of heat that may be deduced therefrom' (1850); translated by W.F. Magie, in Carnot, *Reflections* (1960).
6. Later, Clausius also reworked the problem studied by Clapeyron: a vapour in contact with its liquid.
7. C_V is the specific heat at constant volume for a gas: Regnault and others had experimental measures for this.
8. The fact that the 'enclosed area' is equal to the 'work done' has been explained in Chapter 8 ('Watt') and in Chapter 12.
9. Cardwell, W to C.
10. These examples are for a two-dimensional thermodynamic system—things can become more complicated than this; one can have three thermodynamic parameters, or more.
11. In other words, Carnot's Law could still be upheld.
12. Clausius, 'On the moving force of heat'; this and subsequent quotes.
13. There was some departure from Clapeyron in the cases where the vapours were of high density.
14. Clausius, 'On the moving force of heat'.
15. Clausius, 'On the moving force of heat'.
16. Clausius, 'On the moving force of heat'.
17. Thomson, William, 'On the dynamical theory of heat, with numerical results from Mr. Joule's equivalent of a thermal unit, and M. Regnault's observations on steam', *Transactions of the Royal Society of Edinburgh*, March 1851; this and the following quotation.
18. Carnot, Sadi, Appendix to *Reflections* (1960).
19. Thomson, William, 'Thermo-electric currents, preliminary 97–101, Fundamental principles of general thermo-dynamics recapitulated', *Transactions of the Royal Society of Edinburgh*, vol. xxi, part I, p. 232, read May 1854; also in *Mathematical and Physical Papers*, vol. 1, Cambridge University Press, Cambridge, 1882.
20. Smith, Crosbie and Wise, Norton, *Energy and Empire: a Biographical Study of Lord Kelvin*, Cambridge University Press, Cambridge, 1989.
21. Thomson, William, 'On a universal tendency in nature to the dissipation of mechanical energy', *Philosophical Magazine*, 1852.

22. Clausius' three-temperature ideal engine is given in Appendix III, 'Extras'.
23. An increasing function $f(x)$ is, as its name suggests, a function that always increases as x increases.
24. This is what 'complete' means: starting from any point on the cycle, but then proceeding all the way round and ending back at the same point.
25. Note that the dQs in Equations 16.4 and 16.5 are not any old losses of heat, but they are *ideal* heat transfers within a Carnot cycle: the temperature differences are infinitesimal, the expansion proceeds very slowly, and so on—as described in Chapter 12; see also the quasi-static conditions explained in Chapter 18.
26. Cardwell, W to C, p. 272.
27. Cardwell, W to C, p. 269.
28. The S in Equations 16.5, 16.6, and 16.7 is, in effect, ΔS, in the special case where the start- and end-points are one and the same (i.e. the cycle is *complete*). ΔS is equal to zero for changes within the system, and greater than zero in the surroundings.
29. Clausius, Rudolf, *The Mechanical Theory of Heat: with its Applications to the Steam Engine and to the Physical Properties of Bodies*, translated by John Tyndall, John van Voorst, London, 1867, p. 365.
30. Smith and Wise, *Energy and Empire*.
31. Thomson, William, and Tait, Peter Guthrie, *Treatise in Natural Philosophy*, Macmillan, London, for the University of Oxford, 1867.
32. Smith and Wise, *Energy and Empire*.
33. Hamilton's 'variational' approach was given more prominence on the continent and led to a British/continental divide that, maybe, continues to this day.
34. Clausius, *The Mechanical Theory of Heat*, p. 251.
35. Brush, Stephen, *The Temperature of History: Phases of Science and Culture in the Nineteenth Century*, Lenox Hill Press, New York, 1979.

CHAPTER 17, PAGES 298–330

1. Maxwell, James Clerk, in Gillispie, C.C., *The Edge of Objectivity*, Princeton University Press, Princeton, NJ, 1973.
2. Note that the energy is proportional to the square of the field, whether the field is static or exists as a wave.
3. Feynman, R.P., *Lectures on Physics*, with R. Leighton and M. Sands, Addison-Wesley, Reading, MA, 1966, vol. II.
4. See the section on 'Special and General Relativity'.
5. Grattan-Guiness, Ivor, 'Work is for the workers: advances in engineering, mechanics, and instruction, in France, 1800–1830', *Annals of Science*, 41, 1984, pp. 1–33.
6. Cauchy's idea was itself an advance on Euler's remarkable innovation of an imaginary slice through a material or body of fluid.

7. Einstein, A. and Infeld, L., *The Evolution of Physics*, paperback edition, Simon and Schuster, New York, 1967.

8. Feynman, *Lectures on Physics*, vol. I.

9. Gillispie, C.C., *Essays and Reviews in History and History of Science*, American Philosophical Society, Philadelphia, PA, 2007, ch. 19.

10. Quetelet, Adolphe, *A Treatise on Man and the Development of His Faculties*, William and Robert Chambers, Edinburgh, 1842.

11. Müller, Ingo, *A History of Thermodynamics: the Doctrine of Energy and Entropy*, Springer-Verlag, Berlin, 2007, p. 92.

12. Klein, Martin J., 'The development of Boltzmann's statistical ideas', in Cohen, E.G.D. and Thirring, W. (eds), *The Boltzmann Equation: Theory and Applications*, Springer-Verlag, Vienna, 1973.

13. Maxwell, J. C., *The Scientific Papers of James Clerk Maxwell*, vol. II, edited by W. Niven, Cambridge University Press, Cambridge (1890); reprint New York, 1952, p. 43, referred to in Everitt, C.W.F., *James Clerk Maxwell, Physicist and Natural Philosopher*, Charles Scribner's Sons, New York, 1975.

14. He looked at the time evolution of a gas and proved that Maxwell's distribution was both unique and stable (i.e. that different starting distributions would all tend towards it and, once arrived at, it would keep reproducing itself). These properties are *essential* if the distribution is to truly represent the equilibrium state.

15. Actually, in Maxwell's later work, some of these ideas are explored.

16. Feynman, *Lectures on Physics*, vol. I, Chapter 39.

17. A kind of 'harmonic oscillator'. 'Ideal' means that the spring is not extended beyond its elastic range, and that there is no dissipation—self-evident in this microscopic regime.

18. Cf. Prévost's theory of exchange (see the end of Chapter 6).

19. Feynman, *Lectures on Physics*. At the very end of Chapter 39, Volume I, Feynman adds that it is only approximately true that the internal energy is the sum of vibrational and rotational kinetic energies. More correctly, for an r-atom molecule, the average kinetic energy of the molecule will be $3rkT/2$ joules, of which $3kT/2$ is the energy of translation (in three directions), and the rest, $3(r-1)kT/2$, is internal vibrational and rotational kinetic energy.

20. At room temperature, $T = 300$ K, the mean energy per thermalized unit of energy, $kT/2$, is a tiny 2×10^{-21} J, and so the energy in a litre of gas is about 6 kJ. Note that the peak of the distribution moves to a higher speed for a higher temperature, tying in with Daniel Bernoulli's prediction that T is proportional to the square of the speed (see Chapter 5).

21. This can be seen in various engineering and physics applications (cf. impedance matching in electronics, and acoustics). An analogy also occurs in the flow of traffic: to optimize the number of cars crossing a certain intersection per second, the average car speed mustn't be too low, but also it mustn't be too high, or else the packing density of cars will fall too much.

22. Klein, 'The development of Boltzmann's statistical ideas'; this and subsequent quotes.

23. Provided that 'quasi-static conditions' apply—this is explained further in Chapter 18.

24. Gibbs, Josiah Willard, 'On the equilibrium of heterogeneous substances', in Bumstead, H.A. and Van Name, R.G. (eds), *The Scientific Papers of J. Willard Gibbs*, Longmans, Green & Co., London, 1906; reprinted by Dover Publications, New York, 1961.

25. Klein, Martin J., 'Max Planck and the beginnings of the quantum theory', *Archive for the History of the Exact Sciences*, 1, 1960, pp. 459–79.

26. The metal would feel warmer and warmer to the touch, but no electrons would be emitted if the frequency was below some threshold value.

27. Note that the dimensions of h, and of $\Delta E \Delta t$, are the same as the dimensions of action (see Chapter 7, Part IV).

28. The Bohr–Kramers–Slater theory.

29. Feynman, R.P. and Hibbs, A.R., *Quantum Mechanics and Path Integrals*, McGraw-Hill, New York, 1965.

30. Klein, Martin, J., 'Thermodynamics in Einstein's thought', *Science*, 157, 1967, pp. 509–16.

31. Klein, 'Thermodynamics in Einstein's thought'.

32. In SR, the frames of reference must move at uniform speed and direction, and extend to infinity; in GR, the frames of reference can be accelerating, but are only of finite extent.

33. Readers may be familiar with the fact that, in SR, time intervals depend on the motion of the observer; space intervals depend on the motion of the observer; and space and time no longer exist as independent, absolute quantities, but form a continuum known as spacetime.

34. We have already seen, in Part IV, Chapter 7, how space and time are affected by 'inertia', and vice versa.

35. This moving mass, m, is the product of m_0 and γ, where m_0 is the rest mass, and γ is the Lorentz factor. Strictly speaking, it is more correct to denote 'm_0' as 'm', and then moving mass becomes '$m\gamma$' (Taylor, E.F. and Wheeler, J.A., *Spacetime Physics*, W.H. Freeman and Company, New York, 1992). However, on this occasion we'll stick with the older, less logical, notation, as the equation, $E = mc^2$, then retains its iconic form.

36. Rindler, W., *Special Relativity*, Oliver and Boyd, Edinburgh, 1966, p. 88.

37. Einstein, Albert, in Pais, A., *Subtle is the Lord: the Science and the Life of Albert Einstein*, Oxford University Press, Oxford, 2005.

38. The rest mass can never be negative (Rindler, *Special Relativity*) but it can be zero, as in the case of photons. A faster-than-light particle, the tachyon, has been suggested, and it would have to have an imaginary rest mass.

39. An antimatter particle, such as an antiproton, has exactly the same, *positive*, rest mass as the corresponding matter particle (the proton), but all its other

(quantum-mechanical) properties (electric charge, intrinsic spin, baryon number, etc.) have the opposite signs.

40. A question: can kinetic energy be considered as a 'relativistic correction' to rest energy, in the same way as magnetic energy is a 'relativistic correction' to electrical energy?

41. Perhaps this is no more surprising than the fact that electrical effects were barely known about for millennia, even though the electric force is 10^{40} times stronger than the gravitational force.

42. Inertial mass is the mass in $F = ma$, whereas gravitational mass is the mass in $F_G = Gm_1m_2/r^2$. We have got used to accepting without question a tacit assumption of Newtonian mechanics—that the two kinds of mass are identical. However, Newton the man (as opposed to Newtonian mechanics) took nothing for granted, and he carried out experiments timing pendulums of identical length, but with bobs made from the same mass of different materials ('gold, silver, lead, glass, sand, common salt, water, wood and wheat'). He found that the period of oscillation was always the same, and so the inertial mass was indeed proportional to the gravitational mass (Westfall, R.S., *Force in Newton's Physics*, American Elsevier and Macdonald, London, 1971, p. 462).

43. Einstein and Infeld, *The Evolution of Physics*.

44. Einstein and Infeld, *The Evolution of Physics*.

45. As evidence, note that the lighter or heavier nucleus of an isotope makes no difference to chemical properties, and so in order to purify a sample one must resort to physical means (e.g. diffusion, mass spectrometry, etc.).

46. Muller, I. and Weiss, W., *Energy and Entropy: A Universal Competition*, Springer-Verlag, Berlin, 2005 (note that we are talking of energy and entropy *changes* as the reaction proceeds).

47. Cardwell, D.S.L., *James Joule, a Biography*, Manchester University Press, Manchester, 1989, p. 253.

48. Cardwell, *James Joule, a Biography*, p. 316.

49. Thomson, William and Tait, Peter, G., *Treatise on Natural Philosophy*, Macmillan, London, for the University of Oxford, 1867.

50. Smith, Crosbie and Norton Wise, M., *Energy and Empire: A Biographical Study of Lord Kelvin*, Cambridge University Press, Cambridge, 1989.

51. Maxwell, J.C., *Theory of Heat* (1891), Dover Publications, New York, 2001.

52. Smith, Crosbie, *The Science of Energy: A Cultural History of Energy Physics in Victorian Britain*, The Athlone Press, London, 1998, p. 260.

53. Smith, *The Science of Energy*, p. 258.

54. Smith, Crosbie, *The Science of Energy*.

55. Cercignani, C., *Ludwig Boltzmann: the Man Who Trusted Atoms*, Oxford University Press, Oxford, 1998.

56. Lindley, David, *Boltzmann's Atom: the Great Debate that Launched a Revolution in Physics*, The Free Press, New York, 2001, p. 81.

57. Dyson, Freeman, 'Why is Maxwell's theory so hard to understand?' Available at http://www.clerkmaxwellfoundation.org/DysonFreemanArticle.pdf (accessed 4 December 2014).
58. Dyson, 'Why is Maxwell's theory so hard to understand?'
59. And Lyell did not think that a hot Earth-core contradicted this uniformity.
60. Brush, Stephen, *The Temperature of History: Phases of Science and Culture in the Nineteenth Century*, Lenox Hill Press, New York, 1979, p. 61.
61. Brush, *The Temperature of History*, p. 40.
62. Smith and Norton Wise, *Energy and Empire*, p. 317.
63. Smith and Norton Wise, *Energy and Empire*, p. 500.
64. Müller, Ingo, *A History of Thermodynamics: the Doctrine of Energy and Entropy*, Springer-Verlag, Berlin, 2007.
65. Brush, *The Temperature of History*, pp. 63–4.
66. Brush, *The Temperature of History*.
67. Brush, *The Temperature of History*.
68. Brush, *The Temperature of History*, p. 43.
69. Smith and Wise, *Energy and Empire*, p. 566.
70. Smith, *The Science of Energy*, p. 251.
71. Cercignani, *Ludwig Boltzmann*, p. 84.
72. Introduction, in Lindley, *Boltzmann's Atom*.
73. Cercignani, *Ludwig Boltzmann*, p. 202.
74. Kak, Mark, 'Probability and related topics in physical sciences', Summer Seminar in Applied Mathematics, Boulder Colorado, 1957.
75. Lindley, *Boltzmann's Atom*, p. 102.
76. Lindley, *Boltzmann's Atom*, p. 225.
77. Gibbs, 'On the equilibrium of heterogeneous substances'.
78. Angrist, S. and Hepler, L., *Order and Chaos*, Penguin, London, 1973.
79. DSB on Clausius.
80. Cercignani, *Ludwig Boltzmann*, p. 176.
81. The insight, and the reference to H.G. Wells, comes from my father, Bertie Coopersmith.
82. Cardwell, *James Joule, a Biography*, p. 216.
83. Cardwell, *James Joule, a Biography*, p. 259.
84. Ayrton, W.E., 'The distribution of power', *Nature*, 72, 1905, p. 612.
85. For example, deaths in coal mining, destruction of habitat, radioactive waste, and greenhouse gases.
86. Smith and Wise, *Energy and Empire*.
87. Feynman, *Lectures on Physics*, vol. I, p. 4–8.

CHAPTER 18, PAGES 331–354

1. I *think* I saw this quote in a book by George Gamow in the Baillieu Library, University of Melbourne, around 2005: try as I might, I have not been able to find it again.

2. Pippard, Brian, *The Elements of Classical Thermodynamics*, Cambridge University Press, Cambridge, 1957.

3. Zemansky, Mark, *Heat and Thermodynamics*, 5th edn, McGraw-Hill Kogakusha, Tokyo, 1968.

4. Zemansky, *Heat and Thermodynamics*, p. 1.

5. Zemansky, *Heat and Thermodynamics*, p. 78; and Pippard, *The Elements of Classical Thermodynamics*, p. 16.

6. Zemansky, *Heat and Thermodynamics*, p. 71.

7. Pippard, *The Elements of Classical Thermodynamics*, p. 15.

8. Zemansky, *Heat and Thermodynamics*, p. 76.

9. Pippard, *The Elements of Classical Thermodynamics*, p. 9.

10. I am making Pippard's prescription a little easier to follow by deigning to use the word 'thermometer'.

11. Pippard, *The Elements of Classical Thermodynamics*, p. 12.

12. Referring again to our theme of how the maths and the physics march forward in step, the quasi-static conditions are a *physical* manifestation of 'taking the limit' in calculus.

13. Zemansky, *Heat and Thermodynamics*, p. 148.

14. Zemansky, *Heat and Thermodynamics*, p. 433.

15. Actually, impossible thought experiments must sometimes be allowed. This occurs, for example, in the Principle of Least Action, where the variations happen instantaneously; that is, faster than light (cf. Coopersmith, J., *The Lazy Universe, the Principle of Least Action Explained*, Oxford University Press, to be published in 2015).

16. Feynman, Richard, *Lectures on Physics*, with R. Leighton and M. Sands, Addison-Wesley, Reading, MA, 1966, vol. II.

17. Planck, Max, *Treatise on Thermodynamics*, 1905, translated by Alexander Ogg, reprinted by Dover Publications, New York, 1990, pp. 106–7.

18. Poincaré, Henri, in Doughty, Noel, *Lagrangian Interaction*, Addison-Wesley, Reading, MA, 1990, p. 225.

19. Wolfgang Pauli's letter in Cropper, W.H., *Great Physicists: the Life and Times of Leading Physicists from Galileo to Hawking*, Oxford University Press, Oxford, 2001, p. 335.

20. Pauli, in Cropper, *Great Physicists*.

21. Feynman, *Lectures on Physics*, vol. II, Section 13–16.

22. Note that charge itself is unaffected by the motion of the frame of reference. Feynman shows that the magnitudes of the forces on the exterior electron are still not equal, $F' \neq F$, but the impulse is the same, $F'\Delta t' = F\Delta t$, when the special relativistic effect known as 'time dilation' is taken into account.

23. Maxwell, James Clerk, *Theory of Heat* (1871), Dover Publications, New York, 2001, introduction by Peter Pesic; see p. 301.

24. Maxwell, *Theory of Heat*, p. 302.

25. Landau, L.D. and Lifshitz, E.M., *Course of Theoretical Physics*, vol. 1, *Mechanics*, Pergamon Press, Oxford, 1960.

26. Coopersmith, *The Lazy Universe*; the scalar product uses *two* 'vector spaces', and so the 'dimensionality' of elemental kinetic energy *must* be two.

27. Ron Maimon: former contributor to Physics Forum, on the web; Maimon is a brilliant independent researcher in physics.

28. Assuming that deformation and sound energies are negligible.

29. Incidentally, as Maimon comments, this shows that the kinetic energy can be calibrated by a method—for example, using a thermometer—that does not depend on motion.

30. We are assuming that the mass, m, is unaffected by the motion of the frame of reference.

31. Ehlers, J., Rindler, W. and Penrose, R., 'Energy conservation as the basis of relativistic mechanics II', *American Journal of Physics*, 33(12), 1965, pp. 995–7. Their proof (the non-relativistic version) is explained in Coopersmith, *The Lazy Universe*.

32. The proof of Ehlers, Rindler and Penrose, 'Energy conservation as the basis of relativistic mechanics II' concerns *elastic* collisions, and so it can relate to *particles*, whereas Maimon's proof concerns *inelastic* collisions and so requires *bodies* (a particle cannot act as a 'wall'). Also, of great interest, Ehlers et al. show that there is a zero point of kinetic energy (more on this in the section after next).

33. The wonderful proof of Ehlers, Rindler and Penrose has one defect—it assumes an absolute frame of reference relative to which the energies of other frames can be gauged (see the caveat after their Assumption II). This assumption is unwarranted—the kinetic energy of an isolated free particle (or the kinetic energy of any number of free particles travelling in parallel) *does* depend on the motion of the frame of reference. On the other hand, the authors don't actually need this assumption, as their proof assumes *colliding* particles, and so only *relative* particle speeds are involved (for further discussion, see Coopersmith, *The Lazy Universe*).

34. As we shall comment in the next section, odd scenarios—even unusually simple ones—lead to odd results.

35. An invariant quantity is one that has the same actual value in a different frame of reference; a conserved quantity is one that maintains the same total value in *one* given, isolated frame of reference; a constant is something the value of which doesn't change in time; and a universal constant, such as Planck's constant, is something the value of which doesn't change in time, or in different frames of reference.

36. See Chapter 7, Part IV; Chapter 13; and also Coopersmith, *The Lazy Universe*.

37. Ehlers, Rindler and Penrose, 'Energy conservation as the basis of relativistic mechanics II' also make this point: whereas in Newtonian mechanics this

extra energy term is arbitrary, and usually normalized to zero, in Special Relativity it is *determined* as m_0c^2.

38. And also the 'momentum books'.
39. Jeans, James, *An Introduction to the Kinetic Theory of Gases*, Cambridge University Press, Cambridge, 1940; reissued 2008.
40. As opposed to, say, the function $f = x^3$, where the 'd of f', with respect to x, is given by $3x^2$. From now on, we distinguish exact differentials, dQ, from inexact differentials, đQ, by a tiny horizontal line across the 'd'.
41. Reminder: the tiny horizontal line across the 'd' in đQ indicates that the differential is not exact.
42. Feynman, *Lectures on Physics*, vol. I, ch. 39, page 39–7.
43. Collie, C.H., *Kinetic Theory and Entropy*, Longman, London, 1982, p. 159; also Zemansky, *Heat and Thermodynamics*, p. 82.
44. Granted that we have some primitive notion of a 'scale of hotness', we firm this up by calorimetry: adding 'heat' (ΔQ) causes regular changes in 'hotness' and/or regular changes in some external parameter, typically volume. We define '(dia)thermal contact' and 'equilibrium' and invoke the Zeroth Law. From all the above, we can then develop various empirical scales of temperature (e.g. that ΔT is proportional to ΔV). (It must be admitted, however, that the calorimetry equations invite some circularity between ΔQ and ΔT. This always occurs when two new concepts are introduced at the same time—for example, 'heat' and 'diathermal walls', or m and F in Newton's Second Law of Motion. There is no way out except that the innumerable experimental confirmations make the initial definitions consistent.)

Continuing with the absolute temperature scale, we imagine using an ideal Carnot engine of standard size to map out the degrees: the heat drawn out of or added to a reservoir during a Carnot cycle is proportional to the absolute temperature of that reservoir.

The immediate problem is that we don't have an ideal heat-engine at our disposal. However, we have the next best thing—a gas thermometer. This isn't a cheat: Carnot's proof shows that *any* type of engine will do as long as we aim as closely as possible to the ideal, quasi-static conditions. The real gas thermometer is an excellent approximation to the ideal heat-engine and the approximation becomes better and better the closer the gas is to ideal (negligible intermolecular forces). So the inert gases (helium, neon, argon, etc.) are more ideal than other gases; and the ideal case is reached for any gas in the limit as the density (pressure) of the gas is reduced to zero. Two consequences follow (from this identification of real gas thermometers with ideal Carnot engines). First, the T in the Ideal Gas Laws (the T in $PV = nkT$), coincides with the absolute T. Second, the gas thermometer may be used to calibrate the absolute scale, and then to cross-calibrate it with other types of thermometer that are more convenient in certain ranges of temperature (e.g. mercury-in-glass thermometers, platinum resistance thermometers, radiation pyrometers, adiabatic demagnetization, and so on).

As regards radiation, provided that we consider a system in equilibrium, such as 'black-body radiation' (radiation in a cavity), then we can talk meaningfully about its temperature. We can use electromagnetic theory to find the pressure of the radiation on the walls of the cavity, and proceed with an analysis exactly as if the cavity were a vessel containing a gas (Pippard, *The Elements of Classical Thermodynamics*, p.78). Or, equivalently, we can treat the radiation as an assembly of hot photons, and apply the kinetic theory as if the cavity were a vessel containing gas molecules (Pippard, *The Elements of Classical Thermodynamics*).

Finally, from the kinetic theory, a connection between the macroscopic and microscopic parameters of a gas at one temperature can be shown: $PV \propto <\frac{1}{2}mv^2>$. Comparing this to the ideal gas law, $PV \propto T$, we can then deduce that T is proportional to $<\frac{1}{2}mv^2>$, as Bernoulli and others suspected.

45. Our examples have mostly involved gases, but the astounding fact is that the process of thermalization, and the smallest thermalized unit of energy, is exactly the same, and depends only on T, whatever the system (e.g. a mixture of phases, a mixture of chemicals, a wire, a surface film, an electrical cell, or a paramagnetic solid). However, this doesn't apply in certain extreme regions (e.g. ultra-low temperatures or very high densities), or in non-equilibrium thermodynamics, where the very idea of temperature needs modification.

46. I had a brief correspondence with Professor Sir Brian Pippard on this question.

47. The explanation is adapted from Feynman, *Lectures on Physics*, vol. I, Section 39–4.

48. Assume that all the walls are perfectly insulating, and the moveable wall can slide without friction or leaks.

49. Strictly speaking, we should be equating translational energies in the left–right direction only.

50. For example, in Feynman's demonstration of the two gas compartments, why have we been able to ignore warming of the walls, and friction of the moving wall, when it is the very *microscopic* effects that we are examining?

51. There is no better explanation of entropy, and the Second Law of Thermodynamics, at this level, than that given in Atkins, P.W., *The 2ⁿᵈ Law: Energy, Chaos, and Form*, Scientific American Library, New York, 1994.

52. Even so, apart from a few specialized cases, we are always looking at *changes* in entropy, in other words, we are always looking at ΔS rather than S. (In Equations 16.4, 16.5, 16.6, and 16.7, the 'S' is really the same as ΔS, but applied to a complete cycle—that is, the start- and end-states are one and the same.)

53. This is hypothetical: even drawing from all the languages in the world, there are not nearly enough names; and, also, quantum mechanics does not allow the labelling of individual molecules.

54. The gas molecules have such small masses that we can ignore gravitational clumping on human timescales.

55. Gibbs, Josiah Willard, 'On the equilibrium of heterogeneous substances' (1878), in Bumstead, H.A. and Van Name, R.G. (eds), *The Scientific Papers of J. Willard Gibbs*, Longman, Green & Co., London, 1906.

56. Carathéodory, C., 1909, 'The Second Law of Thermodynamics', in *Benchmark Papers on Energy*, 5, translated by Joseph Kestin, Dowden, Hutchinson and Ross, Stroudsburg, PA, 1976.

57. Actually, there are some rare time-asymmetric microscopic effects (for example, the decay of neutral kaons) and physicists are unsure of the extent to which these influence cosmic time.

58. The discovery, in 1964, of the cosmic microwave background radiation by Arno Penzias and Robert Wilson.

59. Some processes need to be examined carefully. (1) In *photosynthesis*, gaseous molecules of CO_2 and H_2O are tamed and made to reside quietly together in large structures (like cellulose and carbohydrates), with only the release of gaseous oxygen to compensate. Worse still, sunlight is absorbed. According to Ingo Müller, the conversion of sunlight from low-entropy UV to higher-entropy IR in no way compensates for the production of very large, complex molecules. He explains (in Müller, I. and Weiss, W., *Energy and Entropy, a Universal Competition*, Springer-Verlag, Berlin, paperback edition, 2005) how the plant must release almost 500 times as much water vapour for leaf-cooling (i.e. for entropy maximization) as it absorbs just for leaf-growing. (2) *Gravitational clumping* (into planets and stars) leads to higher entropy, while for a gas, clumping implies a lower entropy—see the discussion in the main text.

60. Listen to my talk on ABC Radio National's Science Show, hosted by Robyn Williams on 11 June 2011, available at http://www.abc.net.au/radiona-tional/programs/scienceshow/the-laws-of-thermodynamics-and-the-fate-of-humanity/2915972 (accessed 4 December 2014). I have been unable to discover any research carried out on the connection between efficiency and the ambient temperature of an internal combustion engine (i.c.e.); however, my brilliant car mechanic, Howard Doherty (cf. Chapter 8, end of the section on 'Watt'), assures me that a car travelling from Bendigo to Sydney uses less fuel on winter days (around 10–15°C) than on summer days (35–45°C), other things being equal. The dependence of efficiency on ambient temperature is a small effect, and is obscured by the following facts. (1) The Earth isn't a system in equilibrium, and so it can't, strictly speaking, be assigned one, overall temperature. (2) A higher Earth temperature may occasionally help some 'engines' to reach their starting temperature, even

while it raises T_{sink}. (3) As efficiency is given by $(T_{source} - T_{sink})/T_{source}$ (with all temperatures in Kelvin), then the reductions in efficiency are very small. Nevertheless, over a whole planet, the effects will be noticeable. (4) There is just one process the efficiency of which will actually increase: the planet and 'outer space' comprise *another* system (not an Earth-bound one). In this larger system the 'engine' comprises the 'whole Earth', and as it will start from a higher temperature, it will run more efficiently, and therefore lose heat to 'outer space' at a very slightly faster rate than before global warming.

CHAPTER 19, PAGES 355–361

1. Poussin, Nicolas, *A Dance to the Music of Time*, oil on canvas, *c.* 1640, Wallace Collection, London.

2. I mean this most respectfully—in fact, I am especially interested in philosophical questions in physics.

3. Actually, in fundamental particle physics, there are some properties (baryon number, strangeness, etc.) which do seem to be defined by their property of being conserved, but these 'quantum numbers' all have the feature that they can be simply enumerated (say, 1, 2, 3, etc.), rather than appearing in a multitude of avatars, as for energy.

4. There are also other forms of energy once the gas starts burning; for example, radiant energy. Also, for a truly comprehensive list, we should include the atomic energies, nuclear binding energies, the rest mass energies of the fundamental constituents, and so on.

5. HamiltonW. R., in Hankins, Thomas L., *Sir William Rowan Hamilton*, Johns Hopkins University Press, Baltimore, MD, 1980, p. 177.

6. These discoveries were contemporaneous with the emergence of the idea of energy in the middle of the nineteenth century.

7. Brush, Stephen, *The Kind of Motion We Call Heat*, North-Holland, Amsterdam, 1976, Book I, p. 44.

8. Later, inequalities arose in Special Relativity and in quantum mechanics: in 1905, Einstein's theory of Special Relativity implied that all material particles had to travel at speeds *less* than the speed of light; in 1927, Heisenberg published his Uncertainty Principle, which required that the joint precision of conjugate variables for one particle was *less* than $h/4\pi$.

9. In his theory of General Relativity, 1915.

10. In Germany, the Law of Conservation of Energy is called the Energy *Principle*.

11. One answer is that there will always be small thermal fluctuations, and these *do* give an absolute measure of time (in the same way that atoms bring in an absolute size).

12. If the wave is for a free quantum-mechanical particle, then it will be infinitely long; if the particle is localized, then, strictly speaking, we should be considering a pulse or 'wave packet'.

13. Howard Doherty of Doherty's Garage, St Andrew's Street, Bendigo, personal communication.

APPENDIX II, PAGE 369

1. Adapted from the Wikipaedia Free Encyclopaedia, Wikimedia Commons, Orders of Magnitude for Energy, with gratitude.

Index

Index

Printed and bound by CPI Group (UK) Ltd, Croydon, CR0 4YY